LAND SNAILS AND SLUGS
OF THE PACIFIC NORTHWEST

The John and Shirley Byrne Fund for Books on Nature and the Environment provides generous support that helps make publication of this and other Oregon State University Press books possible. The Press is grateful for this support.

Previously published with the support of this fund:

One City's Wilderness: Portland's Forest Park
by Marcy Cottrell Houle

Among Penguins: A Bird Man in Antarctica
by Noah Strycker

Dragonflies and Damselflies of Oregon: A Field Guide
by Cary Kerst and Steve Gordon

Ellie's Log: Exploring the Forest Where the Great Tree Fell
by Judith L. Li (Illustrations by M. L. Herring)

Land Snails and Slugs of the Pacific Northwest

THOMAS E. BURKE

PHOTOGRAPHS BY WILLIAM P. LEONARD

Oregon State University Press   Corvallis

The author, photographer, and publisher gratefully acknowledge the generous support of the U.S. Bureau of Land Management and the U.S. Forest Service in helping make possible the publication of this book.

The paper in this book meets the guidelines for permanence and durability of the Committee on Production Guidelines for Book Longevity of the Council on Library Resources and the minimum requirements of the American National Standard for Permanence of Paper for Printed Library Materials Z39.48-1984.

Library of Congress Cataloging-in-Publication Data

Burke, Thomas E., 1941-
Land snails and slugs of the Pacific Northwest / Thomas E. Burke; photographs
    by William P. Leonard.
      pages cm
    Includes bibliographical references and index.
    ISBN 978-0-87071-685-0 (alk. paper) -- ISBN 978-0-87071-686-7 (e-book)
    1. Snails--Northwest, Pacific--Identification. 2. Slugs (Mollusks)--
Northwest, Pacific--Identification.   I. Title.
    QL430.4.B876 2013
    594'.317609795--dc23
                          2012044315

Oregon State University Press
121 The Valley Library
Corvallis OR 97331-4501
541-737-3166 • fax 541-737-3170
http://osupress.oregonstate.edu

Contents

Acknowledgments

T. BURKE

My study of the inland mollusks of the Pacific Northwest over the past 35 years has been incidental to my employment as a wildlife biologist. During that time, I have communicated with many individuals regarding the molluscan taxa and have consulted with many agency personnel and consultants regarding specific questions. Every question and every specimen sent to me has provided a learning experience. Personnel of the U.S. Forest Service, U.S. Bureau of Land Management, U.S. Fish and Wildlife Service, and Washington Department of Fish and Wildlife who provided specimens and site information are too numerous to list, but I gratefully acknowledge the contributions from each.

This work was initiated under a Challenge Cost-Share Agreement and associated contracts with the U.S. Department of the Interior, Bureau of Land Management, Oregon State Office, Spokane District, Spokane, Washington. The continued efforts of Neal Hedges and Kelli Van Norman of those offices kept the project on track, and their support is greatly appreciated.

Others with whom I worked particularly closely include Dr. Joseph Furnish, Nancy Duncan, Stephen Dowlan, Dr. John Applegarth, Roger Monthy, Ted Weasma, Darby Hansen, Jo Ellen Richards, Patty Garvey-Darda, Heather Murphy, Susan Piper, Tom Kogut, Mitch Wainwright, Judy Murray-Hoder, Brigitte Ranne, Jim Baugh, John Musser, and Erik Ellis. Paul Hendricks provided invaluable information on Montana mollusks and their distribution, as well as shells to photograph. Casey Richart provided information acquired through his many invertebrate surveys and studies. My special thanks to contract administrators and assistants on various mollusk projects: Neal Hedges, Barb Behan, Joan Ziegltrum, Dale Swedberg, Darci Rivers Pankratz, and Sarina Jepsen.

The late Dr. Terrence Frest of Deixis Consultants lent his expertise in consulting on disjunct populations and undescribed species. He and his colleague Edward Johannes provided specimens and locations for specific taxa and invaluable information from their experience, as well as copies of many reports from their work.

Mr. Bill Leonard, whose skill and patience with the camera provided the photographs used to illustrate the species, the figures of shell characteristics, and the slugs for this work, also lent his invaluable knowledge acquired through his study of slugs.

Special thanks are due to Dr. Barry Roth and Dr. Timothy Pearce for their reviews of the manuscript and their many comments and suggestions for its improvement. I also wish to thank the Academy of Natural Sciences of Philadelphia for permission to use many quotes and citations from Pilsbry's monographs, *Land Mollusca of North America (North of Mexico)*.

Finally, my wife Carol deserves special recognition for her patience as I worked many long days and nights and for her assistance in reviewing the manuscript. I also thank the doctors and nurses of Providence St. Peter Hospital and The Regional Hospital for their patience and indulgence of me as for months I sat by Carol's bed, often working on my laptop and always praying for her recovery from leukemia and the drastic side effects of chemotherapy; thanks to God that she is still with me.

Acknowledgments
W. LEONARD

As a young man I had the incredible good fortune to take a class from acclaimed nature photographers Larry West and John Shaw. Over the past three decades Larry has unselfishly shared macrophotography techniques that he personally pioneered, techniques that enabled me to make the photographs in this book. I am likewise deeply indebted to Tim Pearce, Lyle Chichester, and Tom Wilke for generously sharing their precious time, knowledge, and technical expertise. I also thank my friends and collecting companions Jim Baugh, Casey Richart, Steve Herman, Brad Moon, Kristiina Ovaska, Bob Storm, Kelley Jorgensen, and Joan Ziegltrum; and, especially, my family—Vicki, Nick, and Megan, for their companionship, patience, and assistance on many road trips around the PNW. Without their support, my contribution to this book would not have been possible. Lastly, I thank Tom Burke for opening my eyes to the diversity of mollusks in my adopted home—the Pacific Northwest.

Introduction

This work is intended as a guide for the identification of land snails and slugs of the Pacific Northwest, for use by biologists and other interested individuals who wish to identify most species in the field or laboratory without the necessity of dissecting the animals or sending them to a genetics laboratory. While not all species can be positively identified without the aid of dissection or molecular analysis, with this guide and some practice, most specimens should be recognizable to species. If a more exact identification is required, the specimens can be forwarded to a malacologist or genetics laboratory for further study. However, interested biologists and students of nature will find a large degree of satisfaction in observing and identifying these animals for themselves, and there is a definite need for such guides to identification of the lesser-known animals.

It has long been the practice to dissect gastropods in order to determine their phylogeny and even to confirm certain specific taxa. In today's taxonomy, analysis of the organism's DNA is now complementing the study of anatomy. Both of these disciplines are important to the overall understanding of relationships between species and other taxonomic groups, but they are best left to the specialists. In many instances, even DNA analysis fails to answer all of the questions, and current taxonomy often uses groups of closely related clades instead of classes, orders, or groups of species. Analysis of variations in DNA requires a computer to sort out findings; thus the study is enhanced. But such necessary technology need not constrain the ability or the interest of the naturalist in the fieldwork of conservation.

It is hoped that this guide will be a useful tool, enabling field biologists and other students of nature to continue working in the front lines of natural resource management and conservation. May we never lose the fascination that led us to these explorations, or the curiosity that continues to attract young and old alike to these essential fields of study.

Mollusk Diversity

Mollusca is the second largest phylum of animals on earth, exceeded only by Arthropoda, which contains the insects, arachnids, crustaceans, and other groups of jointed-legged invertebrates. Most of the better-known classes of mollusks occur in marine environments and include such diverse groups as the polyplacophorans

Land and Freshwater Mollusks of the Pacific Northwest

Phylum Mollusca

Class Bivalvia (Pelecypoda)—clams and mussels

Order Heterodonta—includes Corbiculidae and Sphaeridae (Asian clams, pea clams, and fingernail clams)

Order Schizodonta—includes Unionidae and Margaritiferidae (freshwater mussels)

Class Gastropoda—snails, slugs, and limpets

Subclass Eugastropoda—gill-breathing aquatic snails; aperture closed with an operculum

Order Neotaenioglossa—mostly marine and aquatic snails; some salt marsh inhabitants

Subclass Pulmonata—air-breathing snails and slugs; operculum lacking

Order Basommatophora—mostly aquatic snails and limpets; in this work, one salt marsh species and one species of land snail

Order Stylommatophora—land snails and slugs

(chitons), the bivalvia (clams, oysters, mussels), the scaphopods (tooth or tusk shells), the gastropods (snails, abalones, limpets, sea slugs), and the cephalopods (octopus, squid, cuttlefish, nautilus). Although only two of these classes occupy the aquatic and terrestrial habitats of inland North America, most people are surprised at how diverse and abundant these animals are.

The two classes of inland mollusks of the Pacific Northwest include four orders of Gastropoda (snails and slugs) and two of Bivalvia (freshwater mussels or clams). Our study is limited to the terrestrial gastropods, composed of land snails and slugs that inhabit the Pacific Northwest region of the United States, with four salt marsh inhabitants from the Pacific Coast also included. Those four are usually considered marine species, but inhabit areas that are not regularly inundated (see the introduction to the taxonomic list for more details); they are often omitted from books on seashells as well as from those on inland species.

For this work, the Pacific Northwest is defined as the entirety of the states of Washington, Oregon, and Idaho, and the part of Montana west of the Continental Divide. Because the ranges of many species are poorly understood, and because extensions to the known ranges are continually being discovered, other species are included that are not known to occur within that area but which occur near its periphery.

The Importance of Understanding and Identifying Mollusks

Mollusks belong to the second largest phylum in the animal kingdom, which indicates how important they are in the ecology of the earth. Most are prey for a variety of vertebrate and invertebrate predators, and many are eaten by humans as well. They are decomposers of organic matter, thus soil builders; disseminators of spores; predators; herbivores; and scavengers. Others are agricultural, domestic, and even industrial pests; many are hosts for parasites. There is much that remains to be learned about the biological and ecological relationships of the mollusks, and many opportunities remain open for discovery regarding the beneficial or detrimental effects of individual species.

According to the International Union for Conservation of Nature and Natural Resources (IUCN 1993), over 1200 mollusks are known or suspected to be threatened worldwide. Without a thorough knowledge of the biology and ecology of these animals, it will be difficult to preserve the diversity of species they represent.

Threatened Species Conservation

Species threatened, endangered, or of special concern are of particular interest to the public and to government agencies concerned with the management of resources. Under the Endangered Species Act of 1973 (as revised), the National Forest Management Act of 1976, the Organic Act, and other legislation, land management agencies of the United States assess the potential impacts of their proposed activities to ensure that their actions do not jeopardize the continued viability of any threatened or endangered species, or of any species proposed for listing. Under the Northwest Forest Plan (USDA Forest Service and USDI Bureau of Land Management 1994), the viability of all species was incorporated into environmental assessments, and the "Survey and Manage" standards and guidelines in that plan directed the agencies to survey for, and thus identify, certain species of vascular and nonvascular plants and invertebrates (including mollusks) as well as vertebrate animals.

Concern for species survival is valid, as shown in publications of the International Union for Conservation of Nature and Natural Resources (IUCN). The 1994 IUCN Red List of Threatened Animals (IUCN 1993) lists seven categories of threatened species and the nations in which they occur (these categories include extinct, endangered, vulnerable, and rare species, as well as species insufficiently known but suspected to be threatened and those expected to become threatened if commercial exploitation is not regulated).

The IUCN (1993) lists 5,366 threatened animal species worldwide, plus 563 species suspected to be threatened. Twenty percent of all threatened and suspected species on this list are mollusks, and another 20% are insects. In the continental United States, 73% of the 962 IUCN-listed threatened species are invertebrates.

In the Hawaiian and US Virgin Islands, 94% of another 1,651 threatened and suspected species are invertebrates.

Of about 1,130 species of mollusks considered threatened, endangered, rare, and/or recently extinct by the IUCN, 98% are freshwater and terrestrial (in about equal numbers) and 2% are marine. About 61% are from nine families. Geographically these species are mostly restricted to North America (United States), 40%; Australia-New Zealand, 19%; and Europe, 16% (Kay 1995). Within these lists, large, ecologically diverse countries that are species-rich and that have been subject to detailed assessment of their faunas are the ones with the greatest numbers of recognized species of concern (IUCN 1993).

Ecological Balance

Maintaining ecological balance is essential for the proper functioning of ecosystems and for resource production. Ecosystems are composed of biotic communities and of all of the physical, chemical, and biological components that keep them functioning. While systems are resilient because of biodiversity, it is also important to consider that the more common or abundant species exert a greater influence on systems than those that are rare.

Biotic communities are made up of many organisms (e.g., plants, algae, fungi, animals, protozoans, bacteria, viruses), all interacting with one another. An organism's habitat must provide for the basic needs of food, water, and security. Security may take the form of shelter from the elements, maintenance of specific conditions, or escape or concealment from enemies. Security also includes providing eggs or young with proper and safe environmental conditions for their hatching, development, and growth.

How much habitat is required to support a viable population? First we must consider that viable populations do not represent a single species. We need to take into account not only predator-prey and herbivore-plant relationships, but parasite-host and many other competitive and beneficial relationships. Even viruses must use DNA of another organism to live and reproduce, and protozoans require genetic exchange (conjugation) periodically. Therefore, the habitat of any species is commingled with habitats of other species, the niches of which overlap, sometimes extensively but not completely. To support a population of a species, a biotic community must provide all of the habitat components in sufficient quantity to meet the requirements of the organism, its mate, and its offspring (reproduction is not successful unless the offspring survive to reproduce). In the case of animals, the community must also provide sufficient habitats to maintain viable populations of food organisms for each species, and the food organisms for the food species, and enough of all of these habitats to support the families of competing species, and so on.

When disturbed, ecosystems repair themselves. So what is the importance of one or a few individual species? Ecosystems are resilient because they are composed of so many interacting organisms, the functions of which overlap within the system. If one species is lost (or reduced to the point of ineffectiveness), chances are the remaining species will continue to fulfill enough of the functions of the lost species for the system to continue functioning for a time until the lost species has recolonized and/or recovered. However, the loss of each species, with its unique contribution within the system, shifts the balance of that ecosystem. When populations of other species that rely on those lost components or functions are no longer sufficient to fulfill their own functions, a domino effect occurs. How far out of balance can an ecosystem be and still remain healthy and support all of the organisms that rely on it?

Mollusks, specifically, are some of the tools nature employs for the essential tasks of cleaning and recycling. Freshwater clams filter the waters in lakes and streams. Freshwater snails break down organic debris in streams, lakes, and ponds. Land snails and slugs do the same in terrestrial ecosystems. All of these creatures contribute to the recycling of nutrients in forests, grasslands, streams, lakes, and wetlands, helping to reuse organic materials and maintain the balance between production and decomposition. Some terrestrial gastropods are mycophagous, feeding on fungi and dispersing spores of beneficial organisms such as mycorrhizae (McGraw, Duncan, and Cazares 2002). Some of the gastropods are predators, some are scavengers; all function in maintaining the balance in nature. All of the biota working together contribute to the well-being of each other and maintain the balance within the ecosystem. Mollusks are important components of this system.

Mollusks also function in parasite-host relationships (see section below). These relationships, although they may seem to us detrimental, are all part of the natural world, and perform functions that we may not totally understand. The more we study and understand these relationships, the more we are able to make appropriate responses. While we might want to control outbreaks of parasitism, we should not attempt to eradicate native parasites indiscriminately; we are likely to upset a balance that we do not understand and thereby cause more serious problems. "If the biota, in the course of eons, has built something we like but do not understand, then who but a fool would discard seemingly useless parts?" (Aldo Leopold 1949).

Mollusks as Alternate Hosts for Parasites

Mollusks have long been known as alternate hosts for parasites and as vectors for their transmittal to secondary or final hosts. Mollusk identification is important for recognizing alternate host species of parasites of domestic and wild animals and fish, as well as of humans. Following are just a few examples of mollusk parasite-host relationships.

The larvae (glochidia) of the larger freshwater clams (mussels) are parasitic on fish, but the relationship is most often commensalistic; that is, the host fish is not usually harmed unless infestations are excessive. These relationships between mussels and fish are necessary for propagation of the mussels, and some freshwater mussels have specialized organs for attracting fish to receive their glochidia (Barnes 1968; Nedeau et al. 2009; Pennak 1978).

Salmon poisoning is a parasitic fluke, the life cycle of which involves three alternate hosts: freshwater snails of the genus *Juga*, salmonid fish, and carnivorous mammals of the family Canidae. The organism (*Nanophyetus salmincola*) is fatal to dogs, coyotes, and foxes (Simms et al. 1931; USDA 1956). Freshwater snails of the family Lymnaeidae are intermediate hosts for liver flukes of cattle, sheep, and other livestock. Possibly the best-known snail-borne parasites are the schistosome worms that cause the debilitating disease schistosomiasis in up to 300 million humans in tropical countries, where it is contracted from contact with water contaminated with human wastes (Sport Fishing Institute 1981). Land snails and slugs are carriers of such parasites as the bighorn-sheep lungworm (*Protostrongylus* sp.), mule-deer muscle worm (*Parelaphostrongylus odocoilei*), and the white-tailed deer meningeal worm (*Parelaphostrongylus tenuis*), which causes neurologic disease in many other large ungulates (Anderson 1965; Anderson and Prestwood 1981; Severinghaus and Jackson 1970). View a decomposing slug under a microscope and you are likely to see a mass of wriggling roundworms.

Mollusk Ecology (Habitat)

Habitat comprises all the elements within the environment of an organism that are required for the survival and successful reproduction of viable populations of the species (see the discussion on ecological balance, above). While habitats are variable depending on the species, in general we consider that food, water, and cover are required by all organisms.

Food—Terrestrial gastropods may be herbivores, predators, scavengers, or omnivores. Some are considered serious agricultural pests; most such pests in the Pacific Northwest are species introduced from European or other countries.

Most of our native species are found under rocks or woody debris, which they may use for cover while—as the mouths of gastropods are against the substrate—they feed on any organic matter found on its surface. Among forest floor litter and duff they feed on leaf-mold, fungus, or other decaying organic matter. Mushrooms are common sites on which to find many of the slugs. In wet areas, slugs such as *Deroceras laeve, Hemphillia malonei,* and others are commonly found on skunk cabbage. Some of the *Vespericola* are often found on the inner side of fallen tree bark, where scraped trails can be seen from their feeding. The minute species of the family Vertiginidae may be found on the smooth-barked branches or trunks of hardwood trees and shrubs. Some of the snails frequent the droppings of deer or other animals, and small slugs are often seen feeding on the dead bodies of other invertebrates, including gastropods, and sometimes scavenging vertebrate carcasses. Some of our most abundant native forest gastropods, of the family Haplotrematidae, are predators of other invertebrates such as earthworms, other snails and slugs, and even insects. Cannibalism is not uncommon among the gastropods. While most of our native species feed on dead vegetation or animal matter, many will also utilize some living vegetation, at least seasonally.

Water—Gastropods may be found in wet or dry environments, but water is essential for their survival and reproduction. Even those species that are normally found in dry habitats will be most abundant where and when water is present. In arid lands, the best places to find gastropods are at the edges of springs, in riparian habitats, or under talus where moisture can be found. In the Grand Coulee on hot days, about two rocks deep in talus slopes, the surfaces are cool and there may be condensation on some of them, as well as aestivating *Oreohelix junii* sealed to the rocks with dried mucus. In dry open forests, grasslands, or rocky habitats, the best time to search for snails is during rainy periods.

Shells help snails conserve water in their bodies; slugs are generally more dependent on moist cover sites to maintain their body moisture and to survive. When aestivating or hibernating, snails may form a dried mucous membrane, called an epiphragm, over their aperture to help conserve their body moisture. Snails that live among rocks often seal themselves to a rock with mucus, utilizing the added advantage of the rock to form a more solid closure of the aperture.

Other invertebrates also take advantage of the moisture in the body of gastropods. Minute creatures (e.g., mites and/or springtails) can often be seen in the apertures of snails or on the bodies of slugs, especially on hot days. Mites of one family are obligate associates of snails and slugs (B. Roth, personal communication).

Cover—Cover provides security and protection against adverse weather conditions, predation, accidents, and other potentially harmful factors. Cover for gastropods is provided by rocks, large and small woody debris, the bases of shrubs, and other vegetation. Loose bark on the ground shelters *Vespericola*, and *Hemphillia* species may be found on the underside also, as well as inside decaying logs. *Prophysaon andersoni* and *P. foliolatum* may be found on logs under loose slabs of bark or wood, while the smaller *Prophysaon* are more often found in forest floor litter or debris.

Macrosite conditions (the overall site condition: e.g., plant community, talus, riparian habitat) and microhabitat conditions (e.g., specific plants, soil chemistry, moisture) influence the gastropods that inhabit an area. The macrosite may be moist forest with a variety of understory vegetation, dry open forest with grassy understory, scattered shrubland, prairie-steppe, wetland, rocky-grassland, or rock outcroppings, among others. Stage of growth of sites and habitats also apparently influences the mollusk fauna. Mature and old-growth forest appears to be the best habitat for *Hemphillia dromedarius*, while *H. glandulosa* is generally more abundant in closed-canopy younger forests, and *H. malonei* is more common in riparian areas or moist forests, although these are not the only sites in which these species can be found.

The presence of conifer, hardwood, and mixed overstory also influences the mollusk species on sites. For example, *Cryptomastix devia, Monadenia fidelis fidelis, Megomphix hemphilli*, and *Prophysaon dubium* all have a strong affinity for big-leaf maple (*Acer macrophyllum*). *Monadenia fidelis fidelis* can often be found hibernating under mosses growing on root crowns of this maple, or the young of that snail may be found well up the tree trunks under masses of long epiphytic mosses or lichens growing on the trunks. *Cryptomastix devia* are most often found under stands of big-leaf maples, often under sword ferns growing under the crowns of the trees or on the undersides of their logs. *Megomphix hemphilli* are also commonly found under these trees, frequently burrowed into the soft soil under logs on the forest floor; *Prophysaon dubium* use the leaf litter and logs.

The macrosite alone does not necessarily determine the species occurrence. Understory and other microhabitat conditions also influence what fauna occurs on a site. *Vallonia cyclophorella* in Chelan County, Washington, is found among pinegrass under forest stands of scattered trees on gently rolling slopes. In Asotin County, Washington, along the Snake and Grand Ronde Rivers, the same species is found among small rocks on rather steep rocky terrain. In forests along the Columbia River Breaks in Chelan County several undescribed species of *Oreohelix* occur in various site conditions: the Chelan mountainsnail occurs on a rocky ridge and slope with scattered conifers overlooking the southwest side of Lake Chelan, while a close relative (called the Entiat mountainsnail in this volume) is found more extensively in watersheds from the Entiat to the Wenatchee Rivers, in generally dry conifer forests with grass and scattered shrub understories. These snails are generally found in the forest floor litter and duff or under serviceberry shrubs, in small areas of more open canopy where slight depressions trap soil and hold moisture as indicated by the denser and greener ground vegetation. These snails can also be found on mostly open ridgetops in similar habitat.

Within the same areas, other undescribed species (Ranne's and Hoder's mountainsnails) occur only near a ridgetop, apparently separated by the influence of aspect on their habitat conditions. Although their proximities are quite close, the two species do not appear to coexist. While these are anecdotal observations on conditions in which these snails are found, they indicate a need for further study of the varied habitat conditions that influence the occurrence of these and other gastropods.

Reproduction

All of our terrestrial gastropods are hermaphroditic, and although self-fertilization is known to occur within some species, copulation between two individuals is more common.

Most of our terrestrial snails and slugs reproduce oviparously (lay eggs that hatch outside the parent). Terrestrial gastropods need to find places where their eggs can remain cool, moist, and safe throughout the incubation period. Some slugs lay their eggs in the edges of small streams or ponds. Others find cool moist sites under woody debris in closed-canopy forests, under talus, inside rotting logs, or buried in loose soil under logs.

The *Oreohelix* and at least some *Pupilla* species are ovoviviparous (retaining the eggs inside the parent until they develop into juvenile snails). Since these genera generally inhabit drier habitats, carrying the young until they are developed may be their way of solving the problem of finding a suitable incubation site.

Dispersal

Colonization of new habitats or recolonization of recovered habitats following natural or man-made habitat destruction presents particular problems for terrestrial gastropods. Transport of these snails and slugs is generally passive, and in order to survive, they must be deposited within suitable habitat. Without some external means of transportation, the lifelong range of these animals is quite small. Exotic species and others that occur in or near areas inhabited or frequented by humans are often moved about with transplanted vegetation, or with lumber or other commercial shipments. Small snails or slugs can be picked up on clothing when people walk through wet brush, and they are occasionally found stuck to the hair of animals. Shells are found in debris washed up on the shores of rivers or lakes, indicating that these animals can also be transported on floating debris. Other movement of debris (e.g., rolling logs, blowing leaves, etc.) may transport some for short distances. Of course, for successful dispersal, the animals need to be deposited in suitable habitat.

In small areas of suitable habitat, populations can be limited and isolated, as seen in some of our areas of endemism. Some of the *Oreohelix* species are good examples of this. Within a few miles there may be several different varieties (or species), each confined to a specific set of conditions as the plant community, slope, aspect, soil condition, or geology changes. How long does it take for an isolated snail population to evolve into a new species compared to more wide-ranging species?

Distribution and Endemism

Pilsbry (1948), using the work of Binney (1873) and Henderson (1931), delineated North America into ten Terrestrial Mollusk Provinces, three of which are included within the Pacific Northwest (fig. 1). These provinces are roughly defined by specific endemic groups of mollusks, although there is overlap in the ranges of some of the major groups, and others occur much more widely spread. The Oregonian Province encompasses western Washington and Oregon and extends into northwestern California and western British Columbia. The eastern boundary of this province is roughly the Cascade Crest, but some of its mollusk species are also found in the forested areas of the east slope of the Cascades. The Washingtonian Province lies between the Cascade Crest and the Continental Divide, in British Columbia, Washington, Montana, Idaho, and all but the southern (approximately) one-fifth of Oregon east of the Cascades. The Rocky Mountain Province includes south-ernmost Oregon east of the Cascades, southern Idaho, part of southern Montana, most of Wyoming and into South Dakota, western Colorado, Utah, Nevada, and eastern California. Adjacent provinces are the Californian, south of the Oregonian and west of the Rocky Mountain; the Northern, in Alberta and Saskatchewan and extending into northern Montana and eastward; and the Interior, which includes most of Montana east of the Continental Divide and extends through the Great Plains nearly to the eastern seaboard.

Areas of "substantial terrestrial mollusk endemicity" within the Pacific Northwest were delineated by Frest and Johannes (1995). Significant regions by state follow. In Washington: (a) the Eastern Cascades; (b) Northeastern Washington south, includ-ing the Spokane River and extending east through Idaho and into Montana; (c) the Columbia Gorge; and (d) the Blue Mountains, including the Snake River Canyon and extending into Oregon and Idaho. In Oregon: (a) the Columbia Gorge; (b) the lower Deschutes drainage; (c) the Blue Mountains; (d) Hells Canyon, extending into Washington and Idaho; (e) the Umpqua and Rogue River drainages; and (f) the Upper Klamath Lake drainage. In Idaho: (a) Hells Canyon on the Snake River; (b) the adjacent Lower Salmon watershed; (c) the Clearwater drainage; (d) the Coeur d'Alene Basin and Idaho Panhandle to the north into British Columbia and Mon-tana; and (e) southeastern Idaho into adjacent Wyoming and Utah. In Montana: (a) the Purcell Mountains adjacent to the Idaho Panhandle; (b) the Mission Mountains; (c) the Bitterroot and Flathead watersheds; and (d) the Clark Fork drainage.

Environmental Hazards and Mortality

Mortality in snails and slugs occurs from various causes. Some environmental factors include drought, drowning, fire, and mechanical injury. Shells are often found with dried bodies of snails inside, and dried-up slugs are common. In wet areas small slugs may be found drowned in puddles, and some of those dried-up slugs appear to have been desiccated after being killed by drowning. Occasionally a land snail will be found drowned at the bottom of a pool or pond that is too wide or deep for it to escape. Fires, especially controlled burns, kill thousands of snails, sometimes in a single burn. Slash disposal burns may be so devastating to gastropods because the accumulated woody debris provides excellent habitat for

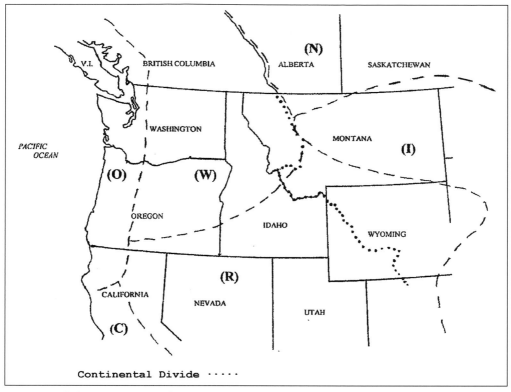

Figure 1. Mollusk Provinces: (C) Californian; (I) Interior; (O) Oregonian; (R) Rocky Mountain; (W) Washingtonian

them. During natural wildfires, it appears that at least some of the snails survive, possibly because they are aestivating underground at the time of the burn, while prescribed broadcast burns are usually done during wet periods when the snails are vulnerable, on or near the surface of the ground. Mechanical injury is caused by vehicles, equipment, tools, or the feet of humans or animals. Chemical poisoning also takes its toll. The use of slug and snail baits is usually limited to areas near human habitation, but the use of other pesticides and fertilizers, the effects of which on gastropods is uncertain, is more widespread. Some other environmental modifications that would be expected to have adverse effects on gastropods and their habitats include acid rain, road salting, forest and range management, agriculture, urbanization, and other development (T. Pearce, personal communication).

Gastropods are prey to a variety of predators, and are also eaten indiscriminately on forage by herbivores, a means by which parasites are transferred from one host to another. Some birds and small mammals are known to feed on snails and slugs. Leonard et al. (1993) and Brown et al. (1995) list known foods of amphibians and reptiles, respectively. Slugs are the primary food of sharptail snakes, and slugs and snails are also eaten by garter snakes and a few others. Pacific giant salamanders and rough-skinned newts eat slugs and other animals, and some other salamanders also include gastropods in their diets. In Chelan County, Washington, introduced wild turkeys were observed to scratch up large patches of forest floor litter in sites known to be inhabited by the yet-to-be-described Entiat mountainsnail.

Some gastropods are cannibalistic if no other food is available, and others are normally predaceous. Members of the family Haplotrematidae attack and feed on other snails and slugs, as well as on earthworms and insects. They will also attack and eat their own species. Some insects are natural predators of gastropods. The larvae of marsh flies (family Sciomyzidae) are predators of aquatic snails (Borror and White 1970; T. Pearce, personal communication). The larvae of lightningbugs (family Lampyridae) and flesh flies (family Sarcophagidae) feed on land snails and slugs as well as on insects (Borror and White 1970; White 1983). Roth (personal communication) has seen the lampyrid larvae feeding on *Monadenia* and said that the "sarcophagid flies account for a lot of mortality in helminthoglyptids in California." *Scaphinotus*, a genus of ground beetle (family Carabidae; tribe Carabini), are specifically adapted for preying on snails. Their long, narrow mouthparts, head, and pronotum fit into the aperture of a shell, which allows them to feed on the snail inside (White 1983). I have seen a dead slug surrounded by these beetles, with their heads toward the center, feeding on the slug, and their large abdomens pointing away from it, thus forming a daisy-petal pattern.

Predation Defenses

As discussed earlier (in the sections Water and Cover) desiccation, environmental conditions, and accidental injury, as well as predation, are hazards faced by gastropods. The primary defense of snails appears to be their shell, which may function as much for water retention as for protection against enemies. The size, shape, rigidity or thickness of the shell, shell structures (e.g., apertural teeth, bristles, reflected and/or constricted peristome, ribs, and carina), and even color may function in snails' defense. Although function of the apertural dentition is not clear since the size and pattern are variable, the denticles appear to further restrict the shell aperture and likely are effective in aiding in the defense of specific species against specific predators. Shells of some snails are adorned with bristles or hair-like structures. I have not read of any specific function of these structures, but they appear to provide some camouflage as they collect and hold bits of fine debris including spiderwebs. Roth (personal communication) believes that the accumulations of debris on these shells "mimic an animal dropping—something unpalatable to predators." The minute pupiform shells of the Vertiginidae appear to be small buds hanging under the branches of shrubs. Ribs and peripheral keels or carina must lend some strength to the shells.

While the snails have their shells for protection, the slugs have had to develop other defenses. The mucus of a slug appears to function partly for defense. A cat that gets an accidental slug in its mouth will quickly spit it out. The mucus of some slugs changes color and consistency when they are irritated. I have seen the predatory snail *Haplotrema vancouverense* drop a *Prophysaon vanattae* after the normally clear mucus of the slug quickly turned white when it was attacked.

Three genera of slugs endemic to the Pacific Northwest have evolved their own unique methods of defense. The jumping-slugs, *Hemphillia* species, rest with their tails curled around to the side, and when attacked or irritated, they writhe and flop around like a fish out of water. This may have a startling effect on the attacker, but it also propels the slug away a short distance, possibly far enough for it to evade its pursuer.

The taildroppers, *Prophysaon* species, are able to drop the ends of their tails as some lizards do. This may not seem very effective for a slug, since the lizard is able to escape while the predator is attracted to its writhing tail. However, it is quite effective against a predatory snail and possibly other small or invertebrate predators. In the instance of *Haplotrema vancouverense* attacking *Prophysaon vanattae* (described above), the snail took the head of the slug into its mouth. The mucus of the slug turned white and the slug was either released by the snail or was able to pull away. The slug then self-amputated its tail, which continued to move in a circle in front of the snail as the slug rather hurriedly retreated. The snail found the tail traveling in a circle in front of it, and ingested that severed organ. Thus the snail had its meal and the slug escaped.

The magnum mantleslug, *Magnipelta mycophaga*, suddenly flares its mantle up when disturbed, similar to quickly popping open an umbrella. I have not found any discussion of anyone observing the results of this action on a specific enemy of the slug, but it appears to be a startling maneuver.

Collecting, Preparing, and Preserving

Collecting terrestrial gastropods is a fascinating pastime, in part because there are so many species and varieties to encounter, and in part because of the many unknowns yet to be researched. Not only are range extensions continually being discovered, but new species are as well.

Random and/or quantifiable surveys can be done, but except for some of the minute snails, many more specimens and more species are likely to be found in a search targeting specific habitats.

Field Equipment—The minimum equipment needed for surveying is a notebook and pencil, a hand lens, containers with labels for specimens, and maps of the area, but certain other items will be of great assistance. Clothing should be appropriate for the weather, and considering that the best time for gastropod surveys is during wet weather, rain gear and waterproof boots are recommended. A small hand cultivator or something to rake through litter and debris is handy, and a headband magnifier or reading glasses help to see the small and minute snails over a wider angle than a hand lens.

Surveys—For simple habitat searches, meandering through an area searching in good microhabitats is the best method. In forested situations, turn over logs when possible, search on the underside of logs and woody debris, and through the forest-floor litter. Lift loose bark and loose slabs of wood from logs where *Prophysaon* species may be found. Pick up a handful of leaves or forest-floor litter, including some damp material if possible, and examine it closely. Search both sides of leaves for minute snails, and look closely at the trunks of smooth-barked hardwood trees and shrubs and on the branches of shrubs. Raise loose hanging mosses and lichens on the trunks of big-leaf maples to look for *Monadenia*. If wet, examine mosses on logs and tree trunks for minute snails and small slugs. Turn over rocks and search through loose rocks and talus. Search through rotten logs or stumps where some *Hemphillia* and small snails hide, and dig through litter and duff under shrubs and sword ferns for some of the larger snails. When done searching each site, replace everything to a state as near natural as possible.

In prairie habitats, look closely at the bases of some grass clumps. Check around the bases of rocks and turn over some of the smaller ones. Check under and on the stems and branches of shrubs, paying special attention to fallen branches, and search around plants with dead fallen leaves. Be sure to check both the upslope and downslope sides of the large ferns and forbs such as *Balsamorhiza*. Look for sites where the vegetation is greener or where forbs are concentrated, indicating more soil moisture.

Riparian areas, springs, and seeps are especially important to land snails and slugs. In forested areas, gastropods will likely be more concentrated in those sites, but in grasslands, they could yield some real surprises. Use the same methods to survey those sites.

Searching through the litter, debris, logs, and rocks disturbs and destroys the habitat. Even replacing the material as it was found does not restore it, because loosened material allows air circulation, which desiccates and raises the temperature of the site. To minimize adverse effects of the search, minimize the area disturbed.

Preparing and Preserving Specimens—Collected shells can be stored in a dry container, but to protect them from damage they should be padded with cotton or tissue paper. Tissue is preferable for shells with hairs because cotton fibers get tangled among the bristles. However, check specimens occasionally for deterioration, as not all materials are acid-free. Storage cabinets should not be made of wood, which gives off an acid that will cause the shells to deteriorate, beginning with the formation of small white dots rising on the surface.

Living snails or slugs may be kept temporarily in collection containers (e.g., film canisters, plastic yogurt cups or other plastic containers, vials, or small jars). The container must be large enough to allow sufficient air for the specimen, so the larger the specimen, the larger the container must be. If the specimens are going to be kept in the collection container for long, it may be more important to maintain the humidity than to provide air holes. If the container is large enough for the animal, and if it is kept cool, opening the container every week will usually allow sufficient air. Moist litter or vegetation from the collection site provides padding during transport or shipping and may provide food for the specimen. Do not use fine material (e.g., crumbling rotten wood) or material with soil on it, especially for small or minute specimens, and do not overfill the container. Soil and other fine particles mixed with the animal's mucus can coat the animal, clog the pneumostome, and smother it. Also, too much material in the container displaces air, and it can cause the examiner to waste a lot of time in the lab searching for tiny specimens. Moss makes good padding for some snails, but it is especially difficult to find tiny specimens among it if too much is used. Back in the laboratory, placing a piece of wet paper towel in the container will maintain the humidity and provide water for the animals. Use distilled or rainwater or at least non-chlorinated water.

Feeding captive specimens may require some experimentation. Exotic species may feed on lettuce, but native animals may not, except sometimes in the spring of the year. Some success has been had feeding raw carrots, and other produce can be tried, but wash it thoroughly first to remove any pesticide residue. Best results for feeding native gastropods is to offer them the organic material on which they were found. For forest species, leaves of big-leaf maples and other native plants are often good, but usually the leaves will need to be dead, moist, and just beginning to decompose. Some species prefer bits of dead bark with cambium exposed. Many will feed on mushrooms. Observe what they are feeding on when collected, and try to offer them the same species, whether leaves, bark, or fungus.

In hot weather it is a good idea to have an ice chest along in which to keep your specimens. However, if blue ice is used, keep something between it and the specimens; if placed in contact with blue ice the specimens may freeze solid. If ice is used, make sure that your containers are sealed or placed in zip-type plastic bags so they don't fill with water and drown the specimens as the ice melts.

To preserve minute snails for their shells (not for dissection), drop them into a vial containing 70% to 80% alcohol (ethyl or isopropyl), which will cause the snail to withdraw into its shell. After a few days or weeks, it should be removed from the alcohol and allowed to dry out.

For larger specimens, drowning is recommended. Place the animal into a small jar filled to the brim with water and close it with as little air inside as possible; this will usually cause the animals to extend, which is needed for snails to protrude from their shells. After about 12 to 24 hours, check whether or not the animal is dead. During hot weather the drowning jar should be moved to a refrigerator after about 12 hours, or in any weather if the animal needs to be left in the water longer than a day, which is not recommended.

Once dead, the snails or slugs should be placed in a 40% solution of alcohol to avoid too-rapid desiccation that may distort the specimens. After about 24 hours they should be moved to 70% to 80% alcohol for storage. For larger snails or slugs, alcohol should also be injected into the body cavity. If the snail is to be removed from the shell, it is better to do it before placing it in the preservative. Specimens or samples that are to be submitted for DNA analysis should be placed in 95% ethyl alcohol as soon as they are dead (B. Roth, personal communication).

To remove a snail from its shell after drowning, place it into a water bath at 60 degrees C (140 degrees F). Allow time for the specimen to attain the temperature of the water, then, using a good pair of forceps, grasp the body of the snail and move it back and forth while pulling firmly but gently. When the mantle breaks loose from the shell, continue the gentle pulling action and, if you are lucky, most of the snail (sometimes all of it) will be extracted. The body can then be preserved. The shell should be cleaned and, if much tissue remains inside it, it should also be placed into alcohol for a few days, and then allowed to dry thoroughly before storing.

How to Use This Book

The first section, Land Snails and Slugs of the Pacific Northwest, is a taxonomical list, with scientific and common names, that includes the species' status (indigenous, peripheral, introduced, etc.) within this study area, as explained in the introduction to the list.

The next section, Keys to the Families and Genera, also has an explanatory introduction. After becoming familiar with the Gastropoda, the user can skip to shortcuts within these keys.For example, one might skip to "Shells Higher than Wide" or "Shells Wider than High" or "Slugs." Genera that have only one or two species within the study area (the Pacific Northwest) will usually have those species keyed with the genus, but the full description and other information for it will still be found in the Species Accounts section. Keys to multiple species and subspecific taxa within a genus are included with the species accounts.

The Species Accounts sections include additional keys to species, subspecies, forms, and varieties. Individual accounts include a description of the species and a discussion of similar species, if any, as well as distribution information and the type of habitat in which the species might be expected. Alternate nomenclature and other comments may also be included, but not for all taxa.

Photographs are included of most species. We did not have specimens of all of the taxa to photograph but intend that, with the available photographs and the descriptions comparing similar species, those taxa not illustrated will be recognizable.

Photographs of shells provided by a specific collection include a code in the caption indicating the collection: Deixis Collection were provided by Dr. Terrence Frest; MNHP are from the Montana Natural Heritage Program collection and were made available by Paul Hendricks; OSAC were from the Oregon State University Arthropod Collection, with access courtesy of Dr. Chris Marshall, Curator. Other specimens were photographed from the collections of the author and of the photographer of this volume, some of which were collected by colleagues noted in the acknowledgments.

Range maps are included for most taxa. In order to include adjacent or peripheral ranges, the maps cover the northwestern and adjacent United States and extend into southwestern Canada. The dotted line on the maps is the Continental Divide, which forms the eastern boundary of the study area in Montana. The one large island shown in Northern Puget Sound, Washington, is actually two: Fidalgo and Whidbey Islands, which are connected to the mainland and to each other by bridges. Many other islands lie in the straits between Vancouver Island and the mainland but are not shown on this map. The San Juan Islands, between the northern Washington coast and Vancouver Island are within our study area, and mollusk surveys are available for federal lands on two of them. Consequently, some of the range maps show occupancy of what appears to be saltwater.

Specific gastropod taxa are represented by patterns drawn around their approximate known or reported ranges or by a specific symbol at the approximate location of the sites at which the species has been reported. An approximation is all that is possible to provide on maps of these scales, so the distribution given within the species accounts for each taxon will sometimes provide more specific locations of the reported sites or ranges.

It should not be thought that a species can be found throughout an area drawn as its range. Some species might be fairly common and be found within most sites of suitable habitat within those areas. Others, however, may rely on specific and often unknown microsite characteristics and be quite rare within their delineated ranges. Some species are currently known from only one or a few very small locations. On the other hand, some species are so widespread across the Pacific Northwest, or North America, that there is no reason to include a range map for them, and most of the exotic species may show up anywhere there is human activity, so range maps are not included for all of them.

Sites shown on these maps are those that are known to me and/or others with whom I have consulted, and records from literature sources. For sources, see the references and acknowledgments sections. Some reports have been considered invalid subsequent to publication because of misidentified species, later splitting or lumping of taxa, or erroneous recording of locations (e.g., Pilsbry [1939, 1940, 1946, 1948] discussed such cases, and other times he simply omitted prior reports that he apparently did not consider valid). Some earlier sites were omitted from this work and these maps for similar reasons.

Those familiar with the environs across the Pacific Northwest will recognize that the lines of occurrence do not necessarily follow the logical areas of habitat in which a species might be expected. A range line that follows a state boundary merely indicates that, to my knowledge, the species has not been reported in the adjacent state, although it might logically be expected to occur there. Much surveying for these animals is still needed, and with all of the ongoing work there are certainly many records for which I do not have access.

Exotic gastropods are continually showing up in new locations, and new sites are constantly being discovered for them and for native species as well. For these reasons, range or site maps for the gastropods will likely never be complete. It should also be recognized that small disjunct populations (e.g., west coast species in northern Idaho or the Blue Mountains of Washington or Oregon) might be introduced by high recreation use in those areas. Many of these populations occur near popular campgrounds.

This work is a compilation of information from many past and recent works and from personal experience, put together for the purpose of providing descriptions and other information to assist in the identification of the land snails and slugs of the Pacific Northwest. For ease of use, literature cited in the text is reduced, and many of the references used in this work are listed following the families or sometimes a group of families. For complete citations, see the references section that follows the glossary. The following sources were used throughout and may not be cited specifically in every case: Bouchet and Rocroi 2005; Burch 1962; Forsyth 2004; Pilsbry 1939, 1940, 1946, 1948; Roth and Sadeghian 2006; and Turgeon et al. 1998. Excerpts from Pilsbry (1939, 1940, 1946, 1948) are reproduced by permission of the Academy of Natural Sciences of Philadelphia.

The following photographs were previously published in *Zootaxa* and are reproduced here with the permission of Magnolia Press: page 301, *Kootenaia burkei*, photograph no. 1 (Leonard et al., 2003); page 302, *Carinacauda stormi*, photograph no. 1; page 304, *Securicauda hermani*, photograph no. 1; and page 307, *Gliabates oregonius* (Leonard et al., 2011).

A glossary of terms is found at the end of the species accounts.

Nomenclature of Shell Characteristics for Use with the Identification Keys

Figure 2.1. Examples of Shell and Aperture Shapes

2.1.A. Low spire; widely lunate aperture, flared basally

2.1.B. Low-moderate spire

2.1.C. Moderate spire; auriculate aperture; reflected lip

2.1.D. Low conic spire; crescentic or narrowly lunate aperture

2.1.E. Very low to nearly flat spire; simple, lunate aperture

2.1.F. Globose shell; simple auriculate aperture

2.1.G. High or domed spire; narrowly lunate with simple peristome

2.1.H. Globose: narrowly flared peristome

2.1.I. Turbinate; obliquely-ovate aperture

2.1.J. Succiniform

2.1.K. Long-ovate or fusiform

Figure 2.2. Pupiform Shells

2.2.A. Ovate

2.2.B. Oblong or cylindrically-ovate

2.2.C. Cylindrical

2.2.D. Elliptically oblong

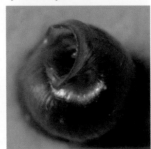

2.2.E. Rimately perforate umbilicus w/crest behind the aperture

Figure 2.3: Shell Characteristics—Whorls

Figure 2.3.A. Whorls expand rapidly

Figure 2.3.B. Whorls expand regularly

Figure 2.3.C. Whorls expand slowly (tightly coiled)

Figure 2.4: Shell Characteristics—Umbilici

2.4.A. Funnelform umbilicate

2.4.B. Umbilicate; last half-whorl expanding rapidly

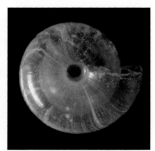

2.4.C. Narrowly umbilicate

2.4.D. Perforate

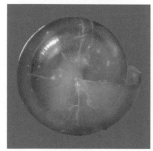

2.4.E. Imperforate

2.4.F. Narrowly umbilicate; umbilicus mostly covered by the peristome

Figure 2.5: Shell Sculpturing

2.5.A1. Ribs (radial or transverse lirae)

2.5. A2. Ribs (radial or transverse lirae)

2.5.B. Ribs and peripheral carina

2.5.C. Spiral lirae and peripheral carina

2.5.D. Low cuticular riblets

2.5.E. High cuticular riblets

2.5.F. Spiral striae and growth-wrinkles

2.5.G. Beaded sculpturing (spiral striae over low ribs)

Land Snails and Slugs of the Pacific Northwest
TAXONOMIC LIST

This is a list of the land snail and slug species of the Pacific Northwest (PNW) region of the United States. Following standard zoological nomenclature, the genera and species are italicised. Many subspecies are included, but subgenera are included only if so doing aids in species identification (i.e., by grouping species that share similar recognizable characteristics). The authority follows the species or subspecies epithet, in parentheses if the name has been modified since the original description. The common name is on the right, followed by a code in brackets indicating the status (indigenous, introduced, etc.) of the species within the Pacific Northwest. A legend of the codes appears in the text box preceding the list, and the terms are further defined below. A few species and most subspecies have not been given a common name, thus that column may be blank. No status code is used for typical subspecies that do not occur within the area of this study (those typical subspecies are presented only to introduce the resident subspecies that follow, e.g., *Columella columella*, where only *C. columella alticola* occurs here).

Taxonomic nomenclature is modified from Turgeon et al. (1998), Bouchet and Rocroi (2005), and Pilsbry (1939, 1940, 1946, 1948), with references to the work of Roth and Sadeghian (2006) and Grimm et al. (2009). Ranges were determined from most of those same publications, as well as others referenced at the ends of the species accounts and in the references section. In keeping with the intent of this work to identify the species by morphological characteristics, taxonomic order was modified to keep the families together with their "informal groups," Aulacopoda and Holopoda (Pilsbry, 1896), which have been used in other keys throughout the twentieth century (superfamilies of Pilsbry, 1946; divisions of Burch, 1962; infraorders by Boss, 1982 [in Bouchet & Rocroi, 2005]; and Burch and Pearce, 1990). All slugs were grouped together and placed at the end of the list.

Some species were retained on the list, appropriately coded {x} or {?}, even though they are not known to occupy the defined study area, because the nearness of their range leads to the possibility of future discoveries of populations within this area. Some undescribed species are included on the list, indicating known taxa for which a formal description has not yet been published. Some of these were included on the list of "Survey and Manage Species" from the Northwest

Forest Plan (USDA Forest Service and USDI Bureau of Land Management 1994). Other undescribed taxa not listed in this work can be found in Roth and Sadeghian (2006) and in unpublished reports by Deixis Consultants (Frest 1999; Frest and Johannes 1995).

The first four species listed are semi-amphibious marine associates, not usually included on lists of land or freshwater snails and often omitted from guides to seashells as well. They are included here because one of them, *Littorina subrotundata*, the Newcomb's littorine snail, has been a candidate for both state and federal listing, and all four species are found in similar habitat. They inhabit the intertidal zone of coastal salt marshes and are usually found in the upper reaches of that zone, which is seldom submerged. Talmadge (1962) said of the Newcomb's littorine snail, "[T]he species is neither a freshwater, nor a true marine gastropod. It was found that the animals could be both smothered and drowned; that it was quite tolerant of both fresh and salt water, yet would climb out of either when immersed."

The following terms and symbols (abbreviated in the legend below) indicate a species' status in the Pacific Northwest: Indigenous {*} indicates the species is native to the study area, defined as Washington, Oregon, Idaho, and the part of Montana west of the Continental Divide; Peripheral {x} indicates that the known range of the species includes area adjacent to the study area, where they may yet be found; Review {?} indicates species for which further research is needed to determine the current status or range (it may also be used when the known range is near to but not immediately adjacent to the defined study area, or if it has been reported from the PNW but is not known to be a current inhabitant). Introduced {I} indicates that the species is not native to the PNW; these are usually native to lands other than North America, but they may also have been introduced from other regions within the United States or North America. A combination of "?" with another code (e.g., {I?} or {*?}) implies some doubt about the validity of the indicated status.

> Legend
>
> {*} Indigenous; {x} Peripheral; {?} Review; {I} Introduced

Land Snails and Slugs of the Pacific Northwest

Phylum: Mollusca
Class: Gastropoda Cuvier, 1795
Subclass: Eugastropoda Shimer & Shrock, 1944
Order: Neotaenioglossa Haller, 1892

Superfamily: LITTORINOIDEA Children, 1834
Family: LITTORINIDAE Children, 1834
Genus: *Littorina* Férussac, 1822
　　　　Littorina subrotundata (Carpenter, 1864)　Newcomb's Littorine Snail　{*}

Superfamily: RISSOOIDEA Gray, 1847
Family: ASSIMINEIDAE H. Adams & A. Adams, 1856
Genus: *Assiminea* Fleming, 1828
　　　　Assiminea californica (Tryon, 1865)　California Assiminea　{*}

Family: POMATIOPSIDAE F. C. Baker, 1926
Genus: *Cecina* A. Adams, 1861
　　　　Cecina manchurica A. Adams, 1861　Manchurian Cecina　{I}

Subclass: Pulmonata Cuvier, 1814
Order: Basommatophora Keferstein, 1865

Family: ELLOBIIDAE L. Pfeiffer, 1854
Genus: *Myosotella* di Monterosato, 1906
　　　　Myosotella myosotis (Draparnaud, 1801)　Saltmarsh Pulmonate　{I}

Family: CARYCHIIDAE Jeffreys, 1829
Genus: *Carychium* Müller, 1774
　　　　Carychium occidentale Pilsbry, 1891　Western Thorn　{*}

Order: Stylommatophora Schmidt, 1855
Suborder: Heterurethra Pilsbry, 1900

Family: SUCCINEIDAE Beck, 1837

Subfamily: CATINELLINAE Odhner, 1950

Genus: *Catinella* Pease, 1871
      Catinella gabbii (Tryon, 1866)   Riblet Ambersnail   {*}
      Catinella rehderi (Pilsbry, 1948)   Chrome Ambersnail   {*}
      Catinella stretchiana (Bland, 1865)   Sierra Ambersnail   {*}
      Catinella vermeta (Say, 1829)   Suboval Ambersnail   {*}

Genus: *Oxyloma* Westerlund, 1885
      Oxyloma decampi (Tryon, 1866)   Marshall Ambersnail   {x}
      Oxyloma hawkinsi (Baird, 1863)   Boundary Ambersnail   {*}
      Oxyloma haydeni (W. G. Binney, 1858)   Niobrara Ambersnail   {*}
      Oxyloma nuttallianum (I. Lea, 1841)   Oblique Ambersnail   {*}
      Oxyloma retusum (I. Lea, 1834)   Blunt Ambersnail   {x}
      Oxyloma sillimani (Bland, 1865)   Humboldt Ambersnail   {*}

Subfamily: SUCCINEINAE Beck, 1837

Genus: *Novisuccinea* Pilsbry, 1948
      Novisuccinea ovalis (Say, 1817)   Oval Ambersnail   {I}
      Novisuccinea strigata Pfeiffer, 1855   Striate Ambersnail   {*}

Genus: *Succinea* Draparnaud, 1801
      Succinea campestris Say, 1817[1]   Crinkled Ambersnail   {I?}
      Succinea oregonensis I. Lea, 1841   Oregon Ambersnail   {*}
      Succinea rusticana Gould, 1846   Rustic Ambersnail   {*}

Suborder: Orthurethra Pilsbry, 1900

Family: CIONELLIDAE Clessin, 1879

Genus: *Cochlicopa* Férussac, 1821
      Cochlicopa lubrica (Müller, 1774)   Glossy Pillar   {*}

Superfamily: PUPILLOIDEA Turton, 1831

Family: PUPILLIDAE Turton, 1831

Genus: *Pupilla* Leach, 1828
      Pupilla hebes (Ancey, 1881)   Crestless Column   {*}
      Pupilla muscorum (Linnaeus, 1758)   Widespread Column   {*}
      Pupilla blandi E. S. Morse, 1865   Rocky Mountain Column   {?}

---

1 *Succinea campestris*, a snail of the southeastern US, is not believed to occur regularly in the Pacific Northwest, but it was once reported on nursery stock in Portland, Oregon (Capizzi, 1962).

Family: VERTIGINIDAE Fitzinger, 1833

Genus: *Columella* Westerlund, 1878

Columella edentula (Draparnaud, 1805)   Toothless Column   {*}
Columella columella (Martens, 1830)   Mellow Column
C. columella alticola (Ingersol, 1875)   Cylindrical Mellow Column   {*}

Genus: *Gastrocopta* Wollaston, 1878

Gastrocopta holzingeri (Sterki, 1889)   Lambda Snaggletooth   {x}

Genus: *Nearctula* Sterki, 1892

Nearctula sp.[2] (=*N. rowelli* of earlier works)   Threaded Nearctula   {*}
Nearctula new sp.[3] Frest, unpublished   Hoko Nearctula   {*}
Nearctula dalliana (Sterki, 1890)   Horseshoe Nearctula   {*}

Genus: *Vertigo* Müller, 1774

Vertigo andrusiana Pilsbry, 1899   Pacific Vertigo   {*}
Vertigo binneyana Sterki, 1890   Cylindrical Vertigo   {*}
Vertigo columbiana Pilsbry & Vanatta, 1900   Columbia Vertigo   {*}
Vertigo concinnula Cockerell, 1897   Mitered Vertigo   {*}
Vertigo elatior Sterki, 1894   Tapered Vertigo   {*}
Vertigo gouldi gouldi (A. Binney, 1843)   Variable Vertigo   {?}
V. gouldi basidens Pilsbry & Vanatta, 1900   {*}
Vertigo idahoensis Pilsbry, 1934   Idaho Vertigo   {*}
Vertigo modesta modesta Say, 1824[4]   Cross Vertigo   {*}
V. modesta corpulenta (Morse, 1865)   {*}
V. modesta parietalis Ancey, 1887   {*}
V. modesta sculptilis Pilsbry, 1934   {*}
Vertigo ovata Say, 1822   Ovate Vertigo   {*}

Family: VALLONIIDAE Morse, 1864

Subfamily: ACANTHINULINAE Steenberg, 1917

Genus: *Planogyra* Morse, 1864

Planogyra clappi (Pilsbry, 1898)   Western Flatwhorl   {*}

Genus: *Zoogenetes* Morse, 1864

Zoogenetes harpa (Say, 1824)   Boreal Top   {x}

---

2 The species previously known as *Vertigo rowellii* Newcomb, 1860, has been determined to be a senior synonym of *Vertigo californica*. This group has recently been included in the genus *Nearctula*, but since *Nearctula rowellii* is now *Nearctula californica*, ours, although an accepted and valid species, is left without a specific epithet and temporarily becomes *Nearctula* sp. (Turgeon et al., 1998). Dr. B. Roth is currently preparing the new description for this species.

3 The *Nearctula* new sp. was discovered by Frest and Johannes and has not yet been formally described. It was included as a "survey and manage" species under the Northwest Forest Plan, USDA Forest Service and USDI Bureau of Land Management (1994).

4 Pilsbry (1948) included several subspecies and forms of *Vertigo modesta* for which the variation and ranges are unclear. The species is quite variable, and although not all that are listed here have been documented from the PNW, the variations within the species found in Washington fit these subspecies.

Subfamily: VALLONIINAE H. Watson, 1920

Genus: *Vallonia* Risso, 1826

    Vallonia albula Sterki, 1893   Indecisive Vallonia   {*}
    Vallonia cyclophorella Sterki, 1892   Silky Vallonia   {*}
    Vallonia excentrica Sterki, 1893   Iroquois Vallonia   {I}
    Vallonia gracilicosta Reinhardt, 1883   Multirib Vallonia   {*}
    Vallonia pulchella (Müller, 1774)   Lovely Vallonia   {I}

Suborder: Sigmurethra Pilsbry, 1900

Superfamily: ACHATINOIDEA Swainson, 1840

Family: SUBULINIDAE P. Fischer & Crosse, 1877

Genus: *Allopeas* H. B. Baker, 1935

    Allopeas mauritianum (Pfeiffer, 1852)   Spike Awlsnail   {I}

Superfamily: RHYTIDOIDEA Pilsbry, 1893

Family: HAPLOTREMATIDAE Baker, 1925

Genus: *Ancotrema* Baker, 1931

    Ancotrema hybridum (Ancey, 1888)   Oregon Lancetooth   {*}
    Ancotrema sportella Gould, 1846   Beaded Lancetooth   {*}
    Ancotrema voyanum Newcomb, 1865[5]   Hooded Lancetooth   {?}

Genus: *Haplotrema* Ancey, 1881

    Haplotrema vancouverense (Lea, 1839)   Robust Lancetooth   {*}

Superfamily: ACAVOIDEA Pilsbry, 1895

Family: MEGOMPHICIDAE H. B. Baker, 1930

Genus: *Megomphix* H. B. Baker, 1930

    Megomphix californicus A. G. Smith, 1960   Natural Bridge Megomphix   {x}
    Megomphix hemphilli (W. G. Binney, 1879)   Oregon Megomphix   {*}
    Megomphix lutarius H. B. Baker, 1932   Umatilla Megomphix   {*}

Genus: *Polygyrella* W. G. Binney, 1863

    Polygyrella polygyrella (Bland & J. G. Cooper, 1861)   Humped Coin   {*}

Superfamily: HELICOIDEA Rafinesque, 1815

Family: HELICIDAE Rafinesque, 1815

Genus: *Helix* Linnaeus, 1758

    Helix aspersa Müller, 1774   Brown Gardensnail   {I}

5 *Ancotrema voyanum* is a species of northern California, and although it has been reported from southern Oregon there is no verification of this. Therefore, for this list its occurrence in the PNW remains questionable.

Genus: *Cepaea* Held, 1837
>*Cepaea nemoralis* (Linnaeus, 1798)   Grovesnail   {I}

Family: HYGROMIIDAE Tryon, 1866
Genus: *Candidula* Kobelt, 1871
>*Candidula intersecta* (Poiret, 1801)   Wrinkled Helicellid   {I}

Genus: *Cernuella* Schlüter, 1838
>*Cernuella virgata* (Müller, 1774)   Maritime Gardensnail   {I}

Family: HELMINTHOGLYPTIDAE Pilsbry, 1939
Genus: *Helminthoglypta* Ancey, 1887
>*Helminthoglypta hertleini* Hanna & Smith, 1937   Oregon Shoulderband   {*}
>*Helminthoglypta mailliardi* Pilsbry, 1927[6]   Del Norte Shoulderband   {*?}

Family: BRADYBAENIDAE Pilsbry, 1939
Genus: *Monadenia* Pilsbry, 1895
>*Monadenia chaceana* S. S. Berry, 1940   Siskiyou Sideband   {*}
>*Monadenia fidelis fidelis* (J. E. Gray, 1834)   Pacific Sideband   {*}
>*M. fidelis celeuthia* Berry, 1937   {*}
>*M. fidelis columbiana* Pilsbry, 1939   {*}
>*M. fidelis flava* Hemphill, 1935   {*}
>*M. fidelis leonina* Berry, 1937   {*}
>*M. fidelis minor* (W. G. Binney, 1885)   {*}
>*M. fidelis semialba* J. Henderson, 1929   {*?}
>*Monadenia* new spp.   (2 species; 1 subspecies)   {*}

Family: POLYGYRIDAE Pilsbry, 1895
Genus: *Allogona* Pilsbry, 1939
Subgenus: *Dysmedoma* Pilsbry, 1939
>*Allogona lombardii* A. G. Smith, 1943   Selway Forestsnail   {*}
>*Allogona ptychophora ptychophora* (A. D. Brown, 1870)   Idaho Forestsnail   {*}
>*A. ptychophora solida* (Vanatta, 1924)   {*}
>*Allogona townsendiana townsendiana* (I. Lea, 1838)   Oregon Forestsnail   {*}
>*A. townsendiana frustrationis* Pilsbry, 1940   {*}

Genus: *Cryptomastix* (Pilsbry, 1939)
>*Cryptomastix devia* (Gould, 1846)   Puget Oregonian   {*}
>*Cryptomastix germana germana* (Gould, 1851)   Pygmy Oregonian   {*}
>*C. germana vancouverinsulae* (Pilsbry & Cook, 1922)   {*}
>*Cryptomastix harfordiana* (W. G. Binney, 1886)   Salmon Oregonian   {*}

---

6 Roth (2003) noted, "Populations in southern Oregon referred to as *Helminthoglypta mailliardi* belong to a new species, being described elsewhere."

Cryptomastix hendersoni (Pilsbry, 1928)   Columbia Oregonian   {*}
Cryptomastix magnidentata (Pilsbry, 1940)   Mission Creek Oregonian   {*}
Cryptomastix mullani mullani (Bland & J. G. Cooper, 1861) Coeur d'Alene
    Oregonian   {*}
C. mullani blandi Hemphill, 1892   {*}
C. mullani clappi Hemphill, 1897   {*}
C. mullani hemphilli W. G. Binney,1886   {*}
C. mullani latilabris Pilsbry, 1940   {*}
C. mullani olneyae (Pilsbry, 1891)   {*}
C. mullani tuckeri Pilsbry & Henderson, 1930   {*}
Cryptomastix populi (Vanatta, 1924)   {*}
Cryptomastix sanburni (W. G. Binney, 1886) Kingston Oregonian   {*}

Genus: *Hochbergellus* **Roth & W. B. Miller, 1992**

Hochbergellus hirsutus Roth & W. B. Miller, 1992   Sisters Hesperian   {*}

Genus: *Trilobopsis* **Pilsbry, 1939**

Trilobopsis loricata (Gould, 1846)
T. loricata nortensis (S. S. Berry, 1933)   Scaly Chaparral   {*}
Trilobopsis tehamana (Pilsbry, 1928)   Tehama Chaparral   {x}

Genus: *Vespericola* **Pilsbry, 1939**

Vespericola columbianus columbianus (Lea, 1838)   Northwest Hesperian   {*}
V. columbianus depressa (Pilsbry & Henderson, 1936)   {*}
V. columbianus latilabrum (Gould, 1846?)   {*}
V. columbianus ssp.[7]   Brushfield Hesperian   {*?}
Vespericola eritrichius Berry, 1939   Velvet Hesperian   {*}
Vespericola euthales Berry, 1939   {*}
Vespericola megasoma (Pilsbry, 1928)   Redwood Hesperian   {*}
Vespericola sierranus (Berry, 1921)   Siskiyou Hesperian   {*}

Family: THYSANOPHORIDAE Pilsbry, 1926

Genus: *Microphysula* **Cockerell & Pilsbry, 1926**

Microphysula cookei (Pilsbry, 1922)   Vancouver Snail   {*}
Microphysula ingersolli (Bland, 1875)   Spruce Snail   {*}

Superfamily: PUNCTOIDEA Morse, 1864 (= PATULOIDEA Tryon, 1866)

Family: OREOHELICIDAE Pilsbry, 1939

Genus: *Oreohelix* **Pilsbry, 1904**

Subgenus: *Oreohelix* **Pilsbry, 1904**

Oreohelix alpina (Elrod, 1901)   Alpine Mountainsnail   {*}
Oreohelix amariradix Pilsbry, 1934   Bitterroot Mountainsnail   {*}
Oreohelix carinifera Pilsbry, 1912   Keeled Mountainsnail   {*}

---

7   This snail was known as *Vespericola columbianus pilosa* (Henderson, 1928) until Roth and Miller (1993) elevated the west-central California populations to full species as *V. pilosus*. In so doing, they pointed out distinct differences between those and specimens from the Columbia River Valley.

Oreohelix elrodi (Pilsbry, 1900)   Carinate Mountainsnail   {*}
Oreohelix eurekensis Henderson & Daniels, 1916   Eureka Mountainsnail   {?}
Oreohelix hammeri Fairbanks, 1984   Seven Devils Mountainsnail   {*}
Oreohelix haydeni haydeni (Gabb, 1869)   Lyrate Mountainsnail   {x}
O. haydeni bruneri (Ancey, 1881)   {*}
O. haydeni corrugata Henderson & Daniels, 1916   {*}
O. haydeni hesperia Pilsbry, 1939   Western Mountainsnail   {*}
O. haydeni hybrida (Hemphill, 1890)   Hybrid Mountainsnail   {*}
O. haydeni oquirrhensis (Hemphill, 1886)   {*}
O. haydeni perplexa Pilsbry, 1939   Enigmatic Mountainsnail   {*}
Oreohelix hemphilli (Newcomb, 1869)   Whitepine Mountainsnail   {*}
Oreohelix howardi Jones, 1944   Mill Creek Mountainsnail   {?}
Oreohelix idahoensis idahoensis (Newcomb, 1866)   Costate Mountainsnail   {*}
O. idahoensis baileyi Bartch, 1916   {*}
Oreohelix intersum (Hemphill, 1890)   Deep Slide Mountainsnail   {*}
Oreohelix jugalis (Hemphill, 1890)   Boulder Pile Mountainsnail   {*}
Oreohelix junii Pilsbry, 1934   Grand Coulee Mountainsnail   {*}
Oreohelix pygmaea Pilsbry, 1913   Pygmy Mountainsnail   {*}
Oreohelix strigosa strigosa Gould, 1846   Rocky Mountainsnail   {*}
O. strigosa berryi Pilsbry, 1915   {x}
O. strigosa buttoni (Hemphill, 1890)   {x}
O. strigosa delicata Pilsbry, 1934   Blue Mountains Mountainsnail   {*}
O. strigosa depressa (Cockerell, 1890)   Depressed Mountainsnail   {*}
O. strigosa fragilis (Hemphill, 1890)   Fragile Mountainsnail   {*}
O. strigosa goniogyra Pilsbry, 1933   Striate Mountainsnail   {*}
Oreohelix subrudis subrudis (Reeve, 1854)   Subalpine Mountainsnail   {*}
O. subrudis apiarium Berry, 1919   {*}
O. subrudis limitaris (Dawson, 1875)   {x}
O. subrudis rugosa (Hemphill, 1890)   {x}
Oreohelix tenuistriata Henderson & Daniels, 1916   Thin-ribbed
   Mountainsnail   {*}
Oreohelix variabilis J. Henderson, 1929   Variable Mountainsnail   {*}
Oreohelix vortex S. S. Berry, 1932   Whorled Mountainsnail   {*}
Oreohelix waltoni Solem, 1975   Lava Rock Mountainsnail   {*}
Oreohelix yavapai Pilsbry, 1905   Yavapai Mountainsnail
O. yavapai extremitatis Pilsbry & Ferriss, 1911   {x}
O. yavapai mariae Bartsch, 1916   {x}

Several new taxa of *Oreohelix* have been found by various researchers in the past several years, many of which have yet to be described. How many actual species occur is controversial. Nine potential species are listed and informally described in this work by proposed common names. Those with numerals following "new sp." are specimens discovered and informally described by Deixis Consultants of Seattle, Washington. In addition to the nine taxa listed below, Deixis Consultants (Frest 1999; Frest and Johannes 1995) have reported at least 32 other undescribed *Oreohelix* from the Pacific Northwest in their unpublished reports.

Undescribed Species of *Oreohelix* **with proposed common names:**

Oreohelix new sp.   Aspen Mountainsnail   {*}

Oreohelix new sp.   Entiat Mountainsnail   {*}

Oreohelix new sp.   Hoder's Mountainsnail   {*}

Oreohelix new sp.   Mad River Mountainsnail   {*}

Oreohelix new sp.   Ranne's Mountainsnail   {*}

Oreohelix new sp. 1   Chelan Mountainsnail   {*}

Oreohelix new sp. 2   Yakima Mountainsnail   {*}

Oreohelix new sp. 20   Sheep Gulch Mountainsnail   {*}

Oreohelix new sp. 30   Pittsburg Landing Mountainsnail   {*}

Family: PUNCTIDAE Morse, 1864 (= PATULIDAE Tryon, 1866)

Genus: *Paralaoma* **Iredale, 1913**

Paralaoma servilis servilis (Shuttleworth, 1852)   Pinhead Spot   {*}

P. servilis alleni (Pilsbry, 1919)   {*}

Genus: *Punctum* **Morse, 1864**

Punctum californicum Pilsbry, 1898   Ribbed Spot   {x}

Punctum minutissimum (I. Lea, 1841)   Small Spot   {*}

Punctum randolphi (Dall, 1895)   Conical Spot   {*}

Family: CHAROPIDAE Hutton, 1884

Genus: *Radiodiscus* **Pilsbry & Ferriss, 1906**

Radiodiscus abietum H. B. Baker, 1930   Fir Pinwheel   {*}

Family: DISCIDAE Thiele, 1931 (= PATULIDAE Tryon, 1866)

Genus: *Anguispira* **Morse, 1864**

Anguispira kochi (Pfeiffer, 1821)   Banded Tigersnail

A. kochi occidentalis (Von Martens, 1882)   {*}

A. kochi eyerdami Clench & Banks, 1939   {*}

Anguispira nimapuna H. B. Baker, 1932   Nimapu Tigersnail   {*}

Genus: *Discus* **Fitzinger, 1833**

Discus brunsoni S. S. Berry, 1955   Lake Disc   {*}

Discus marmorensis H. B. Baker, 1932   Marbled Disc   {*}

Discus rotundatus (Müller, 1774)   Rotund Disc   {I}

Discus shimekii (Pilsbry, 1890)   Striate Disc   {*}

Discus whitneyi (Newcomb, 1864)   Forest Disc   {*}

Family: HELICODISCIDAE H. B. Baker, 1927

Genus: *Helicodiscus* **Morse, 1864**

Helicodiscus salmonaceus Hemphill, 1890   Salmon Coil   {*}

Superfamily: GASTRODONTOIDEA Tryon, 1866

Family: EUCONULIDAE H. B. Baker, 1928

Genus: *Euconulus* Reinhardt, 1883

    Euconulus fulvus fulvus (Müller, 1774)   Brown Hive {*}
    E. fulvus alaskensis (Pilsbry, 1899) {*}

Family: GASTRODONTIDAE Tryon, 1866

Genus: *Striatura* Morse, 1864

    Striatura pugetensis (Dall, 1895)   Northwest Striate {*}

Genus: *Zonitoides* Lehmann, 1862

    Zonitoides arboreus (Say, 1816)   Quick Gloss {*}
    Zonitoides nitidus (Müller, 1774)   Black Gloss {*}

Family: OXYCHILIDAE Hesse, 1927

Subfamily: OXYCHILINAE Hesse, 1927

Genus: *Oxychilus* Fitzinger, 1833

    Oxychilus alliarius (J. S. Miller, 1822)   Garlic Glass-snail {I}
    Oxychilus cellarius (Müller, 1774)   Cellar Glass-snail {I}
    Oxychilus draparnaudi (Beck, 1837)   Dark-bodied Glass-snail {I}
    Oxychilus helveticus (Blum, 1881)   Swiss Glass-snail {I?}

Subfamily: GODWINIINAE Cooke, 1921

Genus: *Nesovitrea* Cooke, 1921

    Nesovitrea binneyana (E. S. Morse, 1864)   Blue Glass
    N. binneyana occidentalis H. B. Baker, 1930 {*}
    Nesovitrea electrina (Gould, 1841)   Amber Glass {*}

Family: PRISTILOMATIDAE Cockerell, 1891

Genus: *Hawaiia* Gude, 1911

    Hawaiia minuscula (A. Binney, 1841)   Minute Gem {*}

Genus: *Pristiloma* Ancey, 1887

Subgenus: *Pristiloma* Ancey, 1887

    Pristiloma arcticum (Lehnert, 1884)   Northern Tightcoil {*}
    Pristiloma crateris Pilsbry, 1946   Crater Lake Tightcoil {*}
    Pristiloma idahoense Pilsbry, 1902   Thinlip Tightcoil {*}
    Pristiloma johnsoni (Dall, 1895)   Broadwhorl Tightcoil {*}
    Pristiloma lansingi (Bland, 1875)   Denticulate Tightcoil {*}
    Pristiloma pilsbryi (Vanatta, 1899)   Crowned Tightcoil {*}
    Pristiloma stearnsi (Bland, 1875)   Striate Tightcoil {*}

Subgenus: *Priscovitrea* Baker, 1931
>   *Pristiloma chersinella* (Dall, 1886)   Blackfoot Tightcoil   {*}
>   *Pristiloma wascoense* (Hemphill, 1911)   Shiny Tightcoil   {*}

Genus: *Ogaridiscus* Chamberlin & Jones, 1929
>   *Ogaridiscus subrupicola* (Dall, 1877)   Southern Tightcoil   {*}

Genus: *Vitrea* Fitzinger, 1833
>   *Vitrea contracta* (Westerlund, 1871)   Contracted Glass-snail   {I}

Superfamily: LIMACOIDEA Lamarck, 1801

Family: VITRINIDAE Fitzinger, 1833

Genus: *Vitrina* Draparnaud, 1891
>   *Vitrina pellucida* (Müller, 1774)   Western Glass-snail   {*}

Family: LIMACIDAE Rafinesque, 1815

Genus: *Lehmannia* Heynemann, 1862
>   *Lehmannia marginatus* Müller, 1774   Tree Slug   {I?}
>   *Lehmannia valentiana* Férussac, 1821   Three-band Gardenslug   {I}

Genus: *Limax* Linnaeus, 1758
>   *Limax flavus* Linnaeus, 1758   Yellow Gardenslug   {I}
>   *Limax maximus* Linnaeus, 1758   Giant Gardenslug   {I}
>   *Limax pseudoflavus* Evans, 1978   Irish Gardenslug   {I}

Family: AGRIOLIMACIDAE Wagner, 1835

Genus: *Deroceras* Rafinesque, 1820

Subgenus: *Deroceras* Rafinesque, 1820
>   *Deroceras hesperium* Pilsbry, 1944   Evening Fieldslug   {*}
>   *Deroceras laeve* (Müller, 1774)   Meadow Fieldslug   {*}
>   *Deroceras monentolophus* Pilsbry, 1944   One-ridge Fieldslug   {*?}

Subgenus: *Malino* Gray, 1855
>   *Deroceras panormitanum* Lessona & Pollonera, 1882   Longneck Fieldslug   {I}

Subgenus: *Agriolimax* Mörch, 1865
>   *Deroceras reticulatum* (Müller, 1774)   Gray Fieldslug   {I}

Family: BOETTGERILLIDAE Wiktor & I. M. Likharev, 1979

Genus: *Boettgerilla* Simroth, 1910
>   *Boettgerilla pallens* Simroth, 1912[8]   Wormslug   {xI}

8 *Boettgerilla pallens* is a European species introduced on Vancouver Island, BC (Reise et al. 2000), thus occurring very near the Pacific Northwestern United States. While natural immigration to the United States from there is impossible, international traffic and commerce provides access from the island to mainland Canada and the United States.

Superfamily: PARMACELLOIDEA P. Fischer, 1856

Family: MILACIDAE Ellis, 1926

Genus: *Milax* Gray, 1855
> *Milax gagates* (Draparnaud, 1801)   Greenhouse Slug   {I}

Superfamily: TESTACELLOIDEA Gray, 1840

Family: TESTACELLIDAE Gray, 1840

Genus: *Testacella* Draparnaud, 1801
> *Testacella haliotidea* Draparnaud, 1801   Earshell Slug   {I}

Superfamily: ARIONOIDEA Gray, 1840

Family: ARIONIDAE Gray, 1840

Genus: *Arion* Férussac, 1819
> *Arion circumscriptus* Johnston, 1828   Brown-banded Arion   {I}
> *Arion distinctus* Mabille, 1868   Darkface Arion   {I}
> *Arion fasciatus* Nilsson, 1823   Orange-banded Arion   {I?}
> *Arion intermedius* (Normand, 1852)   Hedgehog Arion   {I}
> *Arion rufus* (Linnaeus, 1758)   Chocolate Arion   {I}
> *Arion silvaticus* Lomander, 1937   Forest Arion   {I}
> *Arion subfuscus* (Draparnaud, 1805)   Dusky Arion   {I}

Family: ANADENIDAE Pilsbry, 1948

Genus: *Prophysaon* Bland & Binney, 1873
> *Prophysaon andersoni* (J. G. Cooper, 1872)   Reticulate Taildropper {*}
> *Prophysaon coeruleum* Cockerell, 1890   Blue-gray Taildropper   {*}
> *P. coeruleum* ssp.[9]   {*}
> *Prophysaon dubium* Cockerell, 1890   Papillose Taildropper   {*}
> *Prophysaon foliolatum* Gould, 1851   Yellow-bordered Taildropper   {*}
> *Prophysaon humile* Cockerell, 1890   Smoky Taildropper   {*}
> *Prophysaon obscurum* Cockerell, 1890   Mottled Taildropper   {*}
> *Prophysaon vanattae vanattae* Pilsbry, 1948   Scarletback Taildropper   {*}
> *Prophysaon vanattae pardalis*   {*}

Genus: *Kootenaia* Leonard, Chichester, Baugh & Wilke, 2003
> *Kootenaia burkei* Leonard et al., 2003   Pygmy Slug   {*}

Genus: *Carinacauda* Leonard, Chichester, Richart & Young, 2011
> *Carinacauda stormi* Leonard et al., 2011   Cascade Axetail Slug   {*}

Genus: *Securicauda* Leonard, Chichester, Richart & Young, 2011
> *Securicauda hermani* Leonard et al., 2011   Rocky Mountain Axetail Slug   {*}

---

9 Wilke and Davis (2000) found eight genetic clades of *Prophysaon coeruleum* collected in Oregon and Idaho plus two additional clades in California. They are listed here as subspecies because of their similarities, but morphologically they appear to be a composite of at least three species.

Family: ARIOLIMACIDAE Pilsbry & Vanatta, 1898

Genus: *Ariolimax* **Mörch, 1860**

Ariolimax columbianus (Gould, 1851)   Pacific Bananaslug   {*}
Ariolimax sp.[10]   Patos Island Bananaslug   {*}
Ariolimax steindachneri Babor, 1900   (species dubia)   {?}

Genus: *Gliabates* **Webb, 1959**

Gliabates oregonius Webb, 1959   Salamander Slug   {*}

Genus: *Hesperarion* **Simroth, 1891**

Hesperarion mariae Branson, 1991   Tillamook Westernslug   {*}

Genus: *Magnipelta* **Pilsbry, 1953**

Magnipelta mycophaga Pilsbry, 1953   Magnum Mantleslug   {*}

Genus: *Udosarx* **Webb, 1959**

Udosarx lyrata Webb, 1959   Lyre Mantleslug   {*}

Genus: *Zacoleus* **Pilsbry, 1903**

Zacoleus idahoensis Pilsbry, 1903   Sheathed Slug   {*}
Zacoleus leonardi   Ryan Lake Slug   {*}

Family: BINNEYIDAE Cockerell, 1891

Genus: *Hemphillia* **Bland & Binney, 1872**

Hemphillia glandulosa **Group**

Hemphillia burringtoni Pilsbry, 1948   Keeled Jumping-slug   {*}
Hemphillia glandulosa Bland & Binney, 1872   Warty Jumping-slug   {*}
Hemphillia pantherina Branson, 1975   Panther Jumping-slug   {*}

Hemphillia camelus **Group**

Hemphillia camelus Pilsbry & Vanatta, 1897   Pale Jumping-slug   {*}
Hemphillia danielsi Vanatta, 1914   Marbled Jumping-slug   {*}
Hemphillia dromedarius Branson, 1972   Dromedary Jumping-slug   {*}
Hemphillia malonei Pilsbry, 1917   Malone Jumping-slug   {*}

10   The bananaslug, *Ariolimax*, found on Patos Island, is unique in being smaller than *Ariolimax columbianus* from the mainland and other islands of Washington, and in having a monochrome yellow color. *Ariolimax steindachneri* Babor, 1900, was described from "Puget Sound" and has generally been ignored since it has not been rediscovered.

Keys to the Families and Genera

The following is a dichotomous key for the identification of families and genera of the terrestrial gastropods of the Pacific Northwest. Single species in a genus are also included since including a separate key for them under the species accounts would be redundant. However, other redundancies between these keys and the species accounts for other genera have been retained, where repeating the descriptions was advantageous. To use the keys: while examining a specimen, read the first two leads (couplets 1a and 1b) and select the one that best describes the specimen in hand. Then proceed to the indicated couplet following that lead (e.g., [2] or [11]). Continue in this manner until the name of the family, genera, or species is indicated, and then proceed to the species account for that group or species, the page number for which is also given.

These keys are arranged with alternate leads together rather than separated by intervening taxa. Therefore, the indicator at the right must be followed to the next set (or couplet) of leads. The exception occurs when separate sets of genera keys are inserted following the first couplet of a numbered lead. Couplets of keys to families and higher taxa and to monotypic genera and species are numbered, and paired with lowercase "a" or "b" (e.g., 2a and 2b). Couplets of keys to genera or species lead with lowercase letters in parentheses; for example (a) and (aa). Note that numbered indicators to the next couplet use only the first number of the next couplet, enclosed in brackets: "[11]" directs a reader to compare couplet leads 11a and 11b; "[c/cc]" directs the reader to go to couplet leads (c) and (cc).

For some species, size and/or distribution might be the most obvious indicators of the species, so their importance should not be overlooked. However, many specimens will be immature, and range extensions are continually being discovered, so other characteristics also need to be compared.

For size, the relative maturity of the animal is an important factor. These keys were developed primarily for mature specimens, but many or most specimens found are likely to be juvenile, immature, or subadult. As one gains experience in identifying the gastropods, the distinctive forms of the different taxa will become familiar, but it will always be important to know whether the animal being examined is mature, or nearly so, if the size is important to its identification or if the adults have characteristics not seen on immature specimens (e.g., reflected lip or apertural teeth). These are key characteristics as well as indicators of maturity. However, if a peristome (apertural lip) or teeth are not fully developed, their extent of reflection, size, or number could be misleading. The solution to this problem is to be aware of its potential and to examine the other characteristics closely. Try to collect the most mature specimens when possible, and compare specimens with more mature shells or animals from the same site.

Other apparent indicators of maturity include a thickening or slight reflection or flaring around the aperture of species for which a distinctly reflected peristome is not key. A somewhat abrupt deflection (down turning) of the last whorl near the aperture is also an indicator of approaching maturity, as is the obvious shortening of distance between growth-stop lines. The number of whorls is a useful indicator of relative maturity for separating some species when combined with the size of the shell.

Shell characteristics are used in nearly all cases for the snails, but occasionally a lead may refer to characteristics of the animal or, rarely, only to the animal. If the animal isn't available, the succeeding couplets may need to be used to determine the path to follow, keeping in mind the preceding couplets as well.

Keys to the Families and Genera of Land Snails and Slugs of the Pacific Northwest

1a   Shells higher (longer) than wide. One or two pairs of tentacles . . . . [2]

    1b   Shells wider than high or embedded in the mantle so not visible externally. Two pairs of tentacles with the eyespots at the tips of the upper, larger pair. Land snails and slugs. Subclass: Pulmonata (in part); Order: Stylommatophora . . . . [11]

2a   Snails found in intertidal salt marshes along the Pacific Coast . . . . [3]

    2b   Terrestrial snails and slugs, not commonly found in salt marshes (a few species of succiniform snails or small slugs may be semi-amphibious on emergent pond vegetation or around the edges of freshwater wetlands). Subclass: Pulmonata (in part) . . . . [6]

Pacific Coast Salt Marsh Snails

3a   Aperture closed by an operculum when the animal is withdrawn (also includes many freshwater snails not included in this work). Subclass: Eugastropoda; Order: Neotaenioglossa . . . . [4]

    3b   No operculum present. Subclass: Pulmonata (in part); Order: Basommatophora. Shell ellipsoidal with a tightly coiled, conic spire. Aperture long and narrow. Columella encircled by three denticle-like folds. Family: ELLOBIIDAE . . . . *Myosotella myosotis* (=*Phytia myosotis*), p. 70

4a   Shell elongate but little tapered; spire normally truncated and healed as a rounded apex at about three whorls. Aperture small and somewhat teardrop-shaped. Family: POMATIOPSIDAE . . . .*Cecina manchurica*, p. 69

    4b   Shell turbinate, tapering to an acute conic spire (fig. 2.1.I). Aperture teardrop-shaped . . . . [5]

5a   Shell to 8 mm high (usually smaller). Alternating brown and white bands. Normally found on or under *Salicornia* (glasswort or pickleweed) in intertidal salt marshes, associated with *Assiminea* and *Myosotella*. Family: LITTORINIDAE . . . .*Littorina subrotundata* (=*Algamorda subrotundata*), p. 66

    5b   Shell smaller (less than 4 mm high). Color uniformly dark brown. Usually found in salt marshes associated with *Myosotella myosotis*. Family: ASSIMINEIDAE . . . . *Assiminea californica*, p. 68

Land Snails—Shells Higher (Longer) Than Wide

6a   Snails with a single pair of contractile tentacles with eyespots at their bases. Order: Basommatophora. Snails minute (2.25 mm long by 1 mm wide), white, with an elongated shell tapering toward the rounded apex. Aperture lip well-reflected with one or two columellar teeth or lamellae. Family: CARYCHIIDAE . . . . *Carychium occidentale*, p. 72

6b   Snails minute to small (< 2 mm to 20 mm long). Eyespots at tips of the upper pair of retractile tentacles. A second, smaller pair of tentacles usually present on the lower face.
Order: Stylommatophora (in part) . . . . [7]

7a   Shell minute, pupiform, to 4 mm but seldom over 3 mm high. Aperture relatively small and with zero to several apertural teeth. Families: PUPILLIDAE and VERTIGINIDAE . . . . pp. 95, 97

Key to the Genera of Pupiform Snails

PUPILLIDAE: Genus: *Pupilla*

VERTIGINIDAE: Other pupiform genera (see list of terms used to describe these shells under "Pupillid Characteristics and Terminology" on pages 89-91)

(a)   Apertural dentition normally lacking . . . . [b/bb]

    (aa) Apertural dentition normally present . . . .[f/ff]

(b)   Apertural lip thin, not flared or reflected; expanded only at the columellar margin. Shell cylindrical or ovoid, tapering toward the rounded apex, 1.8 to 3 mm high with 5½ to 7 whorls. No crest or sinulus . . . . Genus: *Columella*, p. 97

    (bb) Apertural lip flared or reflected . . . . [c/cc]

(c)   Shell of about 4½ rapidly enlarging whorls, about 2 mm high, and strongly ovately-conic. . . . .*Nearctula dalliana*, p. 102

    (cc) Shell with 5½ to 7 generally cylindrical whorls 2.75 to 4 mm high. Apertural lip reflected following a constriction of the whorl . . . . [d/dd]

(d)   Crest lacking; lip thin and reflected. Shell cylindrical, dull reddish or yellowish-brown, perforate or rimate. Height about 3 mm. Dentition usually absent, but some may have a small parietal cusp. Asotin Co., WA; Salmon River, ID; and south in the Rocky Mountains. . . . . .*Pupilla hebes*, p. 95

    (dd) A prominent crest separated from the reflected apertural lip by a strong constriction of the whorl. Apertural teeth sometimes present . . . .[e/ee]

(e)   Shell ellipsoidally-cylindrical, reddish-brown, rimate. Height 2.75 to 4 mm. A prominent white or light-colored crest separated from the reflected, internally callused lip by a rather wide constriction. Teeth normally lacking in Rocky Mountain varieties but one or two denticles may be present in specimens from farther east. Canada and the northern States west to northeastern Oregon and north and south in the Rocky Mountains.
. . . . *Pupilla muscorum*, p. 96

    (ee) Apertural teeth normally present (seldom lacking) . . . .[f/ff]

(f)   Shell ovoid-cylindrical, light brown, rimate. Height 3.3 mm. Well-developed crest with a deep constriction separating it from the flared or reflected lip, which has a whitish callus encircling the outer rim. Three denticles set well back in the whorl. Rocky Mountains, into the Great Plains.
. . . . *Pupilla blandi*, p. 96

(ff)   Shell, crest, etc. variable, normally with 4 or more denticles in the aperture, but these also vary within a species, occasionally being reduced in size and/or number or lacking altogether . . . .[g/gg]

(g)   Minute, 1.7 mm high by 0.8 mm wide, with about 5 whorls. Shell transparent to whitish, cylindrical. Lip thin and expanded. Apertural teeth 2-1-1-3, parietal and angular, fused to form a reversed lambda (λ) when viewed basally. Columellar and basal teeth large. Eastern Montana eastward and south to New Mexico . . . . *Gastrocopta holzingeri*, p. 98

(gg)   Minute but larger, normally equal to or greater than 1.7 mm high. Shell color light to dark brown . . . . [h/hh]

(h)   Lip flared or reflected. Shell narrowly or widely ovoid, rather strongly tapered above the penultimate whorl to a relatively narrow, rounded apex. Apertural teeth 4 (1-1-2) or 6–7 (1-1-4 or 1-1-5). Without crest or sinulus. Height about 2.4 to 2.8 mm. Perforate or rimately so.
. . . . Genus: *Nearctula,* p. 100

(hh)   Lip not reflected, only slightly if at all flared or expanded. Shell ovoid to cylindrical. Apertural teeth normally 4, 5, or 6— occasionally more or less. Crest, sinulus, and sculpturing variable. Height 1.7 to 2.7 mm in 4½ to 5½ whorls. Perforate or rimate.
. . . . Genus: *Vertigo,* p. 103

7b   Shells other than pupiform and longer (or higher) than 3 mm; without apertural dentition. . . . . [8]

8a   Shells narrowly conic, tapering gently to a rounded apex. Length 7 to 14 mm. Aperture about one-third the shell length or less. Colorless, translucent, and glossy with weak growth-wrinkles. Introduced species seen in greenhouses and on nursery stock.
Family: SUBULINIDAE . . . . *Allopeas mauritianum*, p. 121

8b   Shells either fusiform, succiniform, or somewhat globose, not narrowly conic or straight-sided. Aperture one-third the shell length or greater . . . . [9]

9a   Shell thin, ovoid-conic, somewhat globose; about 3.25 mm high by 2.5 mm wide. Aperture oblique, ovate, simple. Last two whorls with distinct cuticular riblets.
Family: VALLONIIDAE (in part) . . . . *Zoögenetes harpa*, p. 116

9b   Adult shells greater than 5 mm long. . . . . [10]

10a Shell long and narrow, 5.0 to 7.5 mm long with 5½ to 6 whorls, tapering convexly toward the apex and somewhat toward the base. Color glossy brown. Lip not expanded or reflected but thickened within the aperture. Aperture relatively small, about one-third the shell length and less than one-half the shell width.
Family: CIONELLIDAE. . . . . *Cochlicopa lubrica*, p. 88

10b Shells succiniform, with 2½ to 4 whorls, and mostly greater than 7.5
      mm long. Aperture large, greater than one-third the shell
      length and over half of its width. Spire above the body whorl
      conspicuously small. Lip simple. . . . . Family: SUCCINEIDAE

The Genera of SUCCINEIDAE

The family Succineidae is obvious from the shell shape, but there is no sure way of distinguishing the genera or species by using shell characteristics alone. Following Burch (1962), in general, species of *Catinella* are considered to have smaller shells with relatively shorter, rounder apertures. Species of *Succinea* have larger shells with more ovate apertures and swelling around the genital opening, which is located behind the right tentacle. The shells of *Oxyloma* are comparable in size with *Succinea*, or a little larger, but are relatively narrower with long-ovate apertures. The following key characteristics are generally modified from Burch (1962) and Pilsbry (1948), but there is much variation among the western species, and this key should not be wholly relied upon to confirm these genera. Descriptions of genera based on genitalia can be found in Grimm et al. (2009).

Key to the Genera of SUCCINEIDAE

(a) Shell relatively small, 11 mm long or less in 3 to 3½ whorls. Aperture
      roundly ovate, little more than half the shell length.
      . . . . Genus: *Catinella*, pp. 74, 76

   (aa) Aperture distinctly greater than one-half the shell length . . . . . . [b/bb]

(b) Aperture ovate. Shell length mostly 10 mm or greater (*S. oregonensis*
      as small as 6.5 mm) in 2½ to 3½ whorls.
      . . . . Genera: *Succinea* and *Novisuccinea*, pp. 74, 83, 84

   (bb) Aperture relatively long and narrow; shell long-ovate. Shell
          length 7.5 to 20 mm in 2½ to 4 whorls. . . . . Genus: *Oxyloma*, pp. 74, 79

Shells Wider Than High or Enclosed in the Mantle

11a Snails with distinctly coiled shells, wider than high, and into which
      the snail's body can be totally or mostly withdrawn . . . . [12]

   11b Slugs in which the shell is completely or mostly enclosed within
          the mantle or, if exposed, is too small for the animal to withdraw
          into. Shells of *Hemphillia* species can be seen through a slit in the
          back of the mantle, and *Testacella* species carry a small shell on
          their posterior ends . . . . [38]

Snails with Shells Wider Than High

12a Shells minute to small; greater diameter less than 3.8 mm. With
      or without thin cuticular riblets. . . . . [13]

12b Greater diameter usually larger than 3.8 mm (some up to 40 mm or greater). (If 3.5 to 4 mm wide, with 4 whorls or less, and last whorl noticeably wider than adjacent penultimate whorl, see *Nesovitrea* [37b]) .... [18]

13a Spires more elevated, sometimes dome-shaped; light to dark brownish . . [14]

13b Spires quite depressed or low conic. Color variable .... [15]

14a Spire moderately to well elevated. Shell umbilicate; width less than 2.5 mm in about 4 whorls. Color very light to dark or reddish-brown. Cuticular riblets present but sometimes indistinct or microscopic. ....Family: PUNCTIDAE

Key to the Genera of PUNCTIDAE

(a) 1 to 1.9 mm wide. Low, closely-spaced, microscopic riblets, often cuticular and irregular or indistinct. Spiral and/or radial striae may be seen in the interspaces under high magnification..............Genus: *Punctum*, p. 227

(aa) 1.5 to 2.1 mm wide. Rather high cuticular riblets with microscopic radial and spiral striae in the interspaces. Aperture slightly wider than high (compare with *Planogyra clappi*)..... *Paralaoma servilis*, p. 228

14b Spire dome-shaped. Shell perforate or imperforate; diameter about 3 mm. Usually with very closely spaced, microscopic radial striae. Family: EUCONULIDAE .... *Euconulus fulvus*, p. 242

15a Shell about 2 mm diameter with very low or nearly flat spire. Color brownish, with rather high, thin cuticular riblets. Aperture slightly higher than wide, lip simple. Family: VALLONIIDAE (in part) .... *Planogyra clappi*, p. 115

15b Minute (less than 3.8 mm wide); color variable (clear, white, pale greenish, grayish, amber, or brown). With or without riblets, but if riblets are present the shell is white or with a pale greenish tint. Lip simple, thickened, flared or reflected. .... [16]

16a Shells usually less than 3 mm wide. Opaque, white, or with a greenish tint, not glossy. Usually with riblets, but those without riblets with a flared or thickened peristome .... [17]

16b Shells 2 to 3.8 mm wide (rarely to 4 mm but if larger than 3.6 mm, then shell imperforate and with rather distinct sulci). Glossy, transparent or translucent, clear, amber, brownish, grayish, or cloudy-white. Regularly or tightly coiled; imperforate, perforate, or umbilicate. Aperture lip simple, and shell without riblets. .... Family: PRISTILOMATIDAE [36]

17a Shell minute (to 1.8 mm wide), greenish-white, with a low spire and wide umbilicus. Teleoconch sculptured with rather solid, beaded riblets; protoconch with spiral striae. Aperture simple, without dentition. Family: GASTROCOPTIDAE (in part) .... *Striatura pugetensis*, p. 244

17b   Shells white, rather solid, opaque or somewhat translucent.
Diameter 2.5 to 3.0 mm. Aperture lip reflected or thickened
to appear expanded. Cuticular ribs present or not.
Family: VALLONIIDAE (in part) . . . . Genus: *Vallonia*, pp. 114, 117

18a   Shell small to large (greater than 7 mm wide). Species typically with
brown, reddish-brown and/or yellow spiral bands (lacking on some
individuals). Aperture lip usually not reflected, but if reflected only
narrowly, mostly basally, and only on mature specimens . . . . [19]

18b   Shell diameters less than 30 mm. Colored spiral bands lacking.
Aperture lip may or may not be well reflected . . . . [24]

19a   Imperforate as adults, or if greater than 40 mm wide then may be
rimately umbilicate; or immature *Cepaea* may be perforate. Shells more or
less globose, moderately large to quite large. Usually with colored spiral
bands (sometimes lacking). Aperture lip narrowly or not at all reflected in
fully adult specimens. Introduced garden pests in North America.
. . . . Family: HELICIDAE

Key to the Genera of HELICIDAE

(a)   Shells large (attaining a width of 30 to 50 mm), globose, thin and fragile.
Columellar lip margin folds over the umbilicus, making it imperforate to
rimately umbilicate. Colors yellow and/or white with brown- or chestnut-
colored spiral bands of varying widths, and some with radial bands.
Aperture lip narrowly or not at all reflected . . . . Genus: *Helix*, p. 132

(aa)   Shells medium to moderately large (18 to 25 mm or larger), somewhat
globose with an obtuse spire, solid, imperforate to rimately perforate.
Usually yellow with varying numbers of brown spiral bands, but
sometimes all yellow, brown, or pink . . . . *Cepaea nemoralis*, p. 134

19b   Umbilicate. Shells small to large, heliciform to more or less globose.
Normally with colored bands, but the bands may be lacking in some
individuals . . . . [20]

20a   Small to medium (7 to 25 mm wide). Spire moderately high, convex to
rather globose and narrowly umbilicate. Aperture encircled inside by a
thickened ridge. Shell whitish to light yellowish-brown with darker bands
dorsally and basally. Some with fine, closely spaced ribs; some with
irregular, fine, sharp growth lines. Introduced.
. . . . Family: HYGROMIIDAE

Key to the Genera of HYGROMIIDAE

(a)   Small depressed-globose (7 to 13 mm wide by 5 to 8 high), narrowly
and generally symmetrically umbilicate. The periphery obtusely angled
to within a half whorl of the aperture. Sculpturing of fine, closely-spaced
ribs, weaker basally. With multiple darker bands, usually wider dorsally
and narrower basally . . . . *Candidula intersecta*, p. 135

(aa) Small to medium (8 to 25 mm wide by 6 to 19 mm high), narrowly
umbilicate, widening a little more rapidly near the aperture. The
periphery generally rounded. Sculpturing of irregular, fine, sharp
growth lines. With multiple darker bands, usually wider dorsally
and narrower basally. *Cernuella virgata*, p. 135

20b Shell more heliciform than globose, although often with quite high
spire. Without a thickened ridge inside the aperture . . . . [21]

21a Shell 17 to 22 mm wide. Spire rather high, convexly-conic. Periphery
rounded at all ages. Juvenile shells thinner, honey-yellow, with narrow
reddish-brown bands. Shell becoming thicker with age, with higher
spire, darker color, and wider bands. Dorsal and basal surfaces of
subadults slightly lighter than the peripheral bands, with the yellowish
area between the bands retained. Older shells more calcareous, browner,
with markings fainter, the apex often much eroded.
Family: DISCIDAE (in part) . . . . *Anguispira kochi*, p. 232

21b Shells small to large. Heliciform, low to high, conic to convexly-conic
spire, or with a smaller conic spire and a quite capacious body whorl.
Umbilicus very narrow to moderately wide. Bands variable. . . . . . . [22]

22a Shell heliciform with very low to moderately well-elevated spire; mostly
solid and calcareous. Immature shells usually with strongly angled periphery.
Usually light brown or white with two darker bands (lacking in some
species, or with additional, narrower, supernumerary bands also present)
and sometimes flammules or other mottling. Some species with ribs and/or
peripheral carinae. Several species from the eastern Cascades, eastward and
south through the Rocky Mountain states (compare with *Anguispira kochi*
[21]). Family: OREOHELICIDAE. . . . . Genus: *Oreohelix* (in part), p. 183

22b Shells relatively large (18 to 40 mm wide). Heliciform, low to high,
conic to convexly-conic spire, or with a relatively small conic spire
surrounded by a capacious body whorl. . . . . [23]

23a Snails of medium to large size (18 to 40 mm wide) with moderately low to
fairly high conic spire and narrow, open umbilicus. Most are multicolored
(shades of brown and yellow, sometimes white, rarely greenish) with a dark
supraperipheral band, bordered above and below by light yellow or white
bands. Aperture lip narrowly reflected, at least basally, in adults. Periphery of
immature shells shouldered or angled; mostly rounded in adults. Alaska to
southern California, Cascade Mountains, and westward.
Family: BRADYBAENIDAE . . . . Genus: *Monadenia*, p. 139

23b Snails of medium to fairly large size (18 to 33 mm wide), the narrow
umbilicus about half covered by the reflected columellar lip margin.
Narrow dark band above the periphery (hence the common name,
shoulderband). Lip margin simple to narrowly reflected. Periphery
rounded in immature as well as adult shells. Shell sculpturing variable.
Shells generally thinner with a more inflated body whorl than those
of *Monadenia*. A large genus with most species occurring on the Pacific
slope south of Oregon. Southwestern Oregon to southern California.
Family: HELMINTHOGLYPTIDAE . . . . . . Genus: *Helminthoglypta*, p. 137

24a Aperture lip simple, not distinctly reflected. May or may not have apertural dentition . . . .[25]

24b Shells small to large. Apertural lip of adults distinctly reflected. Apertural teeth vary in size and number from zero to three. Periostracum various shades of yellow, tan, or brown, without markings. Shell often with bristles (setae) in most immature specimens. Setae retained by adults of *Cryptomastix germana* and most *Vespericola* species, but rarely, or only as hair-scars or scales in others, if at all.
. . . . .Family: POLYGYRIDAE, p. 150

Key to the Genera of POLYGYRIDAE

(a) Small to medium-sized shell (6 to 18 mm wide), thin and fragile, usually with fairly dense or scattered periostracal hairs (or scale-like sculpturing), rarely lacking. Very narrowly umbilicate to imperforate (or if umbilicus wider, then the shell discoidal and no more than 7 mm wide). Apertural dentition present or not . . . . [b/bb]

(aa) Shells 10 to 35 mm wide. If within the same size range as above, shells are more solid and often with a more depressed spire. Umbilicus relatively larger but sometimes partly or wholly covered by the reflected columellar lip. Dentition variable. Bristles lacking in most as adults . . . .[f/ff]

(b) Shell 6 to 9 mm wide; with apertural dentition . . . . [c/cc]

(bb) Shells larger, 10 to 18 mm wide (*Vespericola sierranus* of northern California as small as 8.4 mm). Usually with periostracal hairs of varying size and density (sometimes lacking). Aperture broadly auriculate. Last whorl constricted before the reflected peristome. With or without apertural dentition. Body whorl relatively capacious; spire conic, low to moderately elevated. Narrowly umbilicate to perforate; umbilicus often partly or wholly covered by the reflected lip margin.
. . . .[e/ee]

(c) Shell depressed-globose, 6 to 8.5 mm wide, normally with scattered or fairly dense, generally long, curved bristles. Last whorl constricted before the reflected peristome. Perforate or imperforate, with a moderate to fairly large parietal tooth. Pacific Coast to the western Cascades.
. . . . *Cryptomastix germana*, p. 158

(cc) Rather low-convex or nearly flat spire, lacking bristles. Aperture with three denticles. Southwest Oregon and northern California. . . . . [d/dd]

(d) Spire low-convexly-conic; shell 6 to 7 mm wide, perforate to narrowly umbilicate. Last whorl constricted behind the thin, narrowly reflected peristome. Small- to moderate-sized parietal, basal, and outer lip teeth. Shell sculptured with collabrally oriented, short, scale-like, raised lines or crescents of about even size and spacing, and fine spiral striae.
. . . . *Trilobopsis loricata*, p. 170

(dd) Shell discoidal, spire nearly flat, about 7 mm wide. Umbilicus rather widely funnelform. Three well-developed apertural teeth. Sculpture of closely-spaced growth-wrinkles and weak spiral striae.
. . . . *Trilobopsis tehamana*, p. 171

(e) Shells tan to brown, measuring (8) 10 to 18 mm wide by 8.5 to 13 mm high with 5½ to 6 whorls. Peristome reflected. Narrow umbilicus often partially covered by the reflected basal lip. Usually sculptured with periostracal hairs of varying size and density (sometimes lacking). A parietal tooth prominent in some, lacking in others . . . . .Genus: *Vespericola*, p. 172

(ee) Shell tan, measuring 13.7 to 17.2 mm wide by 9 to 12.5 mm high in about 5½ to 6¾ whorls. Narrowly umbilicate; partially covered by the reflected basal lip. Peristome white to pinkish-tan, reflected and recurved, somewhat concave. Small, white parietal tooth usually present, and sometimes a thickened basal ridge with a slight cusp at its inner end. Two to three periostracal setae per square millimeter. Curry Co., Oregon . . . . *Hochbergellus hirsutus*,[1] p. 169

(f) Shells variable by species (10 to 25 mm wide). Apertural teeth zero to three, but usually at least a parietal (*C. hendersoni* and *C. populi* typically lack dentition). Bristles often present on the shells of young snails, but retained in adulthood by few species (*C. mullani tuckeri* has flattened, scale-like hairs). Shells usually umbilicate, but the umbilicus may be covered partially or wholly by the reflected lip margin. . . . . Genus: *Cryptomastix*, p. 156

(ff) Shell normally larger (18 to 35 mm in 5 to 6½ whorls). Umbilicate and with neither hairs nor distinct apertural dentition other than sometimes a low callus in the inner basal lip margin . . . . . . Genus: *Allogona*, p. 150

25a Shell heliciform with very low to moderately well-elevated spire; mostly solid and calcareous. Shell ribbed, spirally lirate (with raised threads or ridges), or with very coarse growth-wrinkles. Width 8 to 23 mm. Usually light brown, white, gray, or pinkish (some species with two darker bands [22a]). Several species from the eastern Cascades, eastward and south through the Rocky Mountain states.
Family: OREOHELICIDAE . . . . GENUS: *Oreohelix* (in part), p. 183

25b Shell heliciform with low rounded or nearly flat spire. May or may not be ribbed . . . . [26]

26a Shell medium to large (width 8 to 30 mm). Spire nearly flat to low rounded or roundly elevated, sometimes with a flat or impressed apex. If 11 mm wide or less, then with solid ribs and a large parietal tooth. Lip usually simple but a little thickened in one species. . . . . [27]

26b Shell rather small, width 12 mm or less, with or without solid ribs, cuticular riblets, or spiral lirae. Lip simple. [31]

---

1 Dissection of the genitalia is necessary to confirm this genus, there being no shell characteristics with which to separate it from all species of *Vespericola* in the PNW. For the complete description see Roth and Miller (1992).

27a Shell with solid radial ribs . . . . [28]

  27b Shell without ribs. . . . . [29]

28a Shell 11 to 15 mm wide, low-domed with rather wide umbilicus, angular periphery, and regularly-spaced, medium-fine, sharply angled radial ribs. Idaho Co., Idaho; Wallowa Co., Oregon; and Chelan Co., Washington. Family: DISCIDAE (in part) . . . . *Anguispira nimapuna*, p. 234

  28b Shells 9 to 13 mm wide. Shells yellowish-brown or darker. Spire low-rounded, but first few whorls often flattened or even sunken a little to the apex. Whorls increase slowly in width. Shell smooth basally; dorsal surface ribbed with short crescents distinct to each whorl. Umbilicus large, symmetrical; aperture small, lunate with a relatively large parietal tooth; lip thickened within. Family: MEGOMPHICIDAE (in part). . . . . . *Polygyrella polygyrella*, p. 131

29a Shell 11 to 17 mm wide, narrowly umbilicate, and with a very low convex spire. Whorls somewhat lenticular but with rounded periphery. Light brown dorsally, glossy, with a waxy white area basally around the umbilicus. Aperture widely oval and deeply lunate. Growth wrinkling is only sculpture. Family: OXYCHILIDAE . . . .*Oxychilus draparnaudi*, p. 249

  29b Spire low, conic to nearly flat; umbilicus rather wide, one-quarter to one-third the shell width. . . . . [30]

30a. Medium-sized, shell 8.5 to 20 mm wide. Umbilicus contained 3½ to 4½ times in the shell width. Ivory to amber-white; fresh shells somewhat translucent with slight greenish-yellow tint. Spire very low to nearly flat. Whorls increase more regularly than those of the haplotremes (the difference between the body and penultimate whorls not quite as extreme). Whorls somewhat flattened basally. Shell sculpturing lacking except for smooth growth wrinkling or, in the smaller species, fine, sharp growth lines and faint spiral striae. Family: MEGOMPHICIDAE (in part). . . . . Genus: *Megomphix*, p. 128

  30b Shells medium to large (11 to 30 mm wide), solid, glossy, light greenish, yellowish-brown or brown. Body whorl increases rapidly, is much wider than the penultimate whorl, and is rounded basally. Spire quite depressed (low-conic to nearly flat). Apertural lip not widely reflected, but when fully mature, it may be thickened around the rim and sometimes narrowly reflected, especially basally. Sculpture of either microscopic spiral striae, or low ribs extending around the whorls and cut into beaded sculpture by coarser spiral striae. . . . . Family: HAPLOTREMATIDAE, p. 122

Key to the Genera of HAPLOTREMATIDAE

(a Shell larger, to 29 mm wide in 5 to 5½ whorls (specimens from the eastern part of its range may be smaller). Color light greenish, greenish-brown, or yellowish-brown. Upper (palatal) lip margin flattened or slightly deflected. Sculpturing of growth-wrinkles and very fine, microscopic spiral striae. . . . . *Haplotrema vancouverense*, p. 126

(aa) Shell smaller (11 to 22 mm wide in about 5 whorls). Pale yellowish-green color, dull to glossy. Sculpturing of closely spaced, low transverse ribs, crossed by spiral striae, which give the shell a beaded appearance (sculpturing weak or confined to the first few whorls or the umbilicus in *A. hybridum*). Apertural lip simple, or thickened a little at the edge or very narrowly reflected basally. Palatal lip strongly deflected in adults. . . . . Genus: *Ancotrema*, p. 123

31a Shells small (width 11mm or less) opaque or slightly translucent. Surface dull and sculptured with ribs or spiral lirae. Rarely with colored markings other than bands. Spire flat to moderately elevated . . . . [32]

31b Shell width 12 mm or less. Shell usually translucent or transparent. Surface glossy, without flammules or other markings. Without ribs, but often with microscopic striae, and sometimes with a scalloped corona dorsally. Apertural lip simple. Dentition lacking except in *Pristiloma lansingi*, which has a serrated rib inside the aperture . . . . [34]

32a Diameter 4 to 5 mm, with about 5½ tightly coiled whorls. Discoidal, with a broad, shallow, dish-shaped umbilicus. Raised spiral lines (lirae) run parallel with the whorls. Paired denticles spaced at intervals inside the outer wall of the body whorl. Family: HELICODISCIDAE . . . . *Helicodiscus salmonaceus*, p. 239

32b Major diameter 5 to 11 mm. Shell opaque or slightly translucent, brownish, with solid radial ribs . . . . [33]

33a Shell opaque brown, 5 to 11 mm wide, with low to moderately elevated spire and open umbilicus (about one-third the shell width). First 1½ whorls smooth, followed by regularly spaced solid ribs. Family: DISCIDAE (in part). . . . . Genus: *Discus*, p. 235

33b Shell about 6.5 mm wide; spire very low to nearly flat; umbilicus narrow (about one-sixth the shell width). First two whorls with spiral threads running parallel with the whorls. Following whorls with dense spiral striae visible between distinct ribs. Color a very light olive-buff to brownish. Family: CHAROPIDAE. . . . . *Radiodiscus abietum*, p. 230

34a Shell small (4.5 to 5.5 mm wide by 2.0 to 2.5 mm high in about 5 whorls), thin, glossy, transparent-white. Spire very low-conic to nearly flat, with closely coiled whorls. Aperture simple, narrowly lunate. Umbilicus about one-fourth the shell diameter. Family: THYSANOPHORIDAE . . . . Genus: *Microphysula*, p. 181

34b Shell may be varied combinations of transparent white, nearly flat spire, narrowly lunate aperture, or umbilicus about one-fourth the shell width, but not all characteristics at once. . . . . [35]

35a Shell with about 3 whorls, major diameter 4 to 8 mm. Whorls expanding rapidly. Aperture thin, simple, much larger than the rest of the shell. Shell thin, fragile, transparent, with green tint. Family: VITRINIDAE . . . . *Vitrina pellucida*, p. 268

35b   Shell more regularly whorled; aperture not exceptionally large.
Adults with 3½ or more whorls. Shells normally smooth, most glossy,
transparent or translucent, and uniformly colored; some
with microscopic striae and/or radial grooves . . . . [36]

36a Small snail. Shell 3.5 to 12 mm wide, glossy, with quite low spire.
Umbilicus narrow to moderately wide. . . . . [37]

36b   Minute snails. Shells 2 to 3.8 mm wide. Spire low to moderately high.
Umbilicate, perforate or imperforate.
. . . . Family: PRISTILOMATIDAE, p. 253

Key to the Genera of PRISTILOMATIDAE

Mostly minute snails, shell 2 to 3.8 mm wide. Spire very low to moderately elevated or rarely rather high. Umbilicate, perforate or imperforate.

(a) Umbilicus broad, about one-third the shell diameter, which is 2 to 2.5 mm
in about 4 whorls. Gray to white; sculpture merely of growth-wrinkles.
Aperture quite round or only slightly oblique. . . . . *Hawaiia minuscula*, p. 253

(aa)   Shell imperforate to rimate or narrowly umbilicate. Mostly minute,
but some to 4 mm wide. Shell thin and fragile, generally with very
low conic spire. Regularly spaced, transverse, indented lines (sulci)
are key characteristics on two species, but fainter versions may
appear on dorsal shell surfaces of some other species as well and
can be indicators of this group. . . . . [b/bb]

(b)   Shells imperforate . . . . Subgenus: *Pristiloma* [c/cc]

(bb) Shells narrowly umbilicate, perforate or rimate . . . . [d/dd]

(c)   Shell imperforate, tightly coiled, the last whorl little wider than the
penultimate. Spire conic, low to moderately elevated. Aperture narrowly
lunate; lip simple. Six species ranging from 2 to 3.8 (4) mm wide with
4¾ to 7 whorls (imperforate tightcoils).
. . . . .Genus: *Pristiloma (Pristiloma)*, p. 254

(cc)   Spire low (nearly flat); shell waxy white. Width 2 to 2.5 mm in 3½ to 4
whorls, increasing rather rapidly. Aperture broadly lunate. Imperforate
but may have a fairly deep umbilical depression. Western Washington,
Oregon, and southwestern British Columbia.
. . . . *Pristiloma (Pristiloma) johnsoni*, p. 261

(d)   Shell perforate to very narrowly umbilicate, tightly coiled. Spire low to
moderately elevated, convexly-conic. Diameter 2 to 3.6 mm in 4½ to 5¼
whorls. Aperture narrowly lunate. Two species. Washington, Oregon,
northern California, and Western Montana (perforate tightcoils).
. . . . Genus: *Pristiloma (Priscovitrea)*, p. 254, 262

(dd) Shell width 2 to 3.5 mm (seldom larger). Height/diameter ratio less
than or about equal to 0.5. Spire very low conic . . . .[e/ee]

(e) Shell width 2 to 2.5 mm in about 4 to 4½ whorls. Height less than half the diameter. Umbilicus about one-sixth the diameter. Color clear to cloudy white . . . . *Vitrea contracta*, p. 265

  (ee) Shell width 2.7 to 3.5 mm in about 4 to 4½ whorls. Height about half the diameter. Shell transparent, clear, very narrowly perforate or rimately imperforate by an expansion of the columellar lip margin. Spire strongly depressed, low conic. Umatilla County, Oregon, to Great Salt Lake, Utah, and east-central California.

  . . . . *Ogaridiscus subrupicola*, p. 264

37a Whorls narrower, increasing in size regularly, last not greatly wider than penultimate. Diameter 5 to 7 (8) mm in 4½ to 4¾ whorls. Umbilicus narrow to moderate (about one-fifth the shell width or greater), regularly expanding to the aperture. Aperture roundly to ovately lunate. Family: GASTRODONTIDAE (in part) . . . . Genus: *Zonitoides*, p. 245

  37b Shells 3.5 to 12 mm wide. Whorls expand somewhat faster than those of *Zonitoides*. Either the body whorl conspicuously wider than the adjacent penultimate whorl, or often waxy-white around the umbilicus . . . . Family: OXYCHILIDAE

Key to the Genera of OXYCHILIDAE

(a) Width of adults 5 to 12 mm. Whorls somewhat wider, size increasing moderately and regularly, somewhat compressed to an oval cross-section or somewhat flattened basally. Shells wider with very low to low conic spire. Transparent or translucent, yellowish-brown or greenish-brown, often with a waxy-white sheen basally or around the umbilicus. Umbilicus small to medium (≤ one-sixth the diameter), narrowly funnelform, but expanding more rapidly, but not abruptly so, nearer the aperture. Introduced in North America. . . . . Genus: *Oxychilus*, p. 247

  (aa) Whorls increasing in size regularly to last, which is conspicuously wider than the adjacent penultimate whorl. Diameter 3.5 to 5.5 mm (H/D= 0.55). Microscopic radial grooves on dorsal shell surface. Umbilicus narrow, expanding to about one-fifth the shell width in the last three-quarter whorl . . . . Genus: *Nesovitrea*, p. 250

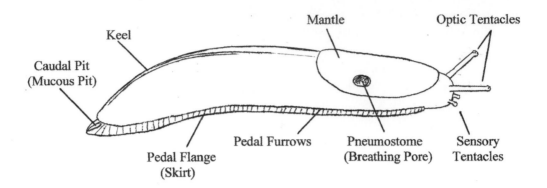

Figure 3: Key characteristics of the slug

Slugs and Slug-Like Gastropods

38a Shell visible, either a shell plate seen through a slit in back of mantle or a small shell at tip of tail . . . . [39]

 38b Shell not visible, completely concealed within the mantle whether solid shell or merely vestigial. . . . . [40]

39a Slug-like animals with a distinct hump, containing the viscera, covered by the mantle. Shell plate visible through a slit in back of mantle. Caudal mucous pit present at convergence of pedal furrows near tip of tail. Superfamily: ARIONOIDEA (in part).
Family: BINNEYIDAE. . . . . Genus: *Hemphillia* [42b]

 39b A burrowing slug up to about 60 mm long, with a small vestigial shell at the end of its tail. Introduced and fairly widespread but seldom seen because of its fossorial habits.
Family: TESTACELLIDAE. . . . . *Testacella haliotidea*, p. 280

40a Fairly large slugs (50 to 70 mm long). Back keeled from posterior edge of mantle to tip of tail. An impressed line or groove delineates a central area of the posterior half of the mantle. Introduced, widespread in California; occasionally found in western Washington and Oregon.
Family: MILACIDAE . . . .*Milax gagates*, p. 279

 40b. Mantle without an impressed line or groove delineating a central portion. Small to large slugs. Keeled or not. [41]

41a A small, slim, wormlike slug (30 to 40 mm long). Mantle with pneumostome located in its right side about midway back or only slightly posterior to its midpoint; without a secondary notch or pore at the posterior right edge of the mantle. Tail with keel extending from back of mantle, but sharp only at posterior end. Pale gray or with yellowish tint. Head bluish-gray. No bands or other markings. Sole tripartite. Superfamily: LIMACOIDEA (in part).
Family: BOETTGERILLIDAE. . . . .*Boettgerilla pallens*, p. 269

41b Small to large slugs. Characteristics variable but not in same combination as for *Boettgerilla*. If colored as *Boettgerilla* and keeled from mantle to tip of tail, then pneumostome placed three-quarters of the way back in the mantle and with a secondary notch in the posterior right edge of the mantle with a pore underneath . . . . [42]

42a Band between pedal furrows and angle of foot relatively narrow and of about even width. Pneumostome located in posterior half of right side of mantle. No caudal pit. Back keeled for about two-thirds of the posterior end of the tail, or sometimes only near tip (not extending forward to the back of the mantle). Sole tripartite (divided into three longitudinal sections). Mantle with concentric wrinkles or folds resembling a fingerprint.
. . . . Superfamily: LIMACOIDEA

Key to the Families and Genera of LIMACOIDEA

(Also containing the snail family Vitrinidae [Bouchet and Rocroi, 2005])

(a) Medium to large slugs, 50 to 120 mm long when extended. Pneumostome approximately two-thirds to three-fourths back in the right side of the mantle. Animal marked with spots or stripes (sometimes faint), or granular-like tubercles of varied colors. Tail profile tapers down to the end, without a truncated appearance. Keel usually extending forward from the posterior end one-third to one-half the length of the tail.
Family: LIMACIDAE. . . . . Genus: *Limax* and Genus: *Lehmannia*, p. 270

  (aa) Small to medium-sized slugs, 15 to 50 (rarely 60) mm long when extended. Pneumostome usually well posterior (approximately three-quarters back) in the right side of the mantle. Generally of one color, the hue or intensity of which may vary; some with irregular speckling or mottling but without distinct patterns of markings. The tail keeled only near the end, often abruptly ending so as to appear truncated in lateral view. Family: AGRIOLIMACIDAE . . . . . Genus: *Deroceras*, p. 274

  42b Band between pedal furrows and angle of foot broader, often forming a flange-like margin widening and most conspicuous in the tail region. Pneumostome may be in anterior or posterior half of the mantle. Size and other characteristics variable (10 to 260 mm long), but see genus keys for distinctive combinations differing from those of the Limacoidea. Superfamily: ARIONOIDEA . . . . p. 281

Key to the Families and Genera of ARIONOIDEA

(a) Shell plate visible through a slit-like opening in the back of the mantle. Visceral pouch under the mantle forms a hump on the back. Caudal mucous pit present. Tail laterally compressed; some species with a high keel.
Family: BINNEYIDAE . . . . Genus: *Hemphillia*, p. 314

  (aa) Shell plate completely enclosed in the mantle (sometimes reduced to granules). No visceral hump present, the visceral cavity contained within the foot. . . . . [b/bb]

(b) Pneumostome in anterior half of right side of mantle, but sometimes nearly halfway back or centered, or if slightly posterior then the slug less than 16 mm long and mantle greater than half its total length . . . . [c/cc]

    (bb) Pneumostome distinctly in posterior half of right side of mantle, which is usually less than half the animal's total length. If mantle greater than half the animal's total length, then the slug greater than 50 mm long. Pneumostome usually well posterior (approximately three-quarters back) in the right side of the mantle . . . .[f/ff]

(c) Caudal mucous pit present at posterior end where pedal furrows meet. Genital orifice below the mantle near the pneumostome. Tail relatively broad, and depressed dorsally. Sole may or may not be tripartite. Family: ARIONIDAE. . . . . Genus: *Arion*, p. 282

    (cc) Caudal mucous pit absent . . . . [d/dd]

(d) Small to medium slug. Caudal mucous pit absent (but excised tail leaves a groove that may appear as such). Genital orifice behind and near the right tentacle. Tail more acute, rounded dorsally rather than depressed, but without a keel. Sole not tripartite. Often with an oblique indentation defining the tail, which may be dropped off when the animal is attacked or irritated. Family: ANADENIDAE . . . . Genus: *Prophysaon*, p. 289

    (dd) Slug small, less than 16 mm long. Mantle longer than half the slug's total length. Tail 27% or less of the animal's length, with a high, sharp mid-dorsal keel and multiple parallel, sharp, lateral ridges . . . .[e/ee]

(e) Pneumostome slightly anterior to midpoint of the right edge of the mantle. Tail 7% to 20% of the animal's total length. Northern Rocky Mountains, Idaho. Family: ANADENIDAE. . . . . *Securicauda hermani*, p. 303

    (ee) Pneumostome about middle of the right edge of the mantle or slightly posterior. Tail 20% to 27% of the animal's length. Northwestern Oregon Cascade Mountains. Family: ANADENIDAE. . . . . *Carinacauda stormi*, p. 302

(f) Caudal mucous pit present where the pedal furrows meet posteriorly . .[g/gg]

    (ff) Caudal mucous pit absent . . . . [h/hh]

(g) Very large slugs (adults attain 185 to 260 mm) when extended. Conspicuous caudal pit. Back distinctly keeled behind the mantle. Sole not tripartite. Family: ARIOLIMACIDAE. . . . . Genus: *Ariolimax,* p. 305

    (gg) Relatively small slugs (adults reach 25 to 45 mm). Conspicuous caudal pit. Tail with narrow, roundly angular ridge; not sharply keeled. Sole indistinctly tripartite. Western Oregon. Family: ARIOLIMACIDAE. . . . . Genus: *Hesperarion*, p. 307

(h) Medium to fairly large slugs (65 to 80 mm long). Mantle elongated, greater than 60% of the total body length and covering the animal when at rest. Pneumostome about 60% back in right side of mantle. Tail short with an angular dorsal ridge. Northeastern Washington to western Montana and north into British Columbia.
Family: ARIOLIMACIDAE. . . . .*Magnipelta mycophaga,* p. 309

  (hh) Small slugs (generally less than 25 mm long). Mantle about one-third to one-half the animal's length . . . . [i/ii]

(i) Tail unkeeled but with longitudinal ridges and grooves somewhat like those of *Prophysaon coeruleum*; however, ridges are wider and grooves less deeply impressed. Mantle about one-half the animal's length. A small slug (9 to 14 mm long), blue-gray in color. Pneumostome slightly posterior to the middle of the right edge of the mantle. Northern Idaho, northwestern Montana, and adjacent British Columbia.
Family: ANADENIDAE. . . . . *Kootenaia burkei,* p. 300

  (ii) Back somewhat to strongly keeled behind the mantle . . . . [j/jj]

(j) Back and tail with strong keel. Mantle relatively large (about one-half the length of the animal). Color gray, mantle darker than sides but otherwise without markings. West-central Oregon.
Family: ARIOLIMACIDAE . . . . *Gliabates oregonius,* p. 306

  (jj) Oval mantle slightly more than one-third the animal's total length. Pneumostome two-thirds to three-fourths back in right side of mantle. Posterior edge of mantle notched just right of center over a pore . . . . [k/kk]

(k) Gray, blue-gray, brown, bluish- or pinkish-white; sometimes with metallic blue head and neck, and/or with thin, darker or lighter line on the keel. Low but distinct keel the full length behind the mantle. Northern Idaho, western Washington, and northwestern Oregon.
Family: ARIOLIMACIDAE . . . . Genus: *Zacoleus,* p. 311

  (kk) Blue-gray or pinkish-brown colored with light mid-dorsal stripe on tail. Mantle with black lateral lines forming a lyre shape with white or light buff patch between. Impressed lines on body darker than adjacent area. Northern Idaho and western Montana.
Family: ARIOLIMACIDAE. . . . .*Udosarx lyrata,* p. 310

Species Accounts

Phylum: Mollusca
Class: Gastropoda
Subclass: Eugastropoda

Gastropods that breathe through gills and have an operculum that closes the aperture when the animal is withdrawn into the shell.

Order: Neotaenioglossa
Superfamily: LITTORINOIDEA
Family: LITTORINIDAE
Genus: *Littorina*

Littorina subrotundata   (Carpenter, 1864)   Newcomb's Littorine Snail
Synonyms: *Algamorda subrotundata; Algamorda newcombiana* (Hemphill, 1877)

Description: A snail of intertidal salt marshes, the shell is higher than wide, turbinate, with a rather globose body whorl and a conical spire tapering to an acute apex. The aperture is teardrop-shaped, about half the height of the shell, and is closed by an operculum when the animal is withdrawn. The shell may attain a height to about 8 mm, but most are smaller. It is brown with narrow lighter or whitish spiral bands. The columellar lip margin is reflected to form a narrow groove under it.

Similar Species: The habitat on which it is found separates this species from most other snails. A similar snail in some of the same areas, *Littorina* sp., has brighter-colored, more distinct and more evenly spaced, alternating yellow and dark brown bands. Its aperture is more bilaterally symmetrical than that of *L. subrotundata*, and it lacks the distinct twist in the columella and the angle at the columellar insertion. It appears to be more numerous than *L. subrotundata* in Coos Bay, Oregon, but much less common in the Washington salt marshes. Of the other salt marsh snails, *Assiminea californica* has a similarly shaped shell, but it is smaller and more narrowly conic than the Newcomb's littorine snail, and it is uniformly brown, lacking the light-colored bands.

Distribution: *L. subrotundata* inhabits the upper intertidal zone, which is inundated only occasionally. It can be found in the salt marshes on *Salicornia* (glasswort or pickleweed) or on the mud under these plants. It occurs in a few bays along the Pacific Coast, having been reported from Humboldt Bay, California, to Neah Bay, Washington. However, there has been some controversy over whether or not these reports are of a single species. *L. subrotundata* occurs in Willapa Bay and Gray's Harbor, Washington; Coos Bay, Oregon; and presumably in Humboldt Bay, California. Its occurrence is questionable and needs to be confirmed in other areas from which it has been reported.

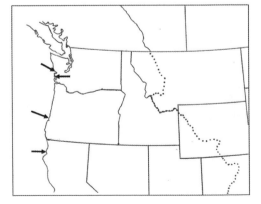

1

2

3

4

1 & 2: *Littorina subrotundata*: Bay Center, Willapa Bay, Pacific Co., WA; scale bar = 1 mm

3. *Littorina subrotundata* (operculum): Bay Center, Willapa Bay, Pacific Co., WA

4: *Littorina* sp. (*L. subrotundata* cf.): Coos Bay, Coos Co., OR; scale bar = 2 mm

Superfamily: RISSOOIDEA
Family: ASSIMINEIDAE
Genus: *Assiminea*

Assiminea californica　(Tryon, 1865)　California Assiminea

Description: A small snail of intertidal salt marshes, the shell is higher than wide, turbinate, with a rather globose body whorl and a conical spire tapering to an acute apex. The aperture is teardrop-shaped, about half the height of the shell, and is closed by an operculum when the animal is withdrawn. The shell, with 5 to 5½ whorls, measures less than 5 mm high by about 3 mm wide. It is dark brown, smooth, with no specific markings. There is no distinct groove under the columellar lip margin.

Similar Species: Within the salt marshes, *Littorina subrotundata* is the most similar species, but its shell is larger, relatively broader, and is marked with alternating brown and whitish bands.

Distribution: *A. californica* can be found along the Pacific Coast and the shores of Puget Sound in intertidal salt marshes, in the upper intertidal zone, under *Salicornia*, and under other plants and debris. Its range extends from southern British Columbia to Baja California.

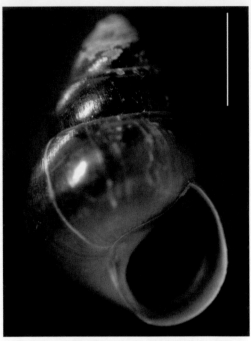

Assiminea californica: South Bend, Willapa Bay, Pacific Co., WA; scale bar = 2 mm

Family: POMATIOPSIDAE
Genus: *Cecina*

Cecina manchurica   A. Adams, 1861   Manchurian Cecina

Description: This is a small snail of intertidal salt marshes with a somewhat cylindrical shell. The periphery of the whorls is only gently rounded, and the sides of the spire are slightly but evenly tapered toward the apex, which is normally truncated and healed to a rounded end at about three whorls. As such, the shell measures about 6 mm in length. The aperture is small, obliquely-ovate or teardrop-shaped, and it is closed by an operculum when the animal is withdrawn.

Similar Species: No other snail of this size and shape occupies the PNW salt marsh habitats.

Distribution: *C. manchurica* is an Asian species introduced in Washington and British Columbia, where it can be found in some salt marshes. Hanna (1966) cited reports of it from Whatcom and Pacific counties, Washington.

Subclass: Pulmonata

Aquatic and terrestrial snails and slugs that lack an operculum and breathe by means of a lung-like modification of the mantle cavity.

Order: Basommatophora

Snails of this order have eyespots at the bases of a single pair of contractile tentacles. They are mostly aquatic, but the order includes the following two families and genera of salt marsh and land snails within our defined study area.

Family: ELLOBIIDAE
Genus: *Myosotella*

Myosotella myosotis   (Draparnaud, 1801)   Saltmarsh Pulmonate
Synonyms: *Phytia myosotis; Ovatella myosotis*

Description: A pulmonate snail of the salt marshes, this snail lacks an operculum. The shell is oval with a conical spire that varies in length. With 5 whorls the shell measures about 7 mm long by 3.5 mm wide. The periostracum is brown (lighter on younger animals, darker on older ones). The aperture is long and narrow, nearly two-thirds the length of the shell. Three denticle-like folds (lamellae) spiral around the columella, but the third one is very small and indistinct. In fresh specimens a row of bristles can be seen spiraling around the shoulder of at least the last whorl. Eyespots are located median to the bases of the rather plump, tapered tentacles.

Similar Species: This snail is unique and easily distinguished from others that occupy the Pacific Northwest salt marsh habitats in the intertidal zone.

Distribution: *Myosotella myosotis* occurs on both the Pacific and Atlantic coasts of North America, although there has been some disagreement as to whether we have one or more species or whether or not it is native (Hanna 1966). It is also found in bays of Puget Sound. It can be found on or under *Salicornia* (glasswort or pickleweed) and other plants and debris in the upper intertidal zone that is not regularly inundated. It is commonly found in association with *Assiminea californica* and with *Littorina subrotundata* where that species occurs.

Selected references for gastropods of Pacific Coast salt marshes: Berman and Carlton 1991; Hanna 1966; Kozloff 1993, 1996; Taylor 1981.

1. *Myosotella myosotis*: Tolmie State Park, Thurston Co., WA (length 7.15 mm)
2. *Myosotella myosotis*: Tolmie State Park, Thurston Co., WA
3. *Myosotella myosotis* (note apertural lamellae): Tolmie State Park, Thurston Co., WA

Family: CARYCHIIDAE
Genus: *Carychium*

Minute white snails, with a narrow, elongated shell tapering toward the narrowly rounded apex. The aperture lip is reflected, and there are one or two lamellae (narrow, raised ridges) spiraling around the columella.

Carychium occidentale   Pilsbry, 1891   Western Thorn

Description: The shell is minute (2.1 to 2.7 mm long by 1 mm wide), white, and narrow, tapering from the body whorl to a narrow, rounded apex. The apertural lip is well reflected. Two long lamellae spiral up the columella. The upper one appears as a small parietal tooth in the aperture; the lower one on the columella is often indistinct. The animal is mostly white. Eyespots are near and in front of the bases of a single pair of contractile tentacles.

Similar Species: *Carychium occidentale* is the one species of this family native to the Pacific Coast states. The white opaque or somewhat translucent shells of these minute elongate snails distinguish *Carychium* from all other genera of land snails in the Pacific Northwest. The single pair of tentacles and the eyespots confirm the order. All other land snails native to this area are of the order Stylommatophora, which have their eyespots at the tips of the upper tentacles rather than at their bases.

   The one snail within the same range that is most similar in appearance to *Carychium* is an aquatic species, *Pristinicola hemphilli* (Pilsbry, 1890). Its shell is generally of a similar size and shape as that of *Carychium*, but it inhabits springs, has an operculum, lacks the apertural dentition or lamellae, and has a lip margin that is thickened but not distinctly reflected. Another species, *Carychium exile* H. C. Lea, 1842, subspecies *canadensis* Clapp, 1906 (the ice thorn), was reported from Vancouver Island by Dr. Hanna (Pilsbry 1948). It has not since been reported from western North America, and its occurrence here is doubtful. It may be distinguished from *C. occidentale* by the thickened, rather than reflected, lip margin and its narrower diameter (about 0.75 mm). The parietal lamella is a flat wing that spirals up the columella. In *C. exile* it expands to become broad and undulating beginning in the penultimate whorl approximately above the aperture; in *C. occidentale* it is narrower and simple.

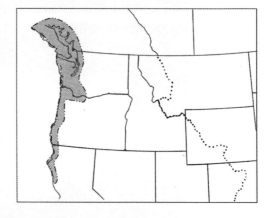

Distribution: *Carychium occidentale* occurs on the Pacific slope from Del Norte Co., California, through western Oregon and Washington to British Columbia, including Vancouver Island. It is found in riparian areas and other moist sites, among leaves and litter collected among and under ferns and shrubs.

Selected references for the family Carychiidae: Burch 1962; Forsyth 2004; Pilsbry 1948.

1. *Carychium occidentale*: Lane Co., OR; scale bar = 1 mm
2. *Carychium occidentale* (minute with apertural lamellae and reflected lip): N of Copalis, Grays Harbor Co., WA; scale bar = 1 mm

Order: Stylommatophora

These snails usually have two pairs of retractile tentacles. The eyes are located at the tips of the upper (larger) pair. This order includes all of the remaining land snails and slugs in the Pacific Northwest.

Suborder: Heterurethra
Family: SUCCINEIDAE

The Succineidae are land or semi-amphibious snails with a distinctive shell shape that has been named for the family. It has a large body whorl with a relatively small spire, which often appears to protrude at an angle. Other than species within this family, the only other snail with the typical succiniform shell that might be found in the Pacific Northwest is *Pseudosuccinea columella* (Say, 1817), an aquatic snail of the family Lymnaeidae that has been introduced with aquatic plants sold for water gardens and ponds. Although both families are Pulmonates, *Pseudosuccinea* is of the order Basommatophora, members of which have their eyespots at the bases of a single pair of tentacles. Therefore, living specimens can readily be separated into their respective families, since the Succineidae are of the order Stylommatophora, members of which have two pairs of tentacles with their eyespots at the tips of the upper pair.

Succineidae is a most difficult family, with no sure way of distinguishing the genera or species from shell characteristics. In general, following Burch (1962), who listed only three genera, species of *Catinella* (*Quickella* of Pilsbry, 1948) are considered to have shorter shells with a rounder aperture and relatively longer spire. Species of *Succinea* have larger shells with more ovate apertures and swelling around the genital opening, which is located behind the right tentacle. The shells of *Oxyloma* are comparable in size to those of *Succinea* or a little larger, but they are relatively narrower with longer, narrower, ovate apertures.

The following keys, species descriptions, and distributions are the product of reviews of literature, examination of published illustrations, and study of specimens from personal collections. In keeping with the general perception that shell characteristics alone are not accurate indicators of species or, in some cases, even of genera, some modifications of sizes and proportions of the shells have been made here. Shell size is used as a major characteristic in these keys, but lengths given are of the larger shells of the species. Growth stage and variation among individual specimens needs to be considered. Specimens that I have collected from the Pacific Northwest are generally smaller than indicated in the literature, and relative proportions (i.e., shell width and aperture length) are not as extreme.

Key to the Species of SUCCINEIDAE

(a)   Width of shell half or less of the length. Genus: *Oxyloma* [j/jj]

   (aa)   Width of shell greater than 55% of the length . . . .[b/bb]

(b)   Width of shell with 3 to 3½ whorls greater than 65% of its length . . . . [c/cc]

   (bb)   Width of shell 57% to 67% of its length. . . . . [d/dd]

(c)   Short and wide, length 10 to 17 mm with nearly 3½ whorls; aperture length about three-fourths the shell length. Width about 70% of length. Shell gray with opaque white or buff streaks. . . . .*Succinea campestris*

   (cc)   Length about 12 mm with 3 whorls; aperture 65% to 70% of shell length. Width about 60% of length. Shell non-glossy, with greenish tint.
. . . . *Succinea rusticana*

(d)   Small (6 to 11 mm long), with 2½ to 3½ whorls; aperture 50% to 70% of shell length, roundly ovate; spire relatively long . . . .[e/ee]

(dd)　Mostly 13 to 19 mm long (some specimens as small as 8 mm);
aperture large and broadly oval . . . . [f/ff]

(e)　Shell small, about 6.5 mm long with 2½ to 3 whorls; translucent with
reddish tint; whorls inflated, rather loosely coiled with well-impressed
sutures. Aperture roundly ovate, nearly symmetrical. Growth-wrinkles
coarse, closely and regularly spaced. . . . . *Succinea oregonensis*

(ee)　With 3 to 3½ whorls, length greater than or equal to 8 mm.
Aperture broadly ovate, obtusely pointed, posteriorly angled at
the columellar insertion. . . . . Genus: *Catinella* [g/gg]

(f)　Shell 8 to 19 mm long, broad, translucent to opaque, reddish or yellowish
with darker and lighter streaks. Average width 65% (57%–74%) of the
length; aperture averages 58% (57%–72%) of the shell length. Aperture
roundly-ovate, mostly symmetrical, with about a right-angle point
posteriorly. Spire rather long, roundly inflated, with a small, nipple-like
apex. Steppe vegetation in Asotin Co., Washington . . . . *Novisuccinea strigata*

(ff)　Shell with 2½ whorls, about 14 to 17 mm long by 9 to 11 mm
wide (width about 65% of length); color greenish-yellow with
pale or reddish apex. Aperture, 70% to 75% of length, roundly
arched anteriorly. . . . . *Novisuccinea ovalis*

(g)　Shells greenish, reddish-orange or reddish-yellow. Length 6 to 11 mm
with about 3 whorls . . . . [h/hh]

(gg)　Shells light yellow or amber; length 9 to 12 mm in 3¼ to 3¾ whorls. . . . [i/ii]

(h)　Shell 6.25 mm long, globose, with deeply impressed sutures, transparent
and glossy with greenish, olive, or orange tint. Aperture greater than 75%
of the shell length, broadly oval, well rounded anteriorly, obtusely pointed
posteriorly. Growth-wrinkles low and irregular . . . . *Catinella stretchiana*

(hh)　Shell 7 to 11 mm long, light yellowish, greenish or reddish-yellow.
Spire narrow and drawn out. Sutures deep, but whorls of the spire
less convex and somewhat flattened laterally. Aperture ovate, about
two-thirds of shell length, the outer margin well inflated, anterior
margin roundly arched or slightly flattened. Growth-wrinkles irregular,
becoming coarse near the aperture. Range throughout the United
States and much of Canada. . . . . *Catinella vermeta*

(i)　Shell about 10 mm long by 6 mm wide. Spire long-conic, light yellowish.
Aperture somewhat reniform, the outer margin well inflated, anterior
margin a little flattened, curving roundly into the columellar, which is
rather straight and forms a slight angle where it joins the columella.
Growth-wrinkles prominent. . . . . *Catinella gabbii*

ii)　Shell about 8.5 to 12.5 mm long. Spire moderately long, conic, amber.
Aperture oval moderately well inflated. The outer margin about parallel
to the columellar margin at the periphery. Anterior margin well rounded,
curving fairly sharply into the columellar, which is slightly curved and
forms a slight angle at the columellar insertion. Growth-wrinkles low
and irregular . . . . *Catinella rehderi*

(j) Shell length 18 to 21 mm. Aperture mostly less than 70% (65%–72%) of the length.
. . . . [k/kk]

(jj)　Shell length less than 18 mm. Aperture mostly about 75% (67%–77%)
of the length . . . . [m/mm]

(k)   Long, rather wide shell, length to 21 mm with 3 whorls. Body whorl well inflated, glossy, amber; the outer aperture lip curved well outward. Spire of moderate length, but narrow and somewhat pointed. Sculptured with close growth-wrinkles and irregular spiral striae. . . . . *Oxyloma haydeni*

     (kk)   Long, narrow shell, aperture narrowly ovate and symmetrical . . . . [l/ll]

(l)   Shell length to 19 mm with 4 whorls; glossy, transparent with reddish tint. Aperture relatively short, well curved anteriorly, acutely angled at columellar insertion. Spire narrow and extended. . . . . *Oxyloma hawkinsi*

     (ll)   Length 14 to 20 mm with 3 whorls. Thin, glossy white or with yellow tint, and with rather coarse growth-wrinkles. Aperture long-ovate, well rounded anteriorly, acutely pointed posteriorly. Spire relatively short, acute . . . . *Oxyloma sillimani*

(m)   Width relatively wide, about half as wide as long. Length, with about 3 whorls, 13 to 17 mm, with a short acute spire. Columellar insertion meets the penultimate whorl at about 90 degrees. Aperture less rounded anteriorly . . . [n/nn]

     (mm) Relatively narrow, width 43% of length. Length 11 to 15 mm in 3 to 3¼ whorls. Aperture long, rather wide, bluntly rounded anteriorly, columellar insertion at right angle or slightly acute. Pale yellow. . . . . *Oxyloma nuttallianum*

(n)   Shell usually 13 to 16 mm long by about 7.5 mm wide; glossy, transparent, light-colored, sometimes with a rose tint. Aperture 67% to 75% of length; spire short, acute. . . . . *Oxyloma decampi*

     (nn)   Shell 15 to 17 mm long by 7 to 9 mm wide, very thin, white or light buff colored. Aperture 71% to 77% of length, often somewhat squared anteriorly; lip slightly dilated . . . . *Oxyloma retusum*

Subfamily: CATINELLINAE
Genus: *Catinella*

Catinella gabbii   (Tryon, 1866)   Riblet Ambersnail
Synonyms: *Succinea gabbi; S. oregonensis gabbi*

Description: With about 3½ whorls, the shell is 9 to 10 mm long by 5.5 to 6.3 mm wide. Length of the aperture is about 70% of the shell length. The shell is thin, semi-transparent, and light yellowish. The spire is relatively long for the family and somewhat conical, terminating with a rather acute apex. The sutures are quite deeply impressed. The aperture is roundly ovate with the anterior edge somewhat flattened. Growth-wrinkles are prominent.

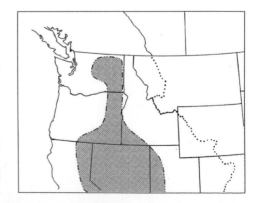

Similar Species: *Succinea oregonensis* has about one less whorl and is a smaller snail with a greater width to length ratio. *Catinella rehderi* has a less acute apex, rather smooth, irregularly spaced growth-wrinkles, and a blunter anterior curve of the aperture.

Distribution: *C. gabbii* has been reported from Walla Walla, Stevens, and Grant counties, Washington; southeastern Oregon; and northeastern California south to Inyo County.

Catinella rehderi (Pilsbry, 1948)   Chrome Ambersnail
Synonyms: *Quickella rehderi* Pilsbry, 1948; *"Succinea oregonensis* Lea of many collections not of Lea" (Pilsbry 1948)

Description: With 3¼ whorls the shell is 8.5 to 12 mm long; the width is about 60% of the length. The aperture is about 65% to 70% of the shell length. The shell is glossy, amber. The spire is relatively long for the family and somewhat conical, terminating with a narrow rounded apex. The sutures are deeply impressed. The aperture is broadly oval to elliptical, the anterior-columellar angle more sharply curved; the outer-anterior curve more smoothly arched. Growth lines are low and irregularly spaced.

Similar Species: *Catinella gabbii* is similar, but with whorls more rounded and more arched laterally, and an aperture less arched anteriorly. The sutures of *C. gabbii* are more deeply impressed, although a rather subtle difference, and its growth-wrinkles are more prominent. *Succinea oregonensis* is smaller and wider relative to the length. It has deeper sutures, and its growth sculpture is sharp and closely spaced.

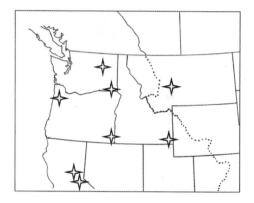

Distribution: The type locality for *Catinella rehderi* is about 5 miles west of Davenport, Lincoln County, Washington. It or very similar specimens have also been reported from southeastern Washington, Montana, Idaho, and throughout much of Oregon and California.

Catinella stretchiana (Bland, 1865)   Sierra Ambersnail
Synonym: *Succinea stretchiana* Bland, 1865

Description: A small succineid, the shell, with 3 whorls, is usually near 6 mm long but up to 8.5 mm by 5 mm wide. The shell is rather capacious, generally more globose than other succineids. It is thin, transparent with a greenish tint, but Pilsbry (1948) says, "They vary in degree of obesity and also color." Some may be more elongate, although still quite small, and the color may vary to olive or orangish. The aperture of the type specimen is about 80% of the shell length, as described, but it appears to be 50% to 60% in those illustrated.

Similar Species: Compare with other *Catinella*. *Succinea oregonensis* may be of similar size, but its shell has a reddish tint instead of the greenish tint of *Catinella stretchiana*.

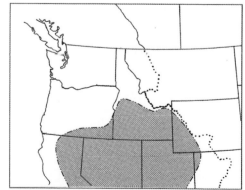

Distribution: *C. stretchiana* has been reported from Big Payette Lake and Cache Valley in southern Idaho, southwestern Wyoming, through Utah, and Nevada, and it is common in the Sierra Nevada of California. Type locality is Washoe Co., Nevada.

Catinella sp.:
Hughes Meadows,
Bonner Co., ID;
scale bar = 1 mm

Catinella vermeta (Say, 1829)   Suboval Ambersnail
Synonyms: *Succinea avara* (Say, 1824); *Mediappendix vermeta* (Say, 1829)
[Grimm et al., 2009]

Description: There are a little more than 3 whorls; the shell is 7 to 11 mm long by 4 to 6.8 wide. The aperture is oval and rather short, about 67% of the shell length or less. The posterior margin of the aperture is curved so the insertion adjoins the penultimate whorl at a near perpendicular or slightly anteriorly directed angle. The shell is narrow, the width about 60% of the length. The whorls are well rounded; the sutures deeply impressed. Color is light yellowish or sometimes pale greenish, reddish-yellow or with a pinkish tint. Growth-wrinkles are irregular and become coarser toward the aperture on the last whorl. The animal has a distinct locomotive disc. The sole is uniformly pale-colored or speckled with black. The mantle, seen through the shell, is pale gray with white and black spots.

Similar Species: Compare this species with other *Catinella* and with *Succinea oregonensis*. "Succineas which do not seem distinguishable from *S. avara* and usually associated in lots with shorter shells formerly referred to *S. oregonensis*, turn up in many places in Washington, Oregon, Idaho and also in Nevada," (Pilsbry 1948).

Distribution: *Catinella vermeta* is known from the Mackenzie River and Field, British Columbia, to California and east, throughout the United States and much of Canada.

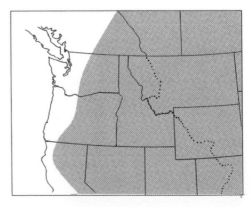

Genus: *Oxyloma*

Oxyloma decampi (Tryon, 1866)   Marshall Ambersnail
Subspecies: *Oxyloma decampi gouldi* (Pilsbry, 1948)

Description: With 3 whorls, the shell may be 13 to 15 (seldom 16) mm long by 7 to 7.5 mm wide. The aperture is about 67% to 75% of the shell length. The shell is ovate, very thin and transparent, sometimes with a rose tint. Growth-wrinkles are faint on a glossy shell. The spire is short but acute. The sutures are moderately impressed but not as deeply as in the *Catinella*. The aperture is long-ovate, widest anterior to the midpoint. Anteriorly the margin is roundly arched; posteriorly the palatal insertion curves inward to meet the penultimate whorl at approximately 90 degrees.

Similar Species: The shell and aperture length of *Oxyloma nuttallianum* is similar. It differs in that its shell is narrower and has a light yellow tint, while the shell of *O. decampi* may have a rose tint.

Distribution: *O. decampi* occurs in eastern Canada and New England to Colorado and Montana. It may be found in "marshy places on and around the aquatic vegetation of muddy pond and river margins and ditches" (Pilsbry 1948).

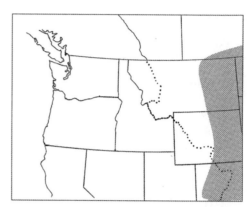

Oxyloma hawkinsi (Baird, 1863)   Boundary Ambersnail

Description: A large succineid, measuring 19 mm long by 8.5 mm wide with 4 whorls. The aperture is 67% of the shell length. A rather long, narrow shell, glossy and transparent with a reddish tint. The spire is long and acute, and the body whorl makes up about two-thirds the shell length. The suture is rather deeply impressed. The growth-wrinkles are moderate. The aperture is long-ovate, symmetrical, well rounded anteriorly, acutely pointed posteriorly.

Similar Species: Other large succineids include *Oxyloma haydeni* and *O. sillimani*, which have 3 to 3½ whorls, and the shells of which are yellowish tinted to amber, compared to the 4 whorls and reddish-tinted shells of *O. hawkinsi*.

Distribution: *O. hawkinsi* has been reported from Colville, Washington, and Lake Osoyoos, British Columbia, to Manitoba, Canada. Henderson (1929) also reported it from Utah.

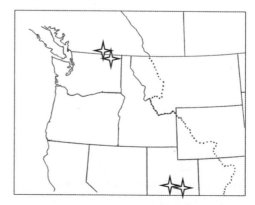

Oxyloma haydeni (W. G. Binney, 1858)   Niobrara Ambersnail

Description: A rather large succineid; with 3 whorls the shell may be as long as 21 mm by 9 mm wide. The shell is long-ovate (distinctly wider than *O. hawkinsi*), thin, glossy, amber-colored. The spire is rather short and acute. The whorls are well inflated, and the suture is moderately impressed. There are closely spaced growth-wrinkles and irregular spiral striae. The aperture is about 70% of the shell length. It is obliquely oval, the anterior end well rounded but not symmetrically. The outer margin is rather widely arched; the columella is straightened posteriorly. The posterior insertion adjoins the penultimate whorl at a barely acute angle.

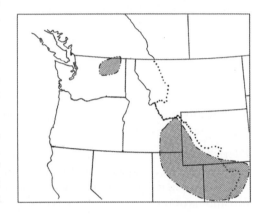

Similar Species: Other large succineids include *Oxyloma hawkinsi,* with 4 whorls and a reddish tint to the shell, and *O. sillimani,* with a wider, lighter-colored shell.

Distribution: Pilsbry (1948) reported *O. haydeni* from Nebraska and northward. Pacific Northwest sites include the vicinity of Colville, Stevens Co., and Park Lake in the Grand Coulee, Grant Co., Washington (Henderson 1929).

Oxyloma nuttallianum (I. Lea, 1841)   Oblique Ambersnail

Description: With 3 whorls the shell is about 15 mm long by 6.4 mm wide. The aperture is about 75% of the shell length. The shell is long-ovate, transparent with a light yellow tint. The spire is short, oblique and somewhat elevated, the sutures moderately impressed. The whorls are rounded, but narrower than those of *O. haydeni*. The aperture is somewhat symmetrical, but the posterior half of the outer margin is more arched; the columellar margin straighter. The anterior margin is well rounded; the posterior rather pointed. Growth-wrinkles are finely striate. The mantle is gray with black spots and stripes around its edge and on the head and sides.

Similar Species: The size of *O. nuttallianum* can be compared with that of *O. decampi gouldi* and *O. retusum*. However, neither of these two species is known to occur west of the Continental Divide, and they are included in this work only because of the proximity of the western extent of their ranges to the Rocky Mountains.

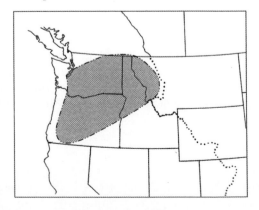

Distribution: *O. nuttallianum* is found west of the Continental Divide from British Columbia to southern Oregon. The type locality is believed to be in the vicinity of the lower Columbia or Willamette River near Portland, Oregon. It has also been recorded from Meadow Lake, southwest of Spokane, Washington, and in southern Idaho.

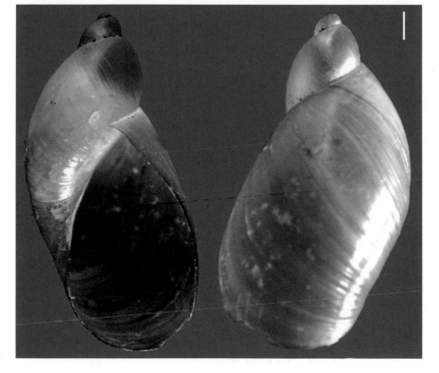

1. *Oxyloma* sp.: Colville River, Stevens Co., WA; scale bar = 1 mm
2. *Oxyloma nuttallianum*: Browns Lk., Stevens Co., WA; scale bar = 1 mm

Oxyloma retusum (I. Lea, 1834)   Blunt Ambersnail

Description: With 3 whorls the shell is 15 to 16.5 mm long by 7 to 9 mm wide. The aperture is about 71% to 77% of the shell length. The shell is oblong-ovate, thin, transparent, light whitish or buff colored. The spire is short, acute, and somewhat elevated. The last whorl is depressed, rounded, and the suture is moderately deep. The aperture is nearly symmetrical; the outer margin is a little more rounded than the columellar margin, which is little arched. There may be a thin parietal fold near the posterior end of the columellar margin, and rarely there is a small, white, parietal denticle. The curve of the anterior apertural lip may be somewhat flattened, and the lip somewhat dilated. The growth-wrinkles are moderately striate.

Similar Species: Other *Oxyloma* of approximately 15 to 16 mm long are *O. decampi gouldi* and *O. nuttallianum. O. decampi gouldi,* with a glossy, transparent shell, which may have a rose tint to it and with faint growth-wrinkles, can be compared with the whitish to light buff tinted shell of *O. retusum* and its moderately strong lines of growth. *O. nuttallianum,* with a yellowish tint to its shell and fine growth-wrinkles, is not likely to occur in the same area as the other two, the range of which would be peripheral to the defined area of this study if occurring within it at all.

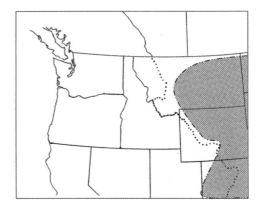

Distribution: *O. retusum* occurs in the northeastern United States, west to the Missouri River on the east slope of the Rocky Mountains in Broadwater Co., Montana. It has also been reported in Alaska, British Columbia, and Riverside Co., California.

Oxyloma sillimani (Bland, 1865)   Humboldt Ambersnail

Description: This rather large succineid with 3 to 3⅓ whorls measures 14 to 20 mm long by 8 to 10 mm wide. The aperture is about 65% to 72% of the shell length. The shell is oblong-ovate, thin, glossy, whitish or with a faint yellowish tint. The whorls are moderately inflated; the sutures are moderately impressed. The spire is short and acute. The aperture is nearly symmetrical; the outer margin is slightly more rounded than the columellar. The growth-wrinkles are rather coarsely striate.

Similar Species: Compare this species with *O. hawkinsi* and *O. haydeni*. Although smaller, the shell of *O. nuttallianum* also appears similar but differs in its slightly longer aperture, more yellow color, and finer growth lines.

Distribution: The type locality of *O. sillimani* is Humboldt Lake, Nevada. It is also known from the Humboldt River, near Carlin, Elko County, Nevada; southern California; and Meadow Lake southwest of Spokane, Washington.

Novisuccinea sp.:
Long Beach,
Pacific Co., WA;
scale bar = 1 mm

Subfamily: SUCCINEINAE

Genus: *Novisuccinea*

Novisuccinea ovalis (Say, 1817) Oval Ambersnail

Description: Fairly large with a well-inflated, oval body whorl, *N. ovalis* is wider than many other members of the family. With 2½ whorls the shell measures 14 to 16.5 mm long by 9 to 11 mm wide; the aperture is 9.8 to 12.2 mm (68% to 76% of the shell length). It is thin, glossy, and translucent, of a greenish-yellow tint, the apex paler or reddish.

Similar Species: *Catinella vermeta* is similarly shaped but smaller, with a half whorl more, and has a relatively longer spire and correspondingly shorter aperture.

Distribution: *Novisuccinea ovalis* is known from eastern Canada to North Dakota and south to Nebraska, southeast to Missouri and Alabama, and east to North Carolina. It has been introduced into Oregon and California (Roth and Sadeghian 2006).

Novisuccinea strigata Pfeiffer, 1855 Striate Ambersnail
Synonym: *Succinea strigata*

Description: The shell is short and broad, thin, translucent or opaque, yellowish or reddish with narrow lighter and darker streaks. With 2¾ to 3½ whorls it measures 8 to 19 mm long (avg. 13.2) by 6.5 to 13 mm wide (avg. 8.6). Average width is 65% (57%–74%) of the length; the aperture is 57% to 72% (avg. 58%) of the length. The whorls are well inflated and convex; the sutures are rather deep. The spire consists of the greatly inflated last half of the penultimate whorl and the very small nipple-like apex containing the earlier whorls. The aperture is broad, symmetrical, and roundly-ovate. The arch of the anterior margin is very slightly if at all flattened; the posterior angle is barely acute.

Similar Species: The strongly inflated shell and roundly-ovate aperture of this shell distinguishes it from other *Succinea* from the same region.

Distribution: An arctic species from Alaska, northern Canada, and Greenland, *Novisuccinea strigata* was found among steppe vegetation in Asotin County, Washington, by Dr. R. Daubenmire and sent to a malacologist (whose name he couldn't recall) for identification. The specimen is #214107 at the US National Museum (R. Daubenmire [1987] letters to Northwest Scientific Association and Washington Department of Wildlife). Also recorded from Long Beach, Pacific Co., Washington.

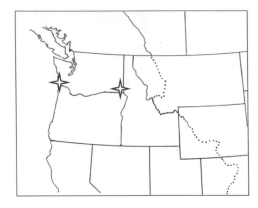

Genus: *Succinea*

Succinea campestris   Say, 1817   Crinkled Ambersnail

Description: This medium to fairly large succineid has a relatively wide shell. With 3⅓ to 3½ whorls it measures 10 to 17 mm long by 7 to 11.5 mm wide; the diameter equals 66% to 74% of the height. The aperture is about three-fourths of the shell length. It has a short conical spire. The shell is streaked with translucent gray and opaque buff, and sculptured with low growth-wrinkles and microscopic granulations. The aperture is ovate and relatively wide for *Succinea*.

Similar Species: The wide shell and aperture may appear more like a *Catinella* species than a *Succinea*. The same characteristics plus the streaks and granulations distinguish it from species native to the Pacific Northwest.

Distribution: *S. campestris* is a snail of the southeastern United States, but Capizzi (1962) reported an introduction on nursery stock in Portland, Oregon (Hanna 1966), although I found no recent records of whether or not it still survives in the area.

1. *Succinea* sp.: Montana; scale bar = 1 mm
2. *Succinea* sp.: First Thought Lk., Stevens Co., WA; scale bar = 1 mm

Succinea oregonensis I. Lea, 1841   Oregon Ambersnail

Description: With 2½ whorls the shell is 6.5 mm long by 4 wide (width 62% of the length). The aperture is about 65% of the shell length. The shell is small but wide, thin and translucent, yellow or reddish with a matte surface. The spire is somewhat elevated. The 2½ to 3 whorls are well inflated and separated by deeply impressed sutures. The aperture is roundly ovate, nearly symmetrical, but the outer-posterior curve is arched more than the columella. The posterior apertural margin curves around to adjoin the penultimate whorl at an obtuse angle. The growth lines are fine, sharp, closely and regularly spaced on the second whorl, becoming more coarse and irregular on the last half-whorl.

Similar Species: Pilsbry (1948) described *Quickella rehderi* (=*Catinella rehderi*) from specimens previously mistaken for *Succinea oregonensis*. He also stated, "Succineas which do not seem distinguishable from *S. avara* and usually associated in lots with shorter shells formerly referred to as *S. oregonensis*, turn up in many places in Washington, Oregon, Idaho and also in Nevada" (1948).

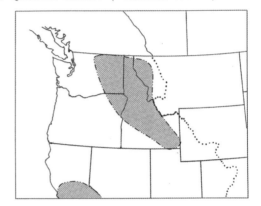

Distribution: The type locality was listed as Oregon. Henderson (1929a) listed *S. oregonensis* from Montana, Idaho, and Washington, to southern California and Vancouver Island, British Columbia.

Succinea oregonensis: Sinlahekin Wildlife Area, Okanogan Co., WA

Succinea rusticana Gould, 1846 Rustic Ambersnail

Description: A small to medium succineid; with 3 whorls the shell is 10 to 12 mm long by about 8 mm wide. Average width is about 67% of the length, and the aperture about 70%. The shell is ovate-conical, thin and non-glossy, with a greenish tint. The spire is short, acute, and elevated. The whorls are moderately inflated; the last is depressed anteriorly; the sutures are moderately well impressed. The aperture is ovate, nearly symmetrical, well rounded anteriorly and pointed posteriorly. Growth-wrinkles are rather coarse.

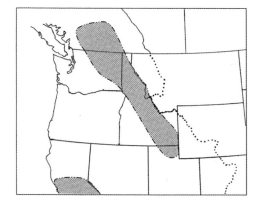

Similar Species: *S. oregonensis* is smaller, yellowish or reddish, and its sutures are more deeply impressed. The body and penultimate whorls of *Novisuccinea strigata* are widely inflated; the aperture is widely ovate.

Distribution: *Succinea rusticana* has been reported from southern British Columbia; the Grand Coulee, Washington; and central California.

Selected references for the family Succineidae: Capizzi 1962; Frest and Johannes 2000; Hanna 1966; Harris and Hubricht 1982; Henderson 1924, 1929a; Hoagland and Davis 1987; Pilsbry 1948; Roth and Sadeghian 2006; Turgeon et al. 1998.

Suborder: Orthurethra
Family: CIONELLIDAE

Also known as Cochlicopidae, Roth (2003) determined Cionellidae to be the correct family name and *Cochlicopa* to be correct for the genus.

Genus: *Cochlicopa*

Cochlicopa lubrica (Müller, 1774)   Glossy Pillar
Synonym: *Cionella lubrica*

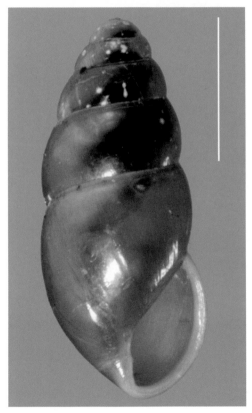

Cochlicopa lubrica: Bellevue, King Co., WA; scale bar = 2 mm

Description: A small snail; shell is 5.0 to 7.5 mm long, narrow, somewhat fusiform and glossy brown. Its apertural lip is neither expanded nor reflected, but it is thickened around the inside. The aperture is about one-third the shell length and less than one-half of the shell width.

Similar Species: Kerney, Cameron, and Riley (1994) describe *Cochlicopa lubricella* from western Europe, which may be difficult to distinguish from *C. lubrica*. It is a similar species but smaller, with a noticeably narrower, more cylindrical shell that is less glossy than *C. lubrica*. Narrower, more cylindrical shells are sometimes found with *C. lubrica* in the Pacific Northwest, but they are mostly longer and do not appear to be *C. lubricella*. Other Pacific Northwest species resemble *Cochlicopa* only superficially. Other small, elongated land snails include those in the families Carychiidae and Pupillidae, all species of which are smaller than *Cochlicopa* and do not otherwise resemble it closely.

Distribution: *Cochlicopa lubrica* is Holarctic in distribution. In the Pacific Northwest it is most often found in lowlands, and it thrives near human habitation under boards, rocks, and other debris.

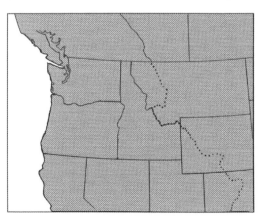

Selected references for the family Cionellidae: Burch 1962; Kerney, Cameron, and Riley 1994; Pilsbry 1948; and Roth 2003.

Superfamily: PUPILLOIDEA

The snails in this group are distinctive. They are minute, and the superfamily is named for the shape of the shells of two of the following three families, which resemble the pupae of some insects (e.g., *Diptera*). Of the snails with pupa-shaped shells, two families, including four genera, occur in the Pacific Northwest, and a fifth genus (*Gastrocopta*) may be found at the eastern edge of this range. The third family, Valloniidae, have minute, mostly heliciform shells; that family is described separately following the discussion of the pupa-shaped species.

Genera of adult Pupillidae and Vertiginidae (the pupillids) can be recognized as summarized below, but to confirm the taxon they should be examined more closely using other characteristics described under each genus or species.

Characteristic Features of the Genus of the Family PUPILLIDAE

Pupilla: Relatively large pupillids with zero to three apertural teeth, depending on the species. Shell cylindrical, about 3 (2.7 to 4.0) mm high, with a distinctly flared or reflected lip margin.

Characteristic Features of the Genera of the Family VERTIGINIDAE

Columella: No apertural teeth; apertural lip simple. Shell less than 2 to about 3 mm high; cylindrical or ovately tapered toward the apex.

Nearctula: Apertural teeth present (except one species); lip flared, reflected, or thickened. Shell ovate or long, ovately oblong, 2 to 2.5 mm high; umbilicus rimate to openly perforate.

Vertigo: Normally with 4 or more apertural teeth (sometimes fewer). Shell usually 1.8 to 2.5 mm high (rarely as small as 1.5 mm). Ovate to cylindrical with a rimate, perforate, or imperforate umbilicus.

Gastrocopta: Usually with 6 or more apertural teeth, parietal and angulars fused. The one species that might occur at the eastern edge of the Pacific Northwest is small (1.75 mm high) and white.

Pupillid Characteristics and Terminology

Shell characteristics used to identify pupillids include size, shape, color, umbilicus, sculpture, sinulus, crest, and apertural dentition.

Size
- Shells are minute, most are 1.7 to 3 mm high. Some of the largest attain a height of about 4 mm.

Shape
- Cylindrical—the sides are relatively straight, with varying degrees of rounding at the ends. Usually the last two or three whorls, which make up the majority of the shell's height, will be approximately the same diameter, and the first two or three small whorls will form the rounded apex. The term is often compounded as cylindrically-ovate, cylindrically-oblong, etc.
- Ovate or ovoid—the shell is somewhat egg-shaped.
- Oval—egg-shaped or elliptical in outline (e.g., shape of the aperture).
- Oblong—the shell is rather cylindrical but tapering somewhat toward the rounded ends.

Umbilicus
- Perforate—as a pinhole.
- Imperforate—with no opening.
- Rimate—as a narrow slit.

These terms may also be compounded, for example, rimately-perforate or rimately-imperforate (used when the umbilical depression terminates in a closed slit).

Sinulus
- A wave in the palatal lip forming an indentation pointing into the aperture. There may also be a groove-like indentation in the shell that extends from the sinulus back over one or both of the palatal denticles.

Crest
- A raised collabral ridge in the last whorl of the shell just behind the aperture.

Apertural dentition
Apertural dentition is composed of denticles or lamellae (formed as layers or folds in the shell nacre extending into the apertural opening), commonly referred to as apertural teeth. These denticles are identified by their location on the apertural wall (fig. 4):
- The parietal is the surface of the penultimate whorl within the aperture and is the location of the parietal and angular denticles or teeth.
- The columella is the central axis of the shell and is the location of the columellar, and sometimes a basal denticle will be located low on the columella at or near the basal-columellar curve.
- The basal wall or lip is the base of the shell when oriented vertically (apex up), and it may contain a basal denticle within it or near the lower end of the columella.
- The palatal is the outer wall of the whorl or aperture (upper in heliciform shells), adjoining the distal portion of the penultimate whorl. There are usually two palatal denticles in species that have these structures, but there may be fewer or more.
- Height of a denticle refers to the distance from its base to its crest.
- Length of a denticle refers to the length of its base from front to back.

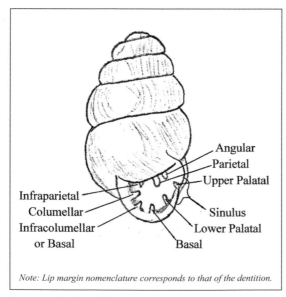

Note: Lip margin nomenclature corresponds to that of the dentition.

Figure 4: The Pupillid Shell

Note: In this work, the dental formula indicates the number of apertural denticles (teeth) beginning with the parietal first, the columellar and the basal (if any) second, and palatals third. Thus, a dental formula of 1-1-2 indicates a parietal, a columellar, no basal, and 2 palatals. The formula 2-2-2 indicates a parietal and an angular on the parietal wall, a columellar and a basal, and 2 palatals, while 1-1-0 indicates a parietal and a columellar only (fig. 5).

In other works the basals and palatals have been combined. However, since there is usually only one columellar, and pupillids of the Pacific Northwest seldom have a basal

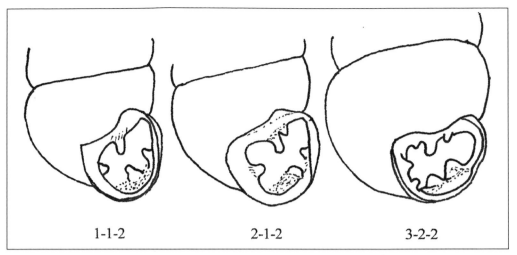

1-1-2   2-1-2   3-2-2

Figure 5: Examples of Dental Formulas (Parietal—Columellar/Basal—Palatal)

without a columellar, combining these two in the formula seems more appropriate. For example, under this method, 1-1-3 nearly always indicates a parietal, no basal, and three palatals, and 1-2-2 indicates a parietal, a columellar and a basal, and two palatals.

Key to the Taxa of Pupiform Snails

Shell minute, usually less than 3 mm high, rarely to 4 mm in the Pacific Northwest. Aperture relatively small and containing zero to several apertural denticles or teeth.

(a)   Apertural dentition normally lacking. If dentition is present, the shell is cylindrical, greater than 2.6 mm high, and with reflected lip margin, or there is a callus ridge around the inside perimeter of the lip . . . .[b/bb]

    (aa)   Apertural dentition normally present. If dentition is lacking, then the width of the last whorl is smaller than that of the penultimate, and/or there is a crest on the last whorl, behind the aperture. . . . . [f/ff]

(b)   Apertural lip thin, not flared or reflected; expanded only at the columellar margin. Shell cylindrical or ovate, and without apertural dentition, crest, or sinulus. Family: VERTIGINIDAE (in part) . . . . Genus: *Columella* [c/cc]

    (bb)   Apertural lip flared or reflected. . . . . [d/dd]

Key to the Species of *Columella*

(c)   Shell ovate or long-ovate, the last whorl about the same width as the penultimate. With 5½ to 6½ whorls the shell is 1.8 to 2.4 mm high. . . . .*Columella edentula*

    (cc)   Shell long-cylindrical, the last whorl may be a little more inflated than the penultimate. Height 2.5 to 3.0 mm with 6 to 7½ whorls
. . . . *Columella columella alticola*

d)   Shell ovately-conic; rapidly expanding whorls well rounded into well impressed
     sutures. Height about 2.1 mm; lip flared. . . . .*Nearctula dalliana* [j/jj]

     (dd) Apertural lip reflected following a constriction of the whorl. Shell relatively
          large, over 2.6 mm high. Color buff to reddish-brown.
          Family: PUPILLIDAE . . . . Genus: *Pupilla* [e/ee]

Key to the Species of *Pupilla*

(e)   Crest lacking; lip thin and reflected. Shell cylindrical, dull reddish or yellowish-
      brown, perforate or rimate. Height about 3 mm. Dentition is usually absent, but
      some may have a small parietal callus. Asotin Co., WA; Salmon River, Idaho; and
      south in the Rocky Mountains . . . . *Pupilla hebes*

      (ee) A prominent crest separated from the reflected apertural lip by a strong
            constriction of the whorl. Apertural teeth sometimes present . . . . [f/ff]

(f)   Shell elliptically-cylindrical, reddish-brown, rimate. Height 2.75 to 4 mm. A
      prominent white or light-colored crest is separated from the reflected, internally
      callused lip by a rather wide constriction. Teeth normally lacking in Rocky
      Mountain varieties but one or two denticles may be present in specimens from
      farther east. Eastern North America, west in Canada and the northern United States
      to northeastern Oregon, and north and south in the Rocky Mountains.
      . . . . *Pupilla muscorum*

      (ff) Apertural teeth normally present (seldom lacking). Peristome, crest,
            and constriction behind the lip variable . . . . [g/gg]

(g)   Shell ovately cylindrical, light brown, rimate. Height about 3.3 mm. Well-developed
      crest with a deep constriction separating it from the flared or reflected lip, which
      has a whitish callus encircling the outer rim. Three denticles are set well back in the
      whorl. Rocky Mountains, east into the Great Plains. . . . . *Pupilla blandi*

      (gg) Shell variable, normally with 4 or more denticles in the aperture, but these
            also vary within a species. Denticles occasionally reduced in size and/or
            number or lacking altogether . . . . [h/hh]

Key to the Species of *Gastrocopta* (one)

(h)   Relatively small, 1.7 mm high by 0.8 mm wide with about 5 whorls. Shell
      cylindrical, transparent to whitish, lip thin and expanded. Apertural teeth 7 (2-2-3),
      parietal and angular fused to form an inverted lambda (λ) when viewed basally.
      Columellar and basal teeth large. Eastern Montana eastward and south to New
      Mexico. . . . .*Gastrocopta holzingeri*

      (hh) Larger, normally equal to or greater than 1.8 mm high. Shell color
           yellowish to dark brown . . . . [i/ii]

(i)   Height about 2.0 to 2.8 mm. Narrowly or widely ovate, rather strongly tapered
      to a rounded apex. Lip more or less flared or reflected. Apertural teeth lacking
      in one species, 4 (1-1-2), or 6 to 7 (1-1-4 or 1-1-5) in others. Crest and sinulus
      lacking . . . . Genus: *Nearctula* [j/jj]

Key to the Species of *Nearctula*

(j) Shell ovately-conic, transparent, greenish-gray with flared apertural lip. Height 2.1 mm with 4½ rapidly enlarging whorls, well rounded into well impressed sutures . . . . *Nearctula dalliana*

    (jj) Sides of spire more rounded. Height about 2.4 to 2.8 mm with 5½ to 6 whorls. Aperture with 4 to 7 denticles . . . . [k/kk]

(k) Height 2.45 to 2.8 mm with 5½ to 6 whorls. Ovately-oblong, tapering to a narrow, rounded apex. Adults rimate; immatures perforate. Teeth strong, white (1-1-2). Thread-like radial striae strongest on the penultimate whorl. West-central California through western Oregon and Washington and into British Columbia . . . .*Nearctula* sp. (formerly *Vertigo rowellii*)

    (kk) Height about 2.5 to less than 3 mm. Shell strongly ovate, buff-colored, openly but narrowly umbilicate. Single high parietal tooth, smaller columellar, and a close-set row of basal-palatals, some of which are fused together. Aperture semi-elliptical; lip flared, slightly reflected basally. Olympic Peninsula . . . .*Nearctula* new sp.

    (ii) Height 1.7 to 2.8 mm in 4½ to 6½ whorls. Shell ovate to cylindrical, perforate or rimate. Lip not reflected, only very slightly if at all expanded. Crest, sinulus, and sculpturing are variable. Apertural teeth normally 4, 5, or 6, but occasionally more, less, or none. . . . . Genus: *Vertigo* [l/ll]

Key to the Species of *Vertigo*

(l) Apertural teeth normally 4 (1-1-2), 1 parietal, 1 columellar, and 2 palatals, but rarely with a small angular next to the parietal . . . .[m/mm]

    (ll) Apertural teeth normally 5 or more (2-1-2), a parietal and a small angular beside it, 1 columellar, 2 palatals and sometimes a basal and additional parietal and palatals . . . . [p/pp]

(m) Height 2.2 mm or more; 5½ to 6 whorls. Cylindrical, tapering elliptically, with a weak to moderate crest complete before the aperture. . . . . [n/nn]

(mm) Height about 2.0 mm (1.9–2.05) with 4½ to 5 whorls . . . . [o/oo]

(n) Height 2.3 to 2.7 mm with 5½ whorls. The four denticles may be in the pattern of a cross. Shell sculpturing is weak. Aperture simple or slightly expanded. Umbilicus of full adults is rimately-imperforate. A variable species, widespread across northern United States and Canada to Alaska. . . . . . *Vertigo modesta*

    (nn) Height 2.2 to 2.4 mm. Dentition variable, typically reduced in size and number, most often lacking the upper palatal. Pattern may be 1-1-1 or 1-1-0 or, less commonly, other combinations including 2-1-2 or an absence of denticles. Shell sculpturing of fine radial striae on the penultimate and preceding whorl. Montana, northeastern Oregon, central Washington . . . .*Vertigo modesta sculptilis*

(o)   Crest low but apparent, usually truncated by the aperture. Sinulus lacking.
      Rimately-perforate. Western Oregon and Washington to Alaska . . . *Vertigo columbiana*

   (oo)   Crest very weak but with a strong sinulus extending back to form
          an indentation over the lower palatal. Rimately imperforate. Vicinity
          of Adams County, Idaho, and northeastern Washington . . . . . *Vertigo idahoensis*

(p)   Normally with 5 denticles; height of shell variable. . . . . [q/qq]

   (pp) Normally 6 or more denticles. Last whorl with a distinct crest and
        normally with a sinulus, which may be strong or weak. . . . .[t/tt]

(q) With 2 parietals (2-1-2). . . . . [r/rr]

   (qq) With 1 parietal (usually 1-2-2) . . . . [s/ss]

(r)   Normally 5 apertural teeth (2-1-2) with a small angular adjacent to the
      parietal. Height 2 to 2.4 mm; cylindrical or long-ovate, roundly tapered
      at the apex. Brown or reddish-brown with sculpture of fine threads,
      a low crest and slightly expanded peristome. Mostly east of the Cascades
      into the Rocky Mountains. . . . . *Vertigo concinnula*

   (rr)   Height 2.4 to 2.7 mm, cylindrically-oblong, dark olive-brown.
          With a distinct crest and weakly striate sculpture. . . . . *Vertigo modesta parietalis*

(s)   Five to seven denticles (1-2-2 or less often 2-2-2 or 2-2-3) with a basal and
      sometimes a small angular. Strong palatal callus. Height 2.2 mm, width 1.2 mm
      in 5 whorls. A weak crest before the aperture and a prominent sinulus
      extending as an impression over the lower palatal. Montana eastward,
      across southern Canada and south to New Mexico . . . .*Vertigo elatior*

   (ss)   Five apertural denticles (1-2-2) including a small basal and 2 strong
          palatals. A small, cylindrically-oblong, chestnut-brown shell, 1.75
          to 2.0 mm high by about 1.0 mm wide, with a low crest and an
          evident sinulus. Rocky Mountains from British Columbia to
          New Mexico. . . . . *Vertigo gouldi basidens*

(t)   Rather large, height 2.3 to 2.5 mm with 5 to 5½ whorls in a brown, cylindrical
      shell. Apertural teeth normally 2-2-2 including a weak basal (however, see
      discussion of the Oswego form in the species account). Low but distinct crest,
      a constriction before the lip, and a slight sinulus extending as an indentation
      over the upper palatal. Western Oregon to southwestern Washington.
      . . . .*Vertigo andrusiana*

   (tt)   Shell slightly smaller, 2.1 to 2.3 mm high in about 5 whorls. Diameter
          greater than 1.0 mm. Five to six or more denticles in varied patterns. . [u/uu]

(u)   Height 2.2 to 2.3 mm by 1.4 mm wide in 5 whorls. Strongly ovate (body
      whorl distinctly larger than the penultimate whorl. Shell reddish-brown with
      light-colored crest and prominent sinulus. Furrows extend over both palatal
      teeth. Six or more denticles (2-2-2 or up to 3-2-4) sometimes with a small
      additional parietal and 2 additional palatals. Widespread. . . . . *Vertigo ovata*

   (uu)   Shell more cylindrical. Penultimate whorl not much smaller than the
          body whorl. Sinulus apparent with only one furrow behind the low but
          distinct crest. Height 2.1 mm in 5 whorls; width 1.1 mm. Normally six
          denticles (2-2-2), including a weak basal. Shell reddish-brown. Eastern
          Montana and southern Canada to Iowa and New Mexico . . . *Vertigo binneyana*

Family: PUPILLIDAE
Genus: *Pupilla*

Relatively large pupillids, 2.75 to 4 mm high. The shell is cylindrical and the apertural lip is distinctly flared or reflected. Pilsbry (1948) warns that distinguishing between the species is sometimes difficult because variations in shell characteristics produce intermediate forms. At least some *Pupilla* species bear living young (Steenberg, cited in Pilsbry 1948)

Pupilla hebes (Ancey, 1881)   Crestless Column

Description: With 6 to 7 whorls this snail measures about 3 mm (2.7 to 4.0) high by 1.5 to 1.9 mm wide. The shell is cylindrical with rounded ends, and is perforate or rimate. The last whorl is usually without a crest, but in some there may be some swelling before a distinct constriction just behind the thin, reflected lip margin. The shell is a reddish or yellowish-brown color; the surface is dull with weak transverse striae. The aperture is small and obliquely ovate. It is usually without apertural teeth (other species are denticulate), but there may be a small raised callus-like parietal cusp.

Similar Species: The other two *Pupilla* species from the Pacific Northwest have prominent crests. *P. muscorum*, with a reddish-brown shell, has a thickened callus around the inner periphery of the aperture, and *P. blandi* is denticulate, as well as having a whitish callus around the outer edge of its peristome. *Columella c. alticola* is darker brown than *Pupilla hebes*. It lacks dentition and has a simple, unreflected lip margin.

Distribution: *P. hebes* is known from Asotin County, Washington; Idaho Co., and southeastern Idaho, south to New Mexico, southeastern California, northern Mexico, and Baja California. Branson (1977) reported it from timberline and above in the Olympic Mountains of Washington, which needs to be confirmed. It is found among talus and under rocks and shrubs and sometimes woody debris.

Pupilla hebes: S of Asotin, Asotin Co., WA; scale bar = 1 mm

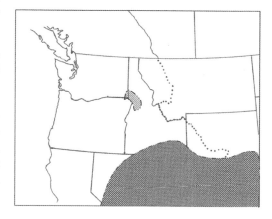

Pupilla muscorum (Linnaeus, 1758)   Widespread Column

Description: Measuring 2.75 to 4 mm high by 1.5 to 1.9 mm wide with about 5¾ to 7 whorls, the reddish-brown shell is rimate, and somewhat elliptically cylindrical. It has a prominent, lighter-colored or white crest separated from the apertural lip by a rather wide constriction. The peristome is reflected and callused inside. The Rocky Mountain varieties normally lack dentition, but populations from farther east may have one or two small teeth, including a parietal and/or lower palatal, or a columellar.

Similar Species: *P. hebes* usually lacks a distinct crest and its reflected peristome is not thickened or callused. *Columella* species lack a reflected lip.

Distribution: *Pupilla muscorum* is native to Europe, Asia, and northern Africa, as well as North America. Its range extends from the northeastern United States and southern Canada, westward to the Rocky Mountains and into northeastern Oregon. In the Rocky Mountains it extends north to Alaska and south to northern Arizona and New Mexico. It may be found in more humid environments, such as wooded sites under woody debris, shrubs, and rocks.

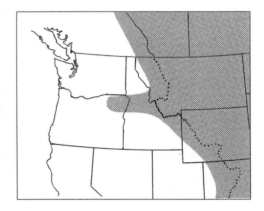

Pupilla blandi   E. S. Morse, 1865   Rocky Mountain Column

Description: The shell with 6½ to 6⅔ whorls measures about 3.3 mm long by 1.5 to 1.75 mm wide. It is light brown in color, somewhat opaque, and ovately cylindrical. The umbilicus is rimate. The last whorl ends with a well-developed crest, then a deep constriction before the flared or reflected lip margin, the rim of which is encircled by a whitish callus. There are normally 3 apertural teeth, all of about the same size: a parietal, a columellar, and a basal or lower palatal set well back in the whorl.

Similar Species: The callused rim of the peristome separates this species from other look-alikes. *P. muscorum* may have some reduced dentition, but its shell is reddish-brown, and it has a callused rim inside the aperture instead of outside around the edge of the peristome.

Distribution: *P. blandi* is a rocky Mountain species known from Alberta, Montana, Nevada, and Kansas, to New Mexico and Texas, occurring in semiarid habitats.

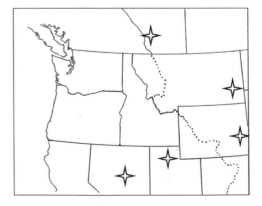

Family: VERTIGINIDAE
Genus: *Columella*

Apertural lip simple, reflected only at the columellar margin. Shell perforate and cylindrical or ovate, lacking apertural dentition. Color glossy, translucent, and cinnamon or darker brown in color. May be distinguished from immature *Vertigo* by its more cylindrical last two to three whorls, which are more squared off at the base, while *Vertigo* usually taper inward a little toward the base.

Columella edentula (Draparnaud, 1805)   Toothless Column
Synonym: *Columella simplex* (Gould, 1840) may be an "unresolved" species (Turgeon et al. 1998)

Description: The shell is usually ovate to ovate-oblong, or larger shells are cylindrical for the last 2 to 3 whorls. The last two whorls are about the same width, above which the spire tapers to the rather broadly rounded apex. With 5½ to 6½ whorls, it measures 1.8 to 2.4 mm high by 1.1 to 1.35 mm wide. It is perforate with a simple, sharp lip and no apertural dentition, sinulus, or crest. The shell is cinnamon brown or darker. Sculpturing is normally of very fine collabral striae, but some specimens from Shoshone County, Idaho, had rather coarse threads.

Similar Species: *C. columella alticola*, *Pupilla hebes*, and some *Vertigo modesta* also lack apertural dentition. Both of these first two similar species are larger and more cylindrical. *Pupilla hebes* shells are also lighter colored and have a reflected lip margin. Also see distribution information for each species. *Vertigo modesta sculptilis* is rimate, has a low to moderate crest, a thickened peristome, and usually some or most specimens in a population will have some apertural dentition. The last whorl of *Vertigo* usually tapers inward a little, while *Columella* and *Pupilla* are more cylindrical to the base.

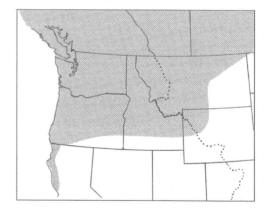

Distribution: The range of *Columella edentula* is from Alaska, across Canada and the northern United States, south through all but the more arid areas of the Pacific Northwest, and into northwestern California. It can be found in moist forested habitats living in forest floor litter, and it can often be found on stems and leaves of sword ferns, other herbaceous plants, and branches of shrubs.

Columella columella alticola (Ingersoll, 1875)   Cylindrical Mellow Column
Synonyms: *Columella columella* (Martens, 1830); *C. alticola* (Ingersoll, 1875)

Description: Larger than *C. edentula*, with 6 to 7 whorls shells of these snails are 2.5 to 3.0 mm high by 1.3 to 1.35 mm wide. The apex is rounded, and the last four whorls are cylindrical. When fully adult, the body whorl is more inflated than the penultimate. It is perforate with a simple lip and without sinulus, crest, or apertural dentition. The shell is chestnut or cinnamon brown and is finely, radially striate.

Similar Species: Of the three Pacific Northwest pupillids that normally lack apertural dentition, *C. edentula* is smaller and more ovate, and the body whorl is of the same width as the penultimate. *C. columella alticola* and *Pupilla hebes* are both larger and more cylindrical. The two species of *Columella* have a simple, sharp lip, except where it flares at the columellar margin. *Pupilla hebes* has a reflected lip margin, often a crest and constriction behind the lip, and, in some areas, may have some reduced dentition.

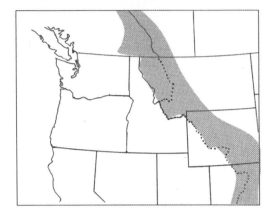

Distribution: *C. columella alticola* occurs in the Rocky Mountain region, from the Kootenay River Valley in British Columbia to Arizona and New Mexico and eastward into Kansas, Iowa, and Illinois. It is most often found among leaf litter, under woody debris, and on low plants in moist forests and riparian zones in mountainous areas.

Genus: *Gastrocopta*

Gastrocopta holzingeri (Sterki, 1889)   Lambda snaggletooth

Description: Not known from within the defined limits of this work, this species is included here as potentially occurring peripherally to this area. It is small, even among the pupillids, measuring about 1.7 mm long by 0.8 mm wide with 5 whorls. The shell is transparent, whitish, cylindrically ovate, slowly tapering to a bluntly rounded apex, and with a rimate umbilicus. The aperture is oval, truncated by the penultimate whorl, and with the lip expanded. Apertural teeth are large, and include a high parietal and angular that join to form a backwards "y" or Greek lambda (λ) when viewed from below (the basis of the common name). There are also large columellar and basal denticles, and 3 palatals, of which the upper one is very small.

Similar Species: No other snail occurring within or near to the Pacific Northwest closely resembles this tiny white pupillid, with its expanded lip and distinctive tooth pattern.

Distribution: *Gastrocopta holzingeri* occurs from Ontario and New York to Helena, Montana, and south to New Mexico. Forsyth (2004) reported it from Alberta and the eastern edge of British Columbia, Canada.

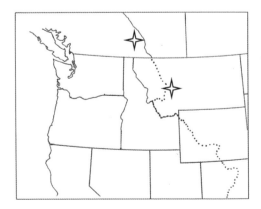

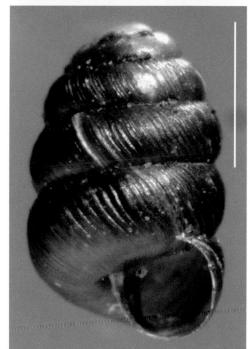

1. *Columella edentula*: Lopez Island, San Juan Co., WA; scale bar = 1 mm

2. *Columella edentula*: Slate Cr., Idaho Co., ID; scale bar = 1 mm

3. *Columella edentula*: Shoshone Co., ID; scale bar = 1 mm

Genus: *Nearctula*

***Nearctula* sp.**   Threaded Vertigo
Synonym: Erroneously known as *Vertigo rowellii*

Description: With 5½ to 6 whorls, this species measures 2.45 to near 3.0 mm long by 1.25 to 1.49 mm wide. The shell is dark olive to light coffee brown. It is rather long-ovate, tapering convexly to a narrow rounded apex, the taper giving it a rather long, narrow appearance. The aperture is oval but truncated by the penultimate whorl; the peristome is flared to narrowly reflected. Immature shells have a round, perforate umbilicus that becomes rimately perforate in adults. There is no apparent crest or sinulus. There are 4 mostly strong, white, apertural teeth, the parietal being quite high. The shell has fine, radial, thread-like sculpturing, most distinct on the penultimate whorl, and indistinct spiral lines can be seen on some specimens.

Note: For many years this species was known as *Vertigo rowellii* (Newcomb), but the original type specimen of *V. rowellii* was instead what has been known as *V. californica* (Rowell, 1861). Correcting the nomenclature for *V. californica* then leaves what has been called *V. rowellii* without a Latin name. "Although this valid species does not currently have a scientific name, we retain this listing because it is the most widespread and one of the most common species of vertiginines in the Pacific States" (Turgeon et al. 1998).

Similar Species: Dentition is similar to that of *V. modesta* and *V. columbiana*, though stronger. The long tapering of this shell and its distinctly flared lip distinguish it from these and other *Vertigo* found in the same areas.

Distribution: The range of this species is from west-central California through western Oregon and western Washington in the Puget Trough and San Juan Islands to the Pacific Ocean and into British Columbia, including Vancouver Island. It is a forest species, often found on branches of shrubs (e.g., ocean spray), and in forest floor litter.

***Nearctula* new sp.**   Hoko Vertigo
Synonym: *Vertigo* **new sp.**

Description: The shell is broadly ovate, tapering roundly from the penultimate whorl to a rounded apex. Height is about 2.5 mm, and it has a narrow, round, open umbilicus. It is buff-colored, and relatively smooth and glossy. The dentition is distinctive. Denticles are whitish including a prominent parietal, a smaller columellar, and multiple basal and palatal cusps, some fused laterally.

Note: This unique pupillid was discovered by Dr. Terrance Frest and Mr. Edward Johannes and was listed by them as *Vertigo* n. sp. 1, the Hoko Vertigo, in Frest and Johannes (1993). It was included as a "Survey and Manage Species" (USDA Forest Service and USDI Bureau of Land Management 1994). It was later included in Kelly et al. (1999) from descriptions

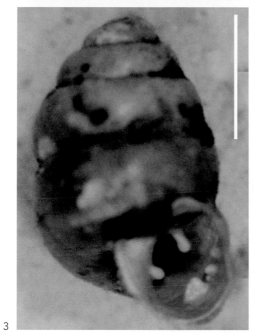

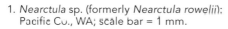

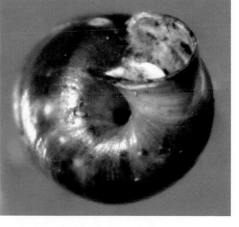

1. *Nearctula* sp. (formerly *Nearctula rowelii*); Pacific Co., WA; scale bar = 1 mm.

2. basal view.

3. *Nearctula* new sp. (Hoko Vertigo): Clallam Co., WA; scale bar = 1 mm.

4. *Nearctula* undescribed sp. (Hoko Vertigo cf.): Scott Cr., Salem District, Bureau of Land Management, OR; scale bar = 1 mm.

5. basal view.

and specimens provided by Frest and Johannes. Dr. Frest (personal communication) later believed it to belong in the genus *Nearctula* rather than *Vertigo*.

Similar Species: The shell most closely resembles that of *Vertigo ovata*, but lacks the sub-columellar denticle, the strong sinulus, and dark reddish-brown color of that species. The Oregon specimen (see distribution) may or may not be of the same animal, but it does not appear to be any other described species.

Distribution: This species was found on the northern Olympic Peninsula, and a single somewhat similar-appearing specimen (TEB 99-066) was submitted by Mr. Stephen Dowlan from the Salem BLM District in Oregon. Frest and Johannes (1996) empha-size its affinity to arboreal habitats, especially the underside of leaves and branches on clumps of deciduous trees and shrubs, some-times among mosses or lichens growing on the trees. It may also be found in leaf litter under such trees.

Nearctula dalliana   Sterki, 1890   Horseshoe Vertigo

Nearctula dalliana

Description: Shell ovately conic, the sides of the spire tapering conically to a rather broad, rounded apex. The shell is transparent, its color greenish-gray or olive, and it is glossy with fine irregular growth-wrinkles. The 4½ whorls are well rounded into the well-impressed sutures. The whorls increase rapidly, the body whorl being obviously the largest. The shell, approximately 2.1 mm high by 1.35 wide, is rimately imperforate. The aperture is edentate, somewhat ovate, and a little oblique; the lip is flared a little, the end insertions ex-tending onto the penultimate whorl.

Similar Species: The edentate, strongly tapered, ovately-conic, pupiform shell with flared lip separates this species from all other land snails of the PNW.

Distribution: *N. dalliana* occurs from Jackson Co., Oregon, to central California (Roth and Sadeghian 2006).

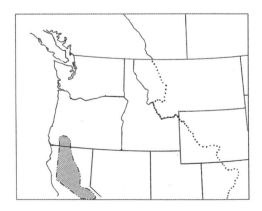

Genus: *Vertigo*

Shell height mostly less than 2.5 (up to 2.8) mm. Shells ovate to sub-cylindrical, tapering somewhat toward the rounded apex. Color brownish, of varying shades. Usually four or more apertural teeth (occasionally fewer). Lip sometimes thickened or expanded; sometimes with a sinulus (wave) in the outer margin. Widespread. Many species, several in the Pacific Northwest (lists below include only PNW species, therefore may not include the species for which the group is titled). In this work, the apertural dentition is relied on heavily in the keys and descriptions of the *Vertigo*. These structures can be quite variable in many species, so other characteristics need to be compared as well. Characteristics may intergrade, so several specimens may need to be examined to determine the most likely taxon of a population.

Groups of Pacific Northwest *Vertigo*

Vertigo gouldi Group: (5 teeth, 2-1-2 or 1-2-2, low crest, weak or no sinulus)
  Vertigo concinnula
  Vertigo gouldi basidens

Vertigo modesta Group: (4 teeth, low crest, no sinulus)
  Vertigo modesta
  Vertigo modesta parietalis
  Vertigo modesta sculptilis

Vertigo ovata Group: (5 or more teeth, low crest, strong sinulus)
  Vertigo binneyana
  Vertigo ovata
  Vertigo elatior

Vertigo pygmaea Group: (4–6 teeth, low crest, with or without sinulus)
  Vertigo andrusiana
  Vertigo columbiana
  Vertigo idahoensis

Vertigo andrusiana   Pilsbry, 1899   Pacific Vertigo

Description: The shell measures 2.3 to 2.46 mm long by 1.3 mm wide in about 5½ whorls. It is rimately-imperforate and cylindrical with a roundly tapered apex. Shell color is cinnamon-brown, but the first whorl is translucent gray. There is a low crest with a shallow constriction before the lip, and a shallow, sometimes indistinct, sinulus with an extended indentation over the upper palatal tooth. Apertural teeth usually 2-2-2, including a weak basal. The columellar ascends inward. There is a thin, light-colored palatal callus. The peristome is reflected at the columella but barely expanded palatally and little so basally. Radial striations are weak if at all apparent.

Variations: In his 1948 monograph, Pilsbry discussed a variety from Oswego, Clackamas County, Oregon, that is smaller (1.9 to 2.25 mm in 4½ to 5 whorls). It is darker colored than typical *andrusiana*. It lacks the angular denticle, and usually the basal also, but it has the palatal callus. "While this Oswego form has characteristics of *V. columbiana*, *V. pygmaea* and *V. a. sanbernardinensis*, it appears most closely related to *V. andrusiana*" (Pilsbry 1948).

Similar Species: *V. columbiana* is similar to Pilsbry's "Oswego" form. It differs in having a rimately perforate umbilicus, its columellar denticle does not ascend inwardly as distinctly, and it lacks a distinct palatal callus. *V. binneyana* is slightly smaller than typical *V. andrusiana*. It is also more cylindrical, reddish-brown in color, and has an indentation over both palatal teeth. They are also separated by geography, except in northwestern Washington and British Columbia where their ranges supposedly merge.

Distribution: *V. andrusiana* occurs in western Oregon through the Puget Trough and Olympic Peninsula of Washington and Vancouver Island, British Columbia. It occurs in forested sites at lower elevations and may be found on trunks and lower branches of deciduous trees and shrubs, as well as among the litter beneath them.

Vertigo binneyana Sterki, 1890   Cylindrical Vertigo

Description: With about 5 whorls *V. binneyana* measures 2.1 mm long by 1.1 mm wide. The shell is reddish-brown, rather cylindrical but with a roundly-tapered spire. There is a low but distinct crest behind the palatal lip. There is a slight sinulus and a distinct impression over the palatal teeth. Apertural teeth, 2-2-2, are generally strong except the angular and basal, which are small. The columellar ascends inward. A palatal callus is evident. Radial striations are weak.

Similar Species: *V. andrusiana*, which is not likely to be found within the same area, is a little larger, and it has an indentation over only the upper palatal tooth. *V. ovata* may sometimes have 6 apertural teeth but with a different pattern and a prominent sinulus. The shell of *V. ovata* is distinctly ovate, while that of *V. binneyana* is cylindrically-ovate.

Distribution: *V. binneyana* has been reported from Manitoba to Vancouver Island, British Columbia; Glendive and Helena, Montana; Seattle, Washington; and New Mexico. However, Forsyth (2004) believed that the Vancouver Island and Seattle, Washington, reports may have been misidentifications. Other than those two sites, the species is known only from east of the Rocky Mountains, although Helena, Montana, is near the Continental Divide.

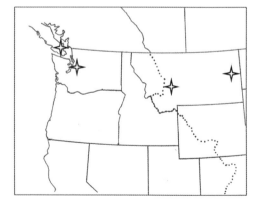

Vertigo columbiana Pilsbry & Vanatta, 1900   Columbia vertigo

Description: Shell 1.9 to 2.05 mm long by 1.1 to 1.2 mm wide in about 5 whorls. The shape is cylindrically-oval tapering to a broad, blunt apex. It is somewhat transparent, glossy, light brown, and is perforate or rimately so. The crest is low but apparent and is truncated by or terminates at the aperture, often without turning inward before the lip. Apertural teeth normally 4 (1-1-2) but occasionally with a very small angular beside the parietal. Parietal and lower palatal are fairly high but not very long. The upper palatal quite small, about half as long as the lower. The columellar ascends little, inwardly. Radial striations are very weak. Apertural lip slightly expanded, if at all.

Similar Species: *V. modesta* has a higher spire, and its apertural teeth are more evenly spaced, so they are more directly opposite, in the pattern of a cross. *V. columbiana* occasionally has a small angular tooth to resemble *V. concinnula* or *V. modesta parietalis*, which are both a little larger, more cylindrical, with a more distinct angular, and a crest that is complete before the aperture.

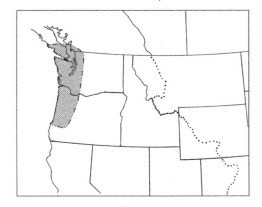

Distribution: *V. columbiana* occurs in Douglas County, Oregon, through western Washington and British Columbia, including Vancouver Island and other offshore islands, to Alaska and the Aleutian Islands. In moist forest habitats it is found among litter, on and under small woody debris, and often on the lower trunks and branches of smooth-barked trees and shrubs.

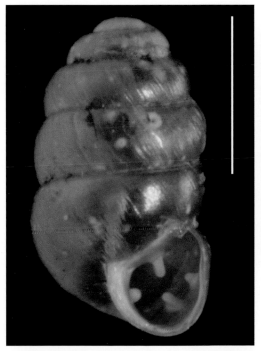

1. *Vertigo columbiana*: Fourmile Cr., Skamania Co., WA; scale bar = 1 mm
2. Basal view: note crest flared to aperture.

Vertigo concinnula Cockerell, 1897   Mitered Vertigo

Description: With about 5 whorls the shell measures 2.0 to 2.4 mm long by 1.1 to 1.25 mm wide. It is cinnamon-brown, rather ovately-cylindrical. The aperture is slightly expanded and with a low to moderate crest behind it. Apertural dentition (2-1-2); a parietal and angular, columellar, and 2 palatals, long but only moderately high, the lower one the longest. The shell sculpture is of radial striae or fine threads, most distinct on the penultimate whorl.

Similar Species: *V. columbiana* sometimes has a small angular tooth, which gives it the same dental pattern, but it is a little smaller, with relatively short palatals, and its crest ends at the aperture. *V. columbiana* is found between the western Cascades and the Pacific Ocean in Oregon, and north to Alaska, while *V. concinnula* is mostly east of the Cascades. *V. modesta parietalis* is most similar, but it is larger, more cylindrical, with short palatals and a less flared peristome.

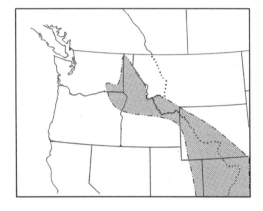

Distribution: *V. concinnula* occurs in southeastern Washington, the Wallowa Valley of Oregon, Idaho, and Montana, to South Dakota, Wyoming, Utah, and New Mexico. It may be found in moist forest sites and riparian zones among litter and under woody debris of deciduous trees.

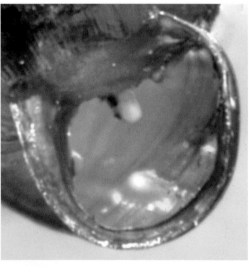

1. *Vertigo concinnula*: Touchet River, Umatilla National Forest, Columbia Co., WA; scale bar = 1 mm
2. Apertural view showing dentition.

1

Vertigo elatior Sterki, 1894 Tapered Vertigo

Description: About 5 whorls; height 2.2 mm; diameter 1.2 mm. Shell oblong-cylindrically-ovate, strongly tapered above the penultimate whorl. There is a weak crest and a prominent sinulus. Apertural teeth 5 to 7 (1-2-2, or less often 2-2-3), strong except for a small but distinct basal, and sometimes a small angular and/or a suprapalatal. There is a prominent palatal callus that connects the two denticles, and a deep impression over the lower palatal.

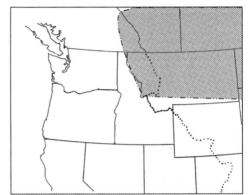

Similar Species: *V. ovata* and *V. idahoensis* also have a strong sinulus but *idahoensis* is smaller, with only four denticles and an indistinct crest. *V. ovata* is larger, and it has a very ovate shell, usually with six or more denticles and a prominent, light-colored crest.

Distribution: *V. elatior* occurs from Minnesota to Hudson Bay; west to Darby and Ward, Montana; and Field, British Columbia.

Vertigo gouldi (A. Binney, 1843) Variable Vertigo

Vertigo gouldi basidens Pilsbry & Vanatta, 1900

Description: A rather small *Vertigo*, 1.75 to 1.95 mm long by 0.95 to 1.05 mm wide. Chestnut brown, cylindrically-ovate. Apertural dentition 1-2-2, a parietal but no angular, a fairly strong columellar, a small basal in the columellar-basal angle, and two relatively strong palatals. There is a rather strong palatal callus above the upper palatal tooth. The crest is low and broad, sometimes prominent, and there is an evident sinulus. Growth striae are fine and distinct, especially on the penultimate and antepenultimate whorls.

Similar Species: Other *V. gouldi* subspecies from outside the Pacific Northwest show a variety of tooth patterns. *V. elatior*, the range of which overlaps that of *V. gouldi basidens*, may have similar dentition, but it is a little larger, and its sinulus and shell taper are much more pronounced.

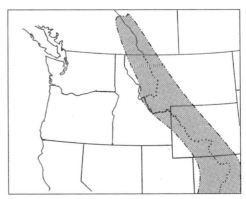

Distribution: Pilsbry (1948) lists *V. gouldi basidens* as a forest dwelling subspecies from Field, British Columbia, and Ward, Montana (both locations being west of the Continental Divide), and south in the Rocky Mountains to Colorado and New Mexico.

Vertigo idahoensis Pilsbry, 1934   Idaho Vertigo

Description: Shell 2.0 mm long by 1.2 mm wide in about 4½ whorls. It is glossy, cinnamon-brown, rimately imperforate, rather ovate, with a roundly conic spire above the penultimate whorl. The crest is very weak, but there is a strong sinulus with an extended indentation over the lower palatal. There are four apertural teeth (1-1-2), the columellar slanting downward toward the aperture. There is no palatal callus. The peristome turns inward above the point of the sinulus, flares outward below it, and is reflected over the columellar margin. Sculpture is of weak growth-wrinkles.

Similar Species: *V. columbiana* and *V. modesta*, also with 4 apertural teeth, are both more cylindrical and lack the distinct sinulus.

Distribution: Type locality of *V. idahoensis* is Meadows, Adams Co., Idaho. A specimen (TEB 577) was also collected in Stevens County in northeastern Washington. It has generally been found in riparian habitats.

Vertigo idahoensis (ovate shell with 4 apertural teeth and sinulus in palatal lip): EF Crown Cr., Stevens Co., WA; scale bar = 1mm

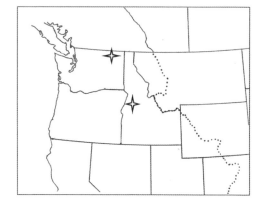

Vertigo modesta modesta Say, 1824   Cross Vertigo

Description: With about 5½ whorls *V. modesta* measures 2.3 to 2.7 mm long by 1.2 to 1.45 mm wide. The shell is glossy, yellowish-brown, rather cylindrical but ovately-tapered toward the apex above the penultimate whorl. Aperture lip simple or slightly expanded

with no apparent sinulus. There is a low crest behind it. There are normally 4 apertural teeth (1-1-2), a parietal, columellar, and 2 palatals that, in the typical variety, form a cross in the aperture. Shell sculpture is weak if present at all.

Similar Species: Apertural teeth in *V. modesta* are especially variable and may range from zero to five in number, so other characteristics need to be examined closely. These variations in dentition have been used to separate out several subspecies, forms, or varieties. An annotated list of subspecies follows, with further descriptions of those that occur in the Pacific Northwest.

   Other species to compare include: *V. columbiana,* generally smaller, less cylindrical and with its crest ending at the aperture; *Nearctula* sp., with a distinctly, tapered, darker brown shell, lacks a crest, and has a flared or reflected lip; *V. concinnula* is normally smaller and has 5 apertural denticles, but some subspecies of *V. modesta* also have 5 denticles (see *V. modesta parietalis*). The palatal teeth of *V. concinnula*, especially the lower one, are longer than that of *V. modesta*.

Vertigo modesta modesta: Adams Co., ID; scale bar = 1 mm

Distribution: *V. modesta* occurs from Labrador to Vancouver Island, Canada; Alaska and across the northern United States; south to Missouri and Kansas; and locally in the northeastern states. It occurs in a variety of forest, shrub, or grassland habitats. It may be found in riparian zones, around springs, or on rock slides.

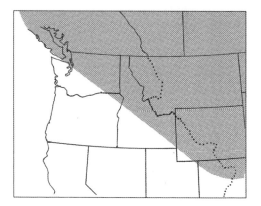

Subspecies of *Vertigo modesta*

Pilsbry (1948) described or discussed the following subspecies, but he said, "The subspecific taxonomy is more or less arbitrary." The recognized varieties include:

- *V. modesta modesta* (see above), from northern United States and southern Canada.
- *V. modesta ultima,* a toothless form from Norton Sound, Alaska.
- *V. m. corpulenta,* a shorter, wider form with a distinct crest and small teeth. Known from Washoe Co., Nevada, and other areas.

- *V. m. parietalis,* larger than *corpulenta* and with larger teeth, which include an angular. From Northern Utah and adjacent Rocky Mountains, the Sierra Nevada of California, and northeastern Washington.
- *V. m. microphasma,* with 2 to 5 very small denticles. From the San Bernardino Mountains of California.
- *V. m. insculpta,* with prominent, sharp striations on the penultimate and antepenultimate whorl. From Arizona, New Mexico, and Utah.
- *V. m. sculptilis,* with prominent, sharp striations on the penultimate and antepenultimate whorl, and with small denticles of 1-1-0 to 2-1-2, but typically 1-1-1; a parietal, columellar, and lower palatal. From Powell Co., Montana, and Wallowa Co., Oregon.
- *V. m. castanea,* toothless or with only very small denticles. From inland central California.
- *V. m. hoppii,* with very small denticles (1-1-0 or 1-1-1). From Greenland.

There appears to be much overlap in the shell characteristics of these different varieties of *V. modesta*. Only those discussed below are documented as occurring in or near the defined area of this work, but other variations are often found. However, unless a population exists that is predominantly of a specific form, that form should not be considered a true subspecies without further study. Where a variety is an occasional occurrence with others of the same species, it might be merely a recurring aberrant form rather than a separate taxon.

V. modesta corpulenta   (Morse, 1865)

Description: The shell is rimately perforate, and with about 4½ whorls it measures 2.2 to 2.45 mm long by 1.35 to 1.4 mm wide. The shell is wider and more ovate than typical *modesta*, and it has a distinct crest behind the lip. There are four teeth but they are reduced in length, and the lower palatal is about the same size or even much smaller than the upper palatal.

Similar Species: This subspecies differs from others by having a more ovate shell, and by the unique characteristic of its lower palatal tooth being about the same size or smaller than the upper.

Distribution: Type locality of *V. modesta corpulenta* is Washoe Co., Nevada, but it is said to also occur with the form *parietalis* over much of its range (Pilsbry 1948), which includes parts of the Pacific Northwest. However, where it occurs with another subspecies, it might best be considered an aberrant form of that subspecies. Habitat is similar to that of *V. modesta modesta*.

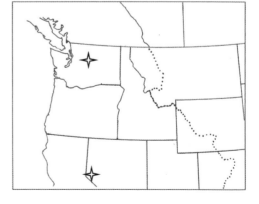

V. modesta parietalis (Ancey, 1887)
Synonyms: *V. modesta corpulenta* form *parietalis* (Pilsbry, 1948); *V. modesta parietalis* (Henderson, 1924); *Pupa corpulenta* var. *parietalis* (Ancey, 1887)

Description: With about 5 whorls the shell is 2.45 mm long by 1.3 mm wide. It is ovately cylindrical, chestnut-brown, and little striate. There are 5 teeth (2-1-2), distinctly longer than those of *corpulenta*, and they include an angular beside the parietal.

Similar Species: Appearance is much like that of *Vertigo concinnula*, but *parietalis* is larger, a little wider, with striations more evenly distributed on the shell, and its peristome is not as flared as that of *V. concinnula*.

Pilsbry (1948) considered this snail *V. modesta corpulenta* form *parietalis* because he felt that they should not be separated from *corpulenta*. The shells of *parietalis*, *corpulenta* and typical *modesta* often occur together, although *parietalis* is usually dominant in such lots. He concluded that "*parietalis* is the more primitive stock, and *corpulenta* a mutation thereof, which has not obtained so wide a distribution." For simplification I have followed Henderson and separated this variety from *corpulenta* because neither its shell shape nor its apertural teeth closely resemble *V. m. corpulenta*. In addition, in some extensive populations of *parietalis*, the subspecies characteristics are essentially uniform.

Distribution: *V. modesta parietalis* is the most common *Vertigo* in northeastern Washington. It also occurs in Idaho (Shoshone, Boise, and Bannock counties); Umatilla Co., Oregon; Glacier National Park and Ravalli Co., Montana; northern Utah (type locality Ogden Canyon); across Colorado and in the Sierra Nevada of California. Habitat is similar to that of typical *V. modesta*.

Vertigo modesta parietalis: tributary of Ruby Cr., Pend Oreille Co., WA; scale bar = 1 mm

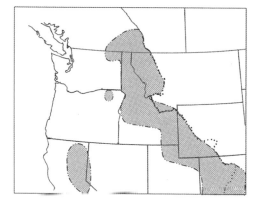

V. modesta sculptilis   Pilsbry, 1934

Description: This subspecies differs from the typical form by distinct, fine radial striae on the penultimate and the antepenultimate whorl. Its shell is about 2.2 to 2.4 mm long. Its dentition is reduced in size and typically lacks the upper palatal, with a formula of 1-1-1. However, variations of 1-1-0 or 2-1-2 also occur, all denticles being small, and the angular and upper palatal being slight, if present at all.

Similar Species: Generally, the size, sculpture, and reduced number and size of the denticles of this subspecies will distinguish it from other pupillids found in the Pacific Northwest, but the species is quite variable, and other forms are often seen which may not yet have been recorded from this region.

Vertigo modesta sculptilis: Swakane Watershed., Chelan Co., WA; scale bar = 1 mm

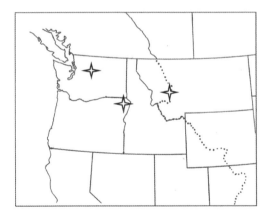

Distribution: *V. m. sculptilis* is known from near Garrison, Powell Co., Montana, and Wallowa Lake, Wallowa Co., Oregon. Specimens have been collected from Chelan Co., Washington, vicinity of Entiat River Valley, its tributaries and adjacent Columbia River Breaks near springs, seeps, and riparian zones.

Vertigo ovata   Say, 1822   Ovate Vertigo

Description: With 5 whorls, *V. ovata* measures up to 2.2 to 2.6 mm long by 1.4 mm wide. Its shell is reddish-brown in color and very ovate. The last whorl is distinctly larger than the penultimate. The palatal lip has a prominent sinulus. There is a strong, light-colored crest behind the palatal lip and a sunken area behind the crest, with a furrow under each of the palatal lamella. Dentition includes 6 to 9 prominent denticles (2-2-2) to (3-2-4). The parietal, columellar, and 2 palatals are quite large, and there are often one or more smaller denticles, including an angular, infraparietal, basal, infrapalatal, and suprapalatal. There is a palatal callus and the apertural lip is expanded but not thickened.

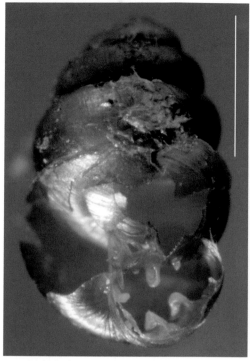

1

2

1. *Vertigo ovata* (ovate shell with sinulus and dental formula 3-2-3): Brown's Lk., Stevens Co., WA; scale bar = 1 mm

2. *Vertigo ovata* (subadult. Ovate shell, lip immature, dental formula 1-3-2); Hoodoo Canyon, Ferry Co., WA; scale bar = 1 mm

Similar Species: *V. ovata* is unique because of its rather large, strongly ovate shell, with a strong crest and sinulus and variable supernumerary denticles.

Distribution: *Vertigo ovata* occurs in the Eastern United States and Canada, west to Puget Sound, Washington, Oregon, and California, with records from Alaska and Arizona. It is most likely found in moist habitats, especially lakeside and riparian areas.

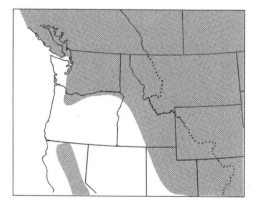

Selected references for the families Pupillidae and Vertiginidae: Branson 1977; Forsyth 2004; Frest and Johannes 1993, 1996; Henderson 1924, 1929a; Kelley et al. 1999; Nekola and Coles 2010; Pilsbry 1948; Roth and Sadeghian 2006; Turgeon et al. 1998.

Family: VALLONIIDAE

Valloniidae is a family of minute, mostly heliciform snails (*Zoogenetes*, the exception in the Pacific Northwest, is ovately-conic). The *Vallonia* are whitish in color; the other two genera are olive or brownish. Most species are ribbed, and all lack apertural dentition.

Key to the Species of VALLONIIDAE

(a)   Shell higher than wide, thin, globosely-ovate, about 3.25 mm high by 2.5 mm wide. Olive-green. Last two whorls with high, widely spaced cuticular riblets . . . . *Zoogenetes harpa*

    (aa)   Shells heliciform or discoidal (wider than high), minute, 3 mm or less in width, with or without thin cuticular riblets on the outer whorls . . .[b/bb]

(b)   Shell brownish in color, with nearly flat spire and high, thin cuticular riblets. Aperture lip thin, or may be thickened on the basal and columellar margins in older individuals . . . .*Planogyra clappi*

    (bb)   Shells whitish, rather solid, although they may be translucent. Aperture lip reflected or thickened to appear expanded. Cuticular ribs present or not . . . . Genus: *Vallonia* [c/cc]

(c)   Riblets present; last whorl descends conspicuously to the aperture . . . .]d/dd]

    (cc)   Riblets lacking. Shells solid and smooth except for lines of growth. Last whorl descends little behind the aperture . . . .[f/ff]

(d)   Lip reflected but thin. Shell diameter 2.7 to 3.0 mm. Last whorl descends sharply behind the aperture so that the lip margin is sometimes free from the penultimate whorl . . . .*Vallonia cyclophorella*

    (dd)   Peristome strong, thickened, expanded or reflected . . . .[e/ee]

(e)   Lip thickened and reflected. Shell diameter 2.5 to 2.8 mm with 3½ to 4 whorls. Umbilicus expands rapidly in last half-whorl . . . . *Vallonia albula*

    (ee)   Lip thickened and strongly expanded except at the palatal insertion. Width 2.5 to 2.9 mm in 3½ whorls. Umbilicus expands regularly in the last half-whorl. . . . . *Vallonia gracilicosta*

(f)   Lip thickened within the aperture and expanded or slightly reflected around most of the margin. Umbilicus increases rapidly but in regular spirals. . . . . *Vallonia pulchella*

    (ff)   Lip thickened but only slightly if at all expanded at the outer margin. Last whorl straightened somewhat in the last one-fourth whorl so the shell and umbilicus are rather oblong. . . . . *Vallonia excentrica*

Subfamily: ACANTHINULINAE

Shells brownish, with simple peristome and rather high cuticular riblets.

Genus: *Planogyra* Morse, 1864

Shell wider than high.

Planogyra clappi (Pilsbry, 1898)   Western Flatwhorl

Planogyra clappi (note spire nearly flat; aperture higher than wide): Patos Island, San Juan Co., WA; scale bar = 1 mm

Description: With about 3½ whorls this minute snail measures approximately 2 mm wide by 1.1 to 1.2 mm high. The shell is light brown in color and has a very low (nearly flat) spire. The umbilicus is fairly wide, funnelform, being about 30% of the shell width. The aperture is roundly oval, slightly but distinctly higher than wide. The peristome is simple. The protoconch of 1½ whorls is microscopically granulose. The teleoconch is sculptured with high cuticular riblets and with microscopic spiral striae in the interspaces.

Similar Species: The most similar appearing species is *Paralaoma servilis* of the family Punctidae. *Paralaoma* differs from *Planogyra clappi* in having an elevated spire and a wider than high aperture. *Vallonia* species differ in being a whitish color with a low spire, and a thickened or reflected peristome.

Distribution: *Planogyra clappi* occurs from the Queen Charlotte Islands and south-western British Columbia through western Washington and Oregon, south to Mendocino Co., California. It occurs primarily within and west of the Cascade Range, but it is also known from northern Idaho. It can be found in forest floor litter, among mosses of moist forest habitats, and around wetlands.

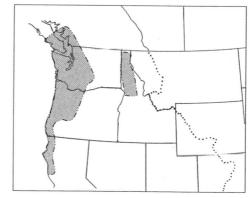

Genus: *Zoogenetes* Morse, 1864
Shell higher than wide.

Zoogenetes harpa   (Say, 1824)   Boreal Top

Zoogenetes harpa (immature): Revelstoke, British Columbia; scale bar = 1 mm

Description: This unique-appearing little snail has a ribbed shell that is higher than wide. With 4 to 4½ whorls, it measures 2.5 mm wide by 3.25 mm high. The shell is ovate-conic, olive green, and narrowly umbilicate. The size of the individual whorls increase rapidly, but because the whorls grow more basally, the width of the shell increases more slowly than the height. The aperture is obliquely oval, a little higher than wide. The two-whorl protoconch is smooth; the following teleoconch is sculptured with fairly high cuticular riblets and microscopic radial lines in the interspaces. The animal is gray dorsally with whitish sides and sole.

Similar Species: The wide top-shaped shell with high riblets and the minute size of *Z. harpa* distinguish it from all other land snails within our area.

Distribution: *Z. harpa* is a boreal species found in northern Europe and Asia, and northern North America (northern United States and Canada from the Rocky Mountains eastward). To my knowledge, it has not been documented within our defined study area, but it occurs at Revelstoke, British Columbia, within the Columbia River watershed, and as far south as Colorado in the Rocky Mountains, so it potentially also occurs within this area. It may be found under hardwood and mixed forest stands in litter and debris, and on herbaceous plants.

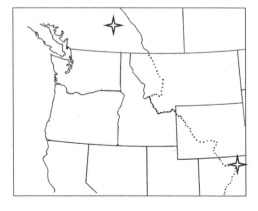

Subfamily: VALLONIINAE

Shells whitish, with low spires and thickened, flared, or reflected peristomes.

Genus: *Vallonia*

Pilsbry (1948) divided *Vallonia* into the following three "groups" that form rather natural breaks based on shell characteristics.

Vallonia costata Group

Shells with ribs, a thickened peristome, at least partly reflected, and with spiral striae on the embryonic whorls. Although *V. costata* itself does not occur in the Pacific Northwest, the following two species of this group are known from within our study area.

Vallonia albula   Sterki, 1893   Indecisive Vallonia

Description: With 3½ to 4 whorls the shell measures 2.5 to 2.8 mm wide and 1.2 to 1.3 mm high. The shell is translucent and whitish colored, with a low conic spire. The whorls increase regularly, the last whorl ascends on the penultimate, then descends to the aperture. The umbilicus is funnelform, wide, a little smaller than one-third the width of the shell, and expands rapidly in the last half-whorl. The aperture is somewhat oval, a little wider than high. The peristome is strong and reflected. The shell is sculptured with fine, close riblets with low cuticular extensions.

Similar Species: *V. gracilicosta* is slightly smaller, and only about 1 mm high. Its umbilicus increases regularly, not accelerating greatly near the aperture. *V. cyclophorella* is larger, and its peristome is reflected, but not thickened.

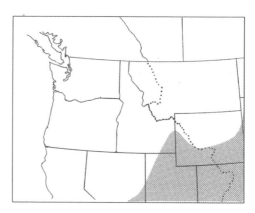

Distribution: *V. albula* is known from the southeastern corner of Idaho (Bannock and Bear Lake counties), south to California and New Mexico, and east through the northern United States and southern Canada.

Vallonia gracilicosta   Reinhardt, 1883   Multirib Vallonia

Description: With about 3½ whorls the shell measures 2.5 to 2.9 mm wide by about 1 mm high. Its color is grayish-white, and it has a low-conic spire. The last whorl is slightly angled around the umbilicus. Its terminal end descends a little to the aperture, which is horizontally oval or a little oblique, but more rounded basally than palatally. The peristome is thickened and expanded except near the palatal insertion. Sculpturing is of distinct cuticular riblets.

Similar Species: This species is somewhat intermediate between *V. albula* and *V. cyclophorella* in size and other characteristics. *V. albula* is about the same diameter but a little higher. *V. cyclophorella* is slightly larger, its reflected lip little if at all thickened, and its last whorl descends more distinctly to the aperture, which may be free from the penultimate whorl.

Distribution: *V. gracilicosta* occurs in the midwestern states to the east slope of the Rockies in Montana, Bannock Co. in southeastern Idaho, and south to Arizona and New Mexico.

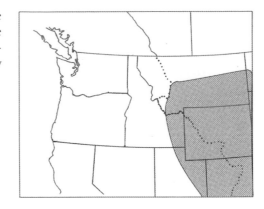

Vallonia cyclophorella Group

Shells ribbed with a thin, reflected peristome.

Vallonia cyclophorella   Sterki, 1892   Silky Vallonia

Vallonia cyclophorella: Grande Ronde River, Asotin Co., WA; scale bar = 1 mm

Description: The shell measures 2.6 to 3.0 mm wide. It is a translucent whitish or grayish color with a very low conic spire. The last one-third whorl ascends then descends rather sharply to the aperture, which is sometimes below and free from the penultimate whorl. The umbilicus is funnelform, large, to greater than one-third of the shell width, and it expands regularly though more rapidly in the last half-whorl. The aperture is roundly-oval, slightly wider than high and more rounded basally than palatally. The peristome is thin and narrowly but distinctly reflected except palatally. Sculpture of the teleoconch is of rather high cuticular ribs; the protoconch has only very faint spiral striae.

Similar Species: *V. albula* and *V. gracilicosta* are slightly smaller, have thickened peristomes, and their body whorls do not descend as deeply to the aperture.

Distribution: *V. cyclophorella* is known from the western half of the United States (North Dakota to Texas and westward). It occurs sporadically throughout the Pacific Northwest east of the Cascade Crest, where it is found in grassy understory of mesic open forests and rocky habitats.

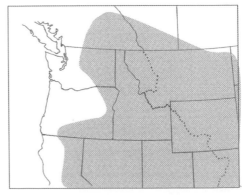

Vallonia pulchella Group

Shell primarily smooth with thickened peristome.

Vallonia excentrica   Sterki, 1893   Iroquois Vallonia

Vallonia excentrica: Thurston Co., WA; scale bar = 1 mm

Description: With 3 to 3½ whorls the shell measures 2.3 mm wide by 1.1 mm high. It is opaque, whitish in color and has a very low spire. Width of the whorls increases regularly, but expands rapidly in about the last one-fourth to one-third whorl. Viewed from above, the periphery of that last one-fourth whorl continues rather straight to the apertural lip. The umbilicus is moderately wide, funnelform, about one-third the width of the shell. It expands rapidly in the last quarter-whorl. The peristome is thickened within but not reflected nor abruptly expanded. The shell is generally smooth except for faint lines of growth.

Similar Species: *V. pulchella*, also unribbed, is rounder in outline and in the umbilicus, and its last whorl ends in a continuous arc to the rather abruptly expanded peristome. *V. excentrica* lacks the appearance of an abruptly flared or reflected peristome. In specimens from east of the Cascades, check the peristome, and for the descent of the last whorl to the aperture, to be sure that a specimen isn't one of the ribbed species with its riblets eroded away.

Distribution: This species is widespread in Europe, Africa, and North America. It is native to New England and into the midwestern states, and has been sporadically introduced in the western states. *V. excentrica* occurs in yards and flower gardens, being spread with nursery stock; it is sometimes found with *V. pulchella*.

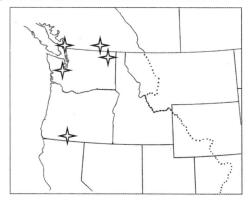

Vallonia excentrica: Thurston Co., WA

Vallonia pulchella   (Müller, 1774)   Lovely Vallonia

Vallonia pulchella: Old Kettle Falls, Stevens Co., WA; scale bar = 1 mm

Description: *V. pulchella* is very similar to *V. excentrica*. With about 3½ whorls it measures 2.4 mm wide by 1.2 mm high. It is opaque, whitish in color, with a very low conic spire. Width of the whorls increases regularly but expands rapidly just before the aperture. Although increasing more rapidly near the end, viewed from above the last quarter-whorl continues in an arc to the thickened and abruptly expanded peristome. The umbilicus is moderately wide, funnelform, a little more than one-fourth the width of the shell, and although expanding more rapidly in the last half-whorl, it is not as much or as straightened as that of *V. excentrica*. The shell is generally smooth except for faint growth-wrinkles.

Similar Species: The outline and umbilicus of *V. excentrica* are more elongated than in *V. pulchella*, and the last whorl of *V. excentrica* ends in a relatively straight line to the apertural lip, which is thickened within but not abruptly expanded, thus showing no appearance of a reflected peristome. See caution regarding ribbed species under *V. excentrica*, above.

Distribution: *V. pulchella* is widespread in Europe, Asia, North Africa, Australia, and North America. Although native in the eastern United States and Canada, it has apparently been introduced into western North America (Pilsbry 1948; Hanna 1966). It is fairly common in the Pacific Northwest, most often in yards and flower gardens, being spread through commerce. It can often be found in or under containers of nursery plants.

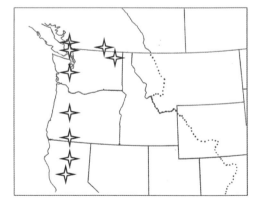

Selected references for the family Valloniidae: Burch 1962; Forsyth 2004; Frest and Johannes 2000; Hanna 1966; Henderson 1924, 1929a; Hendricks, Maxell, and Lenard 2006; Kerney, Cameron, and Riley 1994; Pilsbry 1948; Roth 1985.

Suborder: Sigmurethra Pilsbry, 1900
Superfamily: ACHATINOIDEA Swainson, 1840
Family: SUBULINIDAE
Genus: *Allopeas*

Allopeas mauritianum   (Pfeifer, 1852)   Mauritian Awlsnail
Synonym: *Lamellaxis mauritianus* Pfeifer, 1852

Description: A long, narrowly conic, translucent shell, glossy, with distinct growth-wrinkles and a rounded apex. With about 8 whorls it is approximately 14 mm long. The tapering sides are rather straight, and the individual whorls are little convex. The aperture is small and narrow, the columellar and basal lips flared or thickened, the outer lip thin and simple.

Similar Species: No similar species occurs in the Pacific Northwest. Kerney, Cameron, and Riley (1994) describe *Lamellaxis clavulinus* (Potiez and Michaud, 1938) with *L. mauritianus* as a synonym, but the species they described measures 7.5 to 9 mm long, while Hanna (1966) described the length as 14 mm. The illustration (Kerney, Cameron, and Riley 1994) and photograph (Hanna 1966) appear similar with the same number of whorls. Turgeon et al. (1998) lists *Allopeas mauritianum* and *A. clavulinum* as separate species.

Distribution: *A. mauritianum* is a tropical species, possibly from Africa but widely spread by man. It occurs in moist ground litter in the wild and in greenhouses in temperate climates. Hanna (1966) reported it as *Lamellaxis mauritianus* from Hillsboro, Washington Co., Oregon, in manure being spread around greenhouses in 1955.

Selected references: Hanna 1966; Kerney, Cameron, and Riley 1994; Turgeon et al. 1998.

Superfamily: RHYTIDOIDEA
Family: HAPLOTREMATIDAE

The family Haplotrematidae is composed of medium to large predatory snails. The yellow-ish or greenish to brown shell has a low to nearly flat spire and a wide umbilicus. Whorls enlarge regularly to the last, which is much wider than the penultimate. The aperture is large and without dentition. The shell surface is sculptured with fine to coarse spiral striae and regular growth-wrinkles or beaded ribs. Animals are usually white with gray in the anterior region and brown on the mantle collar.

Key to the Species of HAPLOTREMATIDAE

(a)   Shell large, 20 to over 30 mm wide. Sculpture of fine, closely spaced spiral striae. Palatal lip of the aperture straight or a little extended and/or slightly deflected in adults. . . . . *Haplotrema vancouverense*

   (aa)   Shell 11 to 25 mm wide. Sculpture of beaded ribs formed by relatively coarse spiral striae cutting across closely spaced radial ribs (on inner whorls only of *Ancotrema hybridum*). Palatal lip of adults distinctly deflected and with an oblique crease or dent . . . . Genus: *Ancotrema* [b/bb]

(b)   Beaded sculpturing lacking on the outer whorl, but visible on some earlier whorls or, at least, in the umbilicus. Palatal lip of adult apertures deflected with an oblique dent or channel as typical for *Ancotrema*. . . . . . . *Ancotrema hybridum*

   (bb)   Beaded sculpturing seen on entire teleoconch . . . . [c/cc]

(c)   Size variable, 11 to 22 mm in 5 to 6⅓ whorls. Umbilicus about one-fourth the shell diameter. Palatal lip usually deflected less than half the height of the aperture, and with rather sharply depressed channel running from just median to its center, obliquely outward toward the outer curve of the body whorl. Pacific slope, Alaska southward into northern California, less common east of the Cascade Range. . . . .*Ancotrema sportella*

   (cc)   Size 11 to 15 mm in 5 to 5½ whorls. Umbilicus about one-third the diameter. Palatal lip deflected more than half the height of the aperture with a mid-palatal, more rounded channel. Northern California.
      . . . .*Ancotrema voyanum*

Genus: *Ancotrema*

Medium to moderately large snails with beaded sculpture. Adults have a deflected palatal lip with an oblique crease or channel.

Ancotrema hybridum   (Ancey, 1888)   Oregon Lancetooth

Ancotrema hybridum: Archer Mountain, Skamania Co., WA; scale bar = 1 cm

Description: A medium to fairly large snail, 15 to 25 mm diameter by 7 to 11.5 mm high in 6 to 6¼ whorls. The shell closely resembles that of *A. sportella* except that it may be somewhat larger, and the last whorl lacks the distinct ribbing and spiral striae that form the beaded sculpturing. However, the beaded ribs can usually be seen within the umbilicus and usually on some of the inner whorls. The shell is pale yellowish-green with a very low, slightly rounded or fairly flat spire. The umbilicus is broadly open, about one-fourth the shell diameter. Size of the whorls and the umbilicus increase regularly to the last whorl, which is much larger than the adjacent penultimate whorl. The aperture is wider than high. The lip simple, sharp in immature shells and, in adults, slightly reflected or thickened basally. In mature specimens the deflection and oblique crease in the palatal lip, typical of *Ancotrema*, is evident. The animal is gray dorsally in front of the shell and generally white ventrally and posterior to the shell.

Similar Species: Compare with *Haplotrema vancouverense,* which is larger and lacks beaded sculpturing, and *Ancotrema sportella,* in which the beaded sculpturing extends to the aperture. The typical creased and depressed lip of *Ancotrema* will separate this species from *Haplotrema*, and its lack of beaded sculpturing on the outer whorls distinguishes it from *Ancotrema sportella*.

Distribution: *A. hybridum* occurs from Alaska through western British Columbia, Washington, Oregon, and south to Humboldt Co., California. It was listed as a native species known to occur in Idaho by Frest and Johannes (2000). Some very good specimens have been found in Clackamas and Multnomah counties, Oregon. It is a species of conifer forests, found in litter under sword ferns and other plants, and under rocks, logs, and other woody debris.

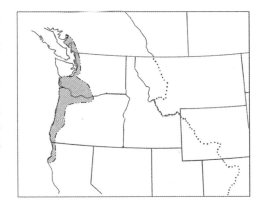

Ancotrema sportella   (Gould, 1846)   Beaded Lancetooth

Description: A medium to fairly large snail, with 5 to 6⅓ whorls it measures 11 to 22 mm diameter by 6 to 10 mm high. The shell is usually a pale yellowish-green with a nearly flat spire. The umbilicus is open, about one-quarter the total shell diameter. The aperture is wider than high. The lip is simple, sharp in immature shells, a little thickened in adults and narrowly reflected basally. The palatal lip is depressed, with a crease-like channel from about the mid-palatal margin running obliquely back to the outer curve of the whorl. The width of the whorls increases regularly to the last whorl which, behind the aperture, is twice or greater the width of the adjacent penultimate whorl. Shell sculpture is of low but usually distinct rib-like radial lirae cut across by spiral striae. This gives the surface a beaded appearance (hence the common name). The size of these snails varies greatly; some populations are of quite small individuals (< 15 mm diameter); others are of medium or large snails (to 22 mm wide). Shell sculpturing is also variable, the ribs being quite

1. *Ancotrema sportella*: Capitol State Forest, Thurston Co., WA; scale bar = 1 cm

2. *Ancotrema sportella*: Patos Island, San Juan Co., WA; scale bar = 1 cm

3. *Ancotrema sportella*: (A) Capitol State Forest, Thurston Co., WA; (B) Woodard Bay, Thurston Co., WA; (C) Miller Sylvania State Park, Thurston Co., WA

coarse and distinct in some populations but fine or indistinct in others. The snail is generally white with gray on the head, neck, and tentacles, and a light brown mantle collar.

Note: *Ancotrema sportella* on Patos Island, San Juan County, Washington, are consistent in size and appearance and are unique from *A. sportella* from the mainland. Their palatal lips are more extremely deflected, for more than half the height of the aperture, so location needs to be considered to avoid confusing them with *A. voyanum* in the species keys.

Similar Species: Other snails having shells of similar shape and color include species of *Haplotrema* and *Megomphix*, as well as other species of *Ancotrema*. *Haplotrema* lacks the radial ribs and the strongly deflected and creased palatal lip, and spiral striae are extremely fine when present. Spiral striae of *Ancotrema sportella* are coarse by comparison. *A. hybridum* appears similar to *A. sportella,* but the beaded sculpturing usually fades out well before the last whorl and may easily be overlooked. *Megomphix hemphilli* lacks any distinct shell sculpturing, having a smooth, glossy shell with flatter basal whorls. Its nearly white, glossy shell gives it a distinctly different appearance from the duller, satiny appearing shells of the haplotremes. Two other *Megomphix* (*M. lutarius* and *M. californicus*) have a somewhat satiny appearance from their fine growth-wrinkles and spiral striae. However, these two species are generally smaller, with noticeably finer shell sculpture than that of *Ancotrema sportella*.

Distribution: *A. sportella sportella* occurs on both sides of the Cascade Mountains in British Columbia, through western Washington and Oregon to Humboldt Co. in northern California, where they are replaced by *A. sportella sinkyonum* (Roth, 1989). In the forests of western Washington and Oregon, *A. sportella* is one of the most common species. The farthest east I have found them is in the town of Colville, Stevens County, Washington, but Frest and Johannes (2000) listed it as a native known to occur in Idaho. They inhabit conifer forests and can be found in litter under sword ferns and other plants, and under rocks, logs, and other woody debris.

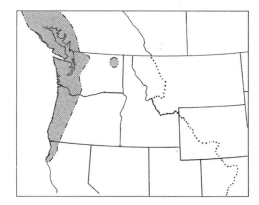

Ancotrema voyanum   (Newcomb, 1865)   Hooded Lancetooth

Description: *A. voyanum* is a medium-sized snail, 11 to 15 mm diameter by 5 to 8 mm high in 5 to 5½ whorls. The shell is pale, somewhat translucent and may have an orange tint to the periostracum. The spire is very low but slightly rounded. The umbilicus is open, funnelform, about one-third the shell diameter. In adults, the lip is narrowly reflected basally and thickened around the margin. The palatal lip is depressed, dipping into a rather deep sinuosity, which gives the aperture a triangular shape when viewed from below. Size of the whorls and the umbilicus increase regularly to the last whorl, which is much wider than the adjacent penultimate whorl. The shell sculpture is of fine but distinct rib-like radial lirae cut across by spiral striae, giving the surface a beaded appearance typical of most *Ancotrema*, but the spiral lines may weaken or fade out in the last half-whorl.

Similar Species: *A. voyanum* is similar to *A. sportella*, but is distinguished from that species by its much greater deflected and distinctly sinuous palatal lip margin. *A. sportella* from

Patos Island, San Juan Co., Washington, might be confused with *A. voyanum* in the species keys, so location needs to be considered when identifying these animals.

Distribution: *A. voyanum* is endemic to northern California (Trinity and Shasta counties), where it occurs with other members of the family. It is included here because of its proximity to the Pacific Northwest and its inclusion on the "Species of Concern" lists of federal land-management agencies.

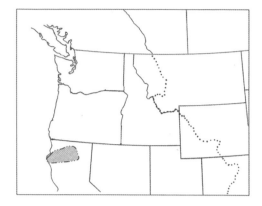

Genus: *Haplotrema*

Haplotrema vancouverense   (Lea, 1839)   Robust Lancetooth

Haplotrema vancouverense: Kraus Ridge, Gifford Pinchot National Forest, Lewis Co., WA; scale bar = 1 cm

Description: This large snail, with a very low, slightly rounded spire, measures 20 to 32 mm wide by 8 to 13 mm high in 5½ to 5¾ whorls. The shell is usually olive or greenish-brown, or occasionally dark brown (form *chocolatum* of Dall, 1905). Live juveniles appear to be mottled brownish from the pigment of the mantle showing through the shell. The umbilicus is open, and measures about one-fourth the shell diameter. The aperture is wider than high. The lip is simple and sharp in immature shells. The basal lip of adults is often reflected or thickened, and the palatal lip is roundly extended and a little deflected, sometimes with a slight thickening or very narrow reflected edge. The width of the whorls increases regularly to the last whorl, which then expands rapidly; just before the aperture it is greater than twice the width of the adjacent penultimate whorl. Shell sculpture is of growth-wrinkles and very fine, closely spaced spiral striae. The striae are diagnostic in the northern part of the range, but may fade on the last whorl of specimens from California and southern Oregon. The snail's body is generally white with gray or light brownish head and tentacles, and irregular dark brown or black blotches on the mantle. In living immature specimens, these blotches show conspicuously through the shells.

Similar Species: Immature *H. vancouverense* can usually be distinguished from *Ancotrema* species by the very fine spiral striae on the shell. Those of the *Ancotrema* are coarser. The radial ribs, typical of *Ancotrema*, are absent from the outer whorls of *Ancotrema hybridum*, so that species can be confused with *Haplotrema*. However, look for the beaded sculpturing of

Ancotrema on the early whorls and within the umbilicus. Mature specimens of *Ancotrema* also show the typical deflection and crease or channel in the palatal lip.

Young *Haplotrema* are sometimes mistaken for *Megomphix*. Close examination of the shell will reveal very fine, close spiral striae on *Haplotrema vancouverense,* giving the shell a satiny appearance. *Megomphix hemphilli* lacks any distinct shell sculpturing, having a smooth, glossy shell, and flatter basal whorls than those of the haplotremes. However, *Megomphix californicus* and *M. lutarius* have weak shell sculpturing, but they are generally smaller with less rapidly increasing whorls, and with apertures flatter basally. The ranges of these two *Megomphix* are very restricted compared to that of *Haplotrema vancouverense.*

Distribution: *H. vancouverense* is common on the west side of the Cascades from Alaska south to Humboldt and Trinity counties, California. Its range extends eastward through the forests of eastern Washington, northeastern Oregon, northern Idaho, and into extreme northwestern Montana. Other species of *Haplotrema* occur locally in California. *H. vancouverense*, the only known *Haplotrema* species in the Pacific Northwest, is common in the conifer forests of western Washington and Oregon. It is found in moist forest habitats in forest floor litter under sword ferns and other plants, shrubs, logs, and woody debris.

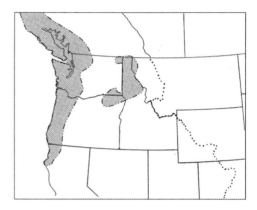

Selected references for the family Haplotrematidae: Brunson and Usher 1957; Frest 1999; Frest and Johannes 2000; Hendricks, Maxell, and Lenard 2006; Hendricks et al. 2007; Kelley et al. 1999; Pilsbry 1946; Roth 1990, 1991; Smith 1970.

Superfamily: ACAVOIDEA Pilsbry, 1895

Family: MEGOMPHICIDAE

Genus: *Megomphix*

Small to medium-sized snails, the shells are white with a faint yellowish-green tint. Shells are discoidal or with very low spires. The apertures are fairly wide, the lips simple and the basal margin somewhat flattened.

Key to the Species of *Megomphix*

(a)   Shell 13 to 20 mm wide. The umbilicus contained about 4½ times in the shell width, increasing very slowly to the last half-whorl. Glossy, without spiral striae. Western Oregon and Washington . . . .*Megomphix hemphilli*

   (aa)   Shell smaller, 9 to 15 mm wide. Umbilicus wider, funnelform. Non-glossy, with faint spiral striae . . . .[b/bb]

(b)   Shell width 9 to 10.5 mm, umbilicus contained about 3⅓ times in the shell width. Southeastern Washington and northeastern Oregon . . . . . *Megomphix lutarius*

   (bb)   Shell width 10 to 15 mm, umbilicus contained about 3½ times in the shell width. Northern California . . . .*Megomphix californicus*

Megomphix californicus   A. G. Smith, 1960   Natural Bridge Megomphix

Megomphix californicus: CA; scale bar = 1 cm

Description: The discoidal shell of *M. californicus* measures 10.3 to 14.8 mm wide by 4.7 to 7.0 mm high with 5⅛ to 5⅝ whorls. The shell is thin, transparent but not glossy, and has a greenish-yellow tint. The umbilicus is broad, funnelform, contained about 3½ times in the shell width. The aperture is somewhat ovate, slightly oblique; the peristome is simple and sharp. The 1¾ protoconch whorls are smooth, followed by fine growth-wrinkles crossed by "subobsolete and hardly noticeable spiral depressions or grooves that are widely spaced, breaking up the growth lines into thin elongated granulations" (Smith 1960).

Similar Species: Most similar to *Megomphix lutarius*, but *M. californicus* is larger, with a slightly wider umbilicus, and its last quarter whorl descends a little. *M. hemphilli* is larger than *californicus* and has a glossy surface and a narrower, more slowly expanding umbilicus. The three species are not known to occur in the same areas.

Distribution: *M. californicus* is known from Siskiyou, Trinity, Shasta, and Napa counties, California. The type locality is in an area of total darkness in a cave; other locations are from springs and along stream corridors, many from inside moist logs (Roth, personal communication).

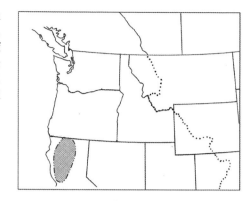

Megomphix hemphilli (W. G. Binney, 1879) Oregon Megomphix

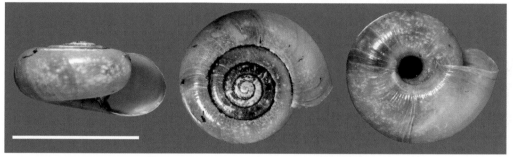

Megomphix hemphilli: Priest Point Pk., Olympia, Thurston Co., WA; scale bar = 1 cm

Description: A medium-sized white snail; the shell is white with a light greenish tint, smooth and glossy without striations, and with low, smooth growth-wrinkles. With 5¼ to 6 whorls its shell measures 13.5 to 20.0 mm across and about 6.8 to 8.5 mm high. Size of the whorls increase regularly but slowly to the last whorl, which is more than twice as wide as the adjacent penultimate whorl. The spire is nearly flat; the aperture is wider than high and somewhat flattened basally; the lip is simple and sharp. The umbilicus is contained about 4½ times in the shell width.

Similar Species: Other *Megomphix* are not found in the same areas. *Haplotrema vancouverense* is much larger, and its fine close spiral striae give it a satiny surface even on juveniles. Both *Haplotrema* and *Ancotrema* species have a wider, more open umbilicus, a more roundly arched basal lip margin, and shell sculpturing that is lacking in *Megomphix hemphilli*. The range of *Microphysula cookei* overlaps that of *M. hemphilli* in Washington, but only juvenile *Megomphix* might be confused with *Microphysula,* which is much smaller, with a thin shell and a relatively narrow aperture.

Distribution: *Megomphix hemphilli* is found at lower elevations in the Willamette Valley and Puget Trough from Douglas Co., Oregon, to Olympia, Thurston Co., Washington. It is generally found in soft forest floor substrate, often in soil under logs that are elevated above the ground, and it is most often associated with big-leaf maple trees.

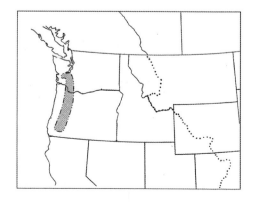

Megomphix lutarius   H.B. Baker, 1932   Umatilla Megomphix

Megomphix lutarius: North Fork John Day River, Umatilla Co., OR; scale bar = 1 cm
(Xerces Society collection)

Description: Smaller than *M. hemphilli, M. lutarius* measures 9.0 to 10.5 mm wide by 4.6
to 5.15 mm high with about 5¼ whorls. It has a very low spire and relatively wide umbilicus, contained about 3⅓ times in the shell width. The whorls expand fairly regularly; the
last is less than twice as wide as the adjacent penultimate. The shell is thin, non-glossy,
and somewhat opaque, with a slight greenish tint. The aperture is widely lunate, slightly
oval and a little oblique. The peristome is simple, the basal area flattened somewhat. The
periphery of the last whorl is rounded. The two protoconch whorls are smooth with the
beginnings of indistinct spiral striae. The teleoconch is sculptured with growth-wrinkles
and very fine, close spiral striae.

Similar Species: Similar species include other *Megomphix*, from which it is separated by
range (see above for descriptions). *Haplotrema* and *Ancotrema* have more rapidly expanding
whorls and more deeply rounded basal lip margins. *Microphysula ingersolli* has a thinner,
more transparent shell with a narrower aperture.

Distribution: Difficult to find, *Megomphix lutarius* was known from Walla Walla,
Washington, and Pine Creek, Umatilla Co., Oregon, on steep lava slope covered with

mosses, ferns, and scattered shrubs with
Douglas-fir, near a small creek. Frest and
Johannes (1995) failed to find this species at
the type locality or at several other sites in
the vicinity. Leonard and Baugh (personal
communication) failed to find the snail, and
found it difficult to find suitable habitat in
which to search because of agriculture and
other development in the area. Jepson and
Carleton, Xerces Society, were the first to dis-
cover an extant population of this species in
recent years, when, in May 2012, they found
it near the NF of the John Day River.

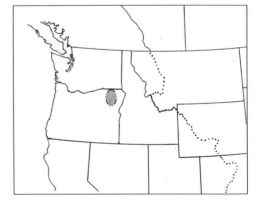

Genus: *Polygyrella*

Polygyrella polygyrella (Bland & J. G. Cooper, 1861) Humped Coin

1

Description: This small to medium-sized snail with a discoidal shell of 6½ to 8½ whorls measures 8.6 to 12.5 mm wide by 3.7 to 6.0 mm high. The spire is low and rounded on the sides but the first three whorls are flattened or depressed a little at the apex. It is glossy, light olive-brown with a pale yellowish-green tint. The umbilicus is wide (one-third the shell width), deep and symmetrical until the last half-whorl, where it expands more rapidly to the aperture. The whorls are tightly coiled, the aperture widely lunate, somewhat triangular, and the peristome is thickened within. A large parietal tooth fills much of the aperture. Beginning in the third whorl, the dorsal surface is sculptured with closely spaced, low but distinct ribs that end above the periphery. The periphery and basal surfaces are rather smooth, as is the protoconch. The ribs fade to nearly obsolete for the last one-sixth whorl.

1. *Polygyrella polygyrella*: O'Hara Campground, Idaho Co., ID; scale bar = 1 cm

2. *Polygyrella polygyrella*: Touchet River, Umatilla National Forest, Columbia Co., WA

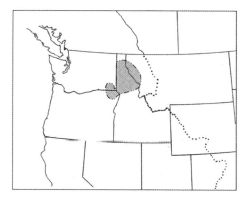

2

Similar Species: This species, with its low rounded, closely ribbed spire, flat apex, large umbilicus and large parietal tooth, is distinct from all other snails in the Pacific Northwest.

Distribution: *P. polygyrella* occurs in Sanders Co., Montana; the eastern slope of the Coeur d'Alene Mountains (type locality); Kootenai, Nez Perce, and Idaho counties, Idaho; Umatilla Co., Oregon; and Walla Walla and Columbia counties, Washington. It may be found in mesic or damp forest habitats, in litter and under rocks, logs, and other woody debris.

Selected references for the family Megomphicidae: Baker 1930, 1932; Forsyth 2004; Frest and Johannes 1995; Henderson 1929a; Hendricks 2003; Hendricks, Maxell, and Lenard 2006; Pilsbry 1939, 1946; Smith 1960.

Superfamily: HELICOIDEA
Family: HELICIDAE

The snails of this family that occur in the Pacific Northwest are all introduced from Europe. They are most often found in or near areas of human habitation.

Key to the Genus *Helix*

(a)   Shell relatively large, to 38 mm wide by 33 high in about 4½ whorls. Thin and fragile, the whorls increasing rapidly with a large, oblique, roundly ovate aperture. Imperforate, the umbilicus normally completely closed by the reflected columellar lip margin. Yellowish with brown bands of varying widths . . . . *Helix aspersa*

    (aa)   Shell larger, to 45 mm or wider. Thin but not quite as fragile. Umbilicus is usually rimately umbilicate, seldom being completely closed by the reflected columellar lip margin. Yellowish with radial and/or spiral brown bands, and sometimes with a white peripheral band. Not established in the PNW—see under Similar Species for *H. aspersa* . . . . *Helix pomatia*

Helix aspersa Müller, 1774   Brown Gardensnail
Synonym: *Cornu aspersa* Born, 1778

1

Description: This exotic snail is rather large, up to 38 mm wide by 33 mm high in about 4½ whorls. The shell is quite thin and obliquely globose. Early whorls are relatively small, but they increase in size very rapidly. In growth the whorls are descending, so the aperture is positioned somewhat basally. The aperture is large, oblique, roundly-ovate, and narrowly reflected except the columellar margin, which is folded back and usually closes the umbilicus. Markings are of varying shades of brown in bands of varied widths, over a yellowish ground color. The bands are broken regularly by narrow, zig-zag yellow streaks, which cross the whorl radially from

1. *Helix aspersa*: Port Townsend, Jefferson Co., WA; scale bar = 1 cm

2. *Helix aspersa* (oblique view): Port Townsend, Jefferson Co., WA

2

(A) *Helix pomatia*: commercial escargot;

(B) *H. aspersa*: Port Townsend, Jefferson Co., WA

near the upper suture to the thin parietal callus. Shells of young are very thin and fragile and are much lighter colored than those of adults. The protoconch is smooth. The following whorls are sculpted with fine regular growth striae. The last whorl is roughened by growth-wrinkles, spiral striae, and irregular wrinkles that form malleations.

Similar Species: *Helix pomatia* Linnaeus, 1758 (escargot), is much larger than *H. aspersa*, attaining a shell width of 45 mm or more. The aperture is not reflected except at the columellar margin, where it is folded back over but does not normally close the umbilicus. The shell appears basically white with spiral bands of varying shades of brown and varying widths, some of which cover most of the shell, but there is usually a wide, light-colored halo around the umbilicus. Radiating dark brown streaks are also often present. The protoconch is smooth; the following whorls have regular growth-wrinkles and weak but distinct spiral striae. Although not currently known to be established in the United States, *H. pomatia* has been found in Michigan and Wisconsin (Pilsbry 1939) and has been intercepted at ports of entry (Hanna 1966; Robinson 1999). Note: Both species of *Helix* have been petitioned for introduction to be raised for escargot, but *Helix aspersa* is already a serious horticultural pest where established, and *H. pomatia*, being a larger snail, may potentially be an even greater threat.

Distribution: Native to western Europe, *H. aspersa* has been introduced and is established in North America and other areas around the world. Pilsbry (1939) reported it from only a few southern states, including California. Hanna (1966) cited records from California, Oregon, Utah, Arizona, and as far north as Pacific Co., Washington. It is well established in Port Townsend, Jefferson Co., Washington, and Forsyth (2004) reported it from Vancouver Island and Vancouver, British Columbia. It inhabits lawns, gardens, and compost piles, and it hides in crevices around houses. At Long Beach, Washington, it was found associated with beach pea plants in the upper, vegetated dune areas.

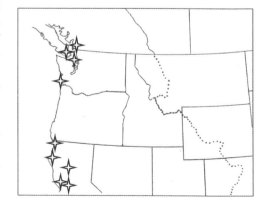

Genus: *Cepaea*

Cepaea nemoralis (Linnaeus, 1798)   Grovesnail

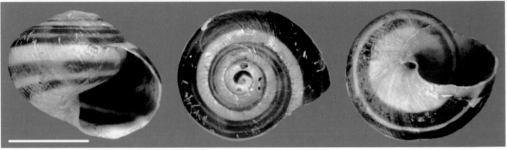

1. *Cepaea nemoralis*: Fife, Pierce Co., WA
2. *Cepaea nemoralis* (immature): Tumwater Canyon, Chelan Co., WA; scale bar = 1 cm

Description: This exotic snail is medium-sized, measuring 18 to 25 mm wide by 12 to 22 high (rarely as large as 32 by 28 mm) with 5½ whorls. The shell is globose or a little more depressed, rather solid and imperforate at maturity. The glossy shell is brightly colored, yellow or brown or variously banded. The face of the flared and thickened apertural lip is white to brown, but is most often dark brown.

Similar Species: The most similar species, although not documented in North America, is *Cepaea hortensis* (Müller, 1774), which is smaller and usually has a white, occasionally brown, peristome. *C. nemoralis* should not be mistaken for any native snail of the Pacific Northwest.

Distribution: Native to central and western Europe, *C. nemoralis* has been introduced into North America, where it has a scattered distribution, occurring in Ontario, Canada; some New England States; Colorado; and California. In Washington it occurs in the east-central and northern Puget Sound vicinity, and one shell was found on the east slope of the Cascades, in Chelan Co., near Leavenworth. Forsyth reported it fairly widely scattered across southern British Columbia, from Vancouver and other islands to Nelson.

Family: HYGROMIIDAE
Genus: *Candidula*

Candidula intersecta (Poiret, 1801)   Wrinkled Helicellid

Candidula intersecta (colors faded): Sisters Rock, Curry Co., OR; scale bar = 1 cm

Description: Shell small, depressed-globose, with 5 to 6½ whorls measuring 7 to 13 mm wide by 5 to 8 mm high. Periphery obtusely angled to the last half-whorl, which is then rounded to the aperture. Width of the whorls increases slowly; the sutures are fairly shallowly impressed. Umbilicus rather narrow, symmetrical and deep (one measured at one-seventh the shell diameter). Aperture oval, wider than high (or obliquely lunate). The lip is thin but there is a distinct thickened ridge just inside it. The fairly thick, opaque shell is sculptured with fine but distinct, closely-spaced, rib-like growth lines. It is whitish to light yellowish-brown, usually with various darker brown bands and/or blotches above and below the periphery.

Similar Species: *Candidula intersecta* and *Cernuella virgata* are quite similar, both recognized by their rather globose shells, the thickened ridges around the inside of their apertures, and their spiral banding. *Candidula intersecta* is generally the smaller of the two species; its whorls increase more slowly and its growth-wrinkles are more coarse and rib-like. *Cepaea nemoralis* (which see) may also be banded, but it is larger and is imperforate as an adult.

Distribution: *Candidula intersecta* occurs in scattered areas of western Europe and the British Isles, where it is found in open grasslands and sand dunes. It has been introduced into North America and as yet is scarce in the PNW.

Genus: *Cernuella*

Cernuella virgata (Müller, 1774)   Maritime Gardensnail

Description: A small to medium-sized shell with a fairly high convexly-conic (somewhat globose) spire. With 5 to 7 whorls it measures 8 to 25 mm wide by 6 to 19 mm high. The periphery is generally rounded, the penultimate whorl being very slightly, if at all, sub-angular at the aperture. The whorls increase rather rapidly; the body whorl, just behind the aperture, is about twice the width of the adjacent penultimate whorl. The sutures are rather shallow. The aperture is a little oblique and wider then high. There is a prominent thickened ridge just inside the aperture, following which the outer lip flares a little and the basal and columellar lips are narrowly reflected. The sculpture is of uneven, fine and closely-spaced growth-wrinkles. The shell is whitish or light yellowish-brown, and usually

Cernuella virgata (colors faded): Seattle, King Co., WA; scale bar = 1 cm

has varying patterns of dark brown bands, lacking in some specimens. The body is usually black (Roth 1987).

Similar Species: *Candidula intersecta* and *Cernuella virgata* are quite similar; both are recognized by their rather globose shells, the thickened ridges around the inside of their apertures, and their spiral banding. *Cernuella virgata* is generally the larger of the two species; its whorls increase more rapidly and its growth-wrinkles are finer and less consistent. *Cepaea nemoralis* (which see) may also be banded, but it is larger and is imperforate as an adult. In California, *Cernuella virgata* appears similar to *Theba pisana* (Müller, 1774), another exotic species and agricultural pest. In contrast, the body of *T. pisana* is tan and has spiral striae and weaker radial ribs (Roth 1987).

Distribution: *Cernuella virgata* is from scattered areas of western Europe and the British Isles, where it is found in open grasslands, sand dunes, and brushy sites. It is established as a serious agricultural pest in Australia. It has been introduced into North America, first found in San Diego Co., California, in 1987 (Roth, Hertz, and Cerutti 1987), and it has been found in the Seattle area, King Co., Washington. Robinson (1999) showed it established in North America and composing 3.5 percent of interceptions by USDA Plant Protection and Quarantine inspectors.

Selected references for the families Helicidae and Hygromiidae: Forsyth 2004; Hanna 1966; Kerney, Cameron, and Riley 1994; Pilsbry 1939; Robinson 1999; Roth, Hertz, and Cerutti 1987.

Family: HELMINTHOGLYPTIDAE
Genus: *Helminthoglypta*

Helminthoglypta is a genus with many species and subspecies, of which only the following two occur in the Pacific Northwest. They are peripheral to our area and are confined to southwestern Oregon and into northern California.

Key to the Species of *Helminthoglypta*

(a)   Shell 18 to 20 mm wide and 75% to 80% as high. Spire high and conic. Umbilicus contained about 14 times in the shell width and more than half covered by the reflected columellar lip. Otherwise the lip is narrowly reflected and a little thickened inside. Sculpture of fine, sharp growth striae dorsally and some weak malleations . . . . *Helminthoglypta mailliardi*

   (aa)   Shell 18 to 23 mm wide by 70% or more as high. Spire fairly low and conic. Umbilicus contained about 9 or more times in the shell width and one-half or less covered by the columellar lip. Otherwise the lip is generally simple but barely flared at the outer and basal margins and a little thickened inside. Sculpture of regular growth-wrinkles and some weak malleations . . . . *Helminthoglypta hertleini*

Helminthoglypta hertleini   Hanna & Smith, 1937   Oregon Shoulderband

Helminthoglypta hertleini: Southwest Oregon; scale bar = 1 cm

Description: The shell of this medium-sized snail is thin and rather fragile. The 5 to 5½ whorls increase in size regularly and fairly rapidly. The shell measures 18 to 23 mm wide and its height is about 70% or more of its width. The narrow umbilicus is contained 9 or more times in the shell width. The body whorl is capacious, the spire fairly low and conic. The color is golden-brown with a narrow, darker brown or reddish band well above the periphery; the band may be bordered by an indistinct area that is a lighter shade of the shell color. The aperture is roundly lunate; the lip is generally simple but thickened a little inside. The outer and basal lip margins are barely flared, but the columellar is reflected to cover about half of the umbilicus. Sculpturing is merely growth-wrinkles, although on some shells weak malleations may be seen.

Similar Species: There are many species of *Helminthoglypta* in California, but only the two discussed here are known from southern Oregon. The large body whorl with round periphery and with a single dark band well above the periphery separate the *Helminthoglypta* from other snails within our study area.

Distribution: *H. hertleini* occurs in the Klamath Mountains in Jackson County, Oregon, and south into Tehama County, California. It inhabits open forests and rocks and talus in prairie habitats.

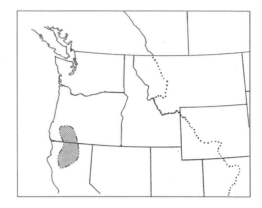

Helminthoglypta mailliardi   Pilsbry, 1927   Del Norte Shoulderband

Description: The shell of this medium-sized snail is thin and rather fragile. The 5⅓ to 5½ whorls increase in size regularly and fairly rapidly. The shell measures 18 to 20 mm wide, and its height is about 76% to 85% or more of its width. The narrow umbilicus is contained about 14 or more times in the shell width, and it is more than half covered by the reflected columellar lip margin. The body whorl is capacious, the spire fairly high and conic. The color is yellowish-tan with a narrow, chestnut-brown band above the periphery; the band may be bordered by an indistinct area that is a lighter shade of the shell color. The aperture is widely lunate; the lip is narrowly flared and thickened a little inside. Sculpturing is strong, fine growth striae dorsally, which fades out below the periphery, and irregular weak malleations.

Similar Species: Of many species of *Helminthoglypta* from California, only the ranges of *H. mailliardi* and *H. hertleini* extend into our area of study. However, Roth and Sadeghian (2006) stated in an endnote that southern Oregon populations "formerly referred to *Helminthoglypta mailliardi* belong to a new species, being described elsewhere."

Distribution: The species referred to as *H. mailliardi* occurs in Douglas and Jackson counties, Oregon. It inhabits dry rocky sites in forest openings, prairie, and shrub habitats.

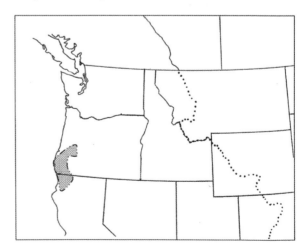

Family: BRADYBAENIDAE

Genus: *Monadenia*

Monadenia, a fairly large and diverse genus, is endemic to the Northwest Pacific Rim. The majority of the species and subspecies are found farther south, in California. However, *Monadenia fidelis*, with several subspecies, occurs in the Cascades and westward from Alaska to west-central California, and several others occur in northern California and southwestern Oregon. Some color variations that occur among the Pacific Northwest *Monadenia*, other than the recognized subspecies, are included in the keys and discussions to alert the reader of their existence. There are also some new, undescribed species discussed below under *Monadenia* New to the Pacific Northwest Fauna.

Key to the Species and Subspecies of *Monadenia* of the Pacific Northwest

(a)   Shells rather large (30 to 38 mm wide by 19 to 26 mm high); spire
      usually fairly high . . . .[b/bb]

  (aa)   Shells 18 to 30 mm wide by 10 to 15 mm high; spire low to
         moderately elevated . . . .[g/gg]

(b)   Base, below the yellowish or white peripheral band, normally dark brown
      to chestnut colored. Dorsal coloration a mixture of the base and peripheral
      band colors, but may be mostly creamy or mostly brown. Umbilicus contained
      9 to 11 times in the shell diameter. Cascades Range to the Pacific Ocean.
      . . . . *Monadenia fidelis fidelis*

  (bb)   Color not as above or if same pattern then not as sharply contrasted . . . [c/cc]

(c)   Color mostly creamy yellowish, if bands are apparent they are relatively faint,
      blending into the general color of the shell. Occasionally found among many
      populations . . . . *Monadenia fidelis* (white or yellow form)

  (cc)   Shell other than all creamy-yellow, although a significant portion
         of it may be that color. . . . . [d/dd]

(d)   Shell mostly dark brownish or black. . . . .[e/ee]

  (dd)   Shells other than mostly dark brownish or black . . . .[f/ff]

(e)   Shell all black or nearly so, with a very faint peripheral band.
      . . . . *Monadenia fidelis* (melanistic form)

  (ee)   Shell mostly blackish-brown; spire little if any lighter than the base or
         chestnut brown. Dark supraperipheral band wide, to 4 mm or more.
         Lighter bands much narrower, the upper one sometimes indistinct.
         . . . . *Monadenia fidelis columbiana*

(f) Color pattern similar to that of typical *M. fidelis*, except the base has a dark green
      cast; there may also be a green tint over the spire. In addition, the light-colored
      bands may be more tan than yellow. Curry County, Oregon.
      . . . . *Monadenia fidelis flava* (= *M. f. beryllica*)

  (ff)   Color similar to *M. fidelis* except the light-colored peripheral band
         extends down under the base to a dark brown patch around the
         umbilicus. Fidalgo Island, Washington . . . . *Monadenia fidelis semialba*

(g)   Dorsal color darker than average for typical *M. fidelis* and with additional bands
      of varied shades of brown and buff on the spire. Shell width smaller than typical
      (23) 27 to 30 mm wide by 19 mm high with 6½ whorls. Umbilicus nearly

half-covered by the reflected basal-columellar lip margin.
Rogue River watershed, Jackson Co., Oregon . . . .*Monadenia fidelis celeuthia*

(gg)   Dorsal color similar to that of *M. fidelis fidelis*, or lighter. Umbilicus
       less then one-quarter covered by the lip margin . . . . [h/hh]

(h)   Southern Oregon into northern California. Spire and basal colors not greatly
      different . . . . [i/ii]

(hh)   Northern Oregon and southern Washington in Columbia Gorge and
       Deschutes River watershed, on east side of the Cascade Mountains. Shells
       small, 20 to 28 mm wide. Umbilicus contained about 7 to 9 times in the
       shell diameter. Color patterns as in typical *M. fidelis* except sometimes
       with one or two light brown spiral bands between the supraperipheral
       band and the suture. Spiral striae rather distinct . . . . . .*Monadenia fidelis minor*

(i)   Shell small, rather thin; color phases of light creamy-buff or a fairly moderately
      dark reddish- or yellowish-brown. Papillose sculpture of protoconch merges
      into nearly rib-like growth lines of the succeeding whorls. Basal aperture
      somewhat flattened, the reflected lip squared off by curving abruptly upward
      at its columellar end . . . .*Monadenia fidelis leonina*

(ii)   Shell moderately heavy; color varies around lighter or darker shades of
       yellowish-brown. Sculpture of protoconch progresses from smooth to
       weak, irregular radial wrinkles, to crowded, fine granules, which do not
       extend onto the variable growth wrinkling and weak, irregular spiral
       striae of the teleoconch. Basal aperture little flattened, if at all; the
       columellar lip extends across the umbilicus at an ascending angle.
       . . . . *Monadenia chaceana*

Monadenia chaceana   Berry, 1940   Siskiyou Sideband

Monadenia chaceana: Medford BLM District, OR

Description: *M. chaceana* with about 5½ to 6 whorls measures 18 to 26 mm diameter
by 10 to 15 mm high (average H/D ratio 0.57). The umbilicus, which is one-seventh
to one-tenth the total shell width, is covered less than one-fourth by the columellar
lip margin. The spire is low to moderately elevated and is convexly-conic. The base
is medium to fairly dark brown; the spire is somewhat lighter, diffused with the light
yellowish color of the peripheral band. Above the light cream-colored peripheral band
is a dark brown to reddish-brown band, which is bordered above by another narrower
cream-colored one.

Similar Species: The basal color of *Monadenia fidelis* is darker than that of *M. chaceana*. Its
spire is relatively higher and its umbilicus is more covered (one-third to nearly one-half)

Monadenia chaceana (immature; atypical color): South Umpqua, Jackson Co., OR (OSAC #PGC07-004)

by the basal-columellar lip, but these characteristics require mature shells to be apparent. Some subspecies (e.g., *M. fidelis minor*) may closely resemble *M. chaceana*, often having a lower spire and sometimes a lighter basal coloration. Sculpturing on the protoconch can be helpful if it isn't too severely eroded; that of *chaceana* is fine granules, the rows of which are not distinct. On *M. fidelis* the granules are finer and more sharply cut by the crossing of oblique, curved, microscopic striations. *M. fidelis leonina* is even more similar, and it occurs in the same vicinity as *M. chaceana*. *M. fidelis minor* is generally found in northern Oregon and the eastern Columbia Gorge, while *M. chaceana* and *M. fidelis leonina* are confined to southwestern Oregon and northern California. The reflected columellar lip of *M. f. leonina* ascends more abruptly to the umbilical insertion, while that of *M. chaceana* angles upward more gently.

Monadenia churchi Hana and Smith, 1933, is somewhat similar to *M. chaceana* and may be found in the same near vicinity in California, but it has not been confirmed in Oregon. *M. churchi* can be distinguished by its distinctive, regularly-spaced, elongated granules arranged spirally but not in distinct rows, its less glossy periostracum, and slightly less covered umbilicus.

Distribution: *Monadenia chaceana* is found in the Umpqua River watershed, Oregon, and southward into northern California.

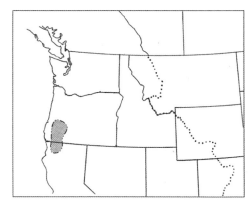

Monadenia fidelis fidelis   (J. E. Gray, 1834)   Pacific Sideband

Description: *Monadenia fidelis* is the largest native land snail in the Pacific Northwest. Its shell of 6½ to 7½ whorls measures 31 to 38 mm wide by 20 to 27 high (see the discussion of *M. fidelis minor*, which is smaller). The spire is moderately to fairly high, conic or a little convex. The last whorl descends rather abruptly to the aperture, which is widely oval, its lip narrowly flared and narrowly reflected basally. The shell is white inside, but the brown color of the base and the peripheral band shows through. The umbilicus is narrow, about one-tenth the shell width. The color pattern varies but is typically dark brown or chestnut basally with a light yellow band at the periphery, a dark brown to black band just above the periphery, and a narrow yellow area above the dark band followed by a blending of

1. *Monadenia fidelis fidelis*: western Washington; scale bar = 1 cm

2. (A) *Monadenia fidelis fidelis*, Des Moines, King Co., WA; (B) *M. f. fidelis*, San Juan Island, San Juan Co., WA; (C) *M. f. columbiana*, Hood River R.D., Multnomah Co., OR

the yellow and brown colors on the spire. The resulting color depends on the dominance of those two individual colors. The yellow is often replaced by white, and there are usually radial streaks of brown on the spire behind growth-stop lines.

Color variations occur. Some have shells that are all creamy-white or nearly all dark brown, chestnut, or black (melanistic), and greenish tints appear in some. These forms are uncommon, but they often occur among typical populations, and some local populations are composed of individuals of unique color. Subspecies have been named for populations that predominantly display variations in color, size, or other unique characteristics, some of which are discussed below.

Not only is *M. fidelis* a large and striking species, the shell sculpture of immature specimens is interesting to examine under a microscope. Mature shells are glossy with a protoconch of about two whorls, sculpted with striations in curving lines running radially and obliquely forward and back to leave curving rows of very fine granules. Whorls following the protoconch show growth-wrinkles and regular or weak spiral striae, which begin on the third or fourth whorls and continue through the body whorl of adults. Immature specimens with from three to five whorls have a unique pattern in their periostracum. Very fine parallel lines of what appear to be overlapping layers of periostracum run forward and obliquely outward from the suture, around the periphery and onto the basal whorl. These lines are generally straight, but their edges are irregular or ragged-appearing. The animals are orange, pinkish, or rose colored, and the grooves between the tubercles are gray. The sole is light gray.

Similar Species: Although other similar *Monadenia* occur in California, only *Monadenia chaceana*, which occurs in southwestern Oregon and northern California, should be mistaken for *Monadenia fidelis* within the range of this study, once they develop their distinct color patterns. As juveniles, the angled shoulder and the spreading lines in the periostracum distinguish *Monadenia* from other genera of snails within its range. *Oreohelix strigosa* is also a banded snail but is smaller and has a more sharply angled periphery when immature. Its range overlaps that of *Monadenia fidelis* in the eastern Washington Cascade foothills. However, the color and banding patterns are distinct for each of the two genera, and oreohelices generally have a dull surface with little or very thin periostracum, while *Monadenia* have a distinct, glossy periostracum.

As stated above, *Monadenia fidelis* is variable to some extent in color, size, and shape. See the discussion of subspecies, below.

Distribution: *Monadenia fidelis* occurs from southern Alaska to northern California, in the Cascade Mountains and west to the Pacific Ocean. It is a species of moist forests, usually of mixed hardwoods and confers but often predominantly hardwoods and/or shrubs. It may be found under logs and other debris; in litter under ferns, shrubs, and other vegetation; and among scattered boulders and rock outcroppings in humid, non-forested areas. In late fall and winter it can often be found hibernating under mosses on the bases of big-leaf maple trees.

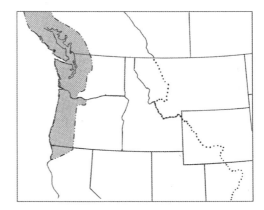

Subspecies of *Monadenia fidelis*

Monadenia fidelis beryllica   Chase, 1935

See *Monadenia fidelis flava*

Monadenia fidelis celeuthia   Berry, 1937

Monadenia fidelis celeuthia: SW Oregon; scale bar = 1 cm

Description: This subspecies has a shell of medium size for the species. With 6 to 6½ whorls it measures 27 to 30 mm wide by 17 to 19 mm high. Its umbilicus is one-ninth to one-tenth the shell width and is nearly half-covered by the reflected columellar lip margin. Its spire is moderate to fairly high, and quite dark brown. Its base is "from chestnut to a dark liver brown" (Berry). The aperture is widely lunate; the lip is narrowly reflected from the outer margin to the columellar insertion, strongly so basally. It has the typical peripheral banding, plus its distinguishing characteristic—multiple bands of varying shades of brown and buff on the spire.

A shell in hand that was labeled as this subspecies by the collector, and which appears to be it, is smaller. It is an adult shell with 6⅛ whorls; it measures 22.9 mm wide by 19.3 mm high, and its umbilicus is contained 12½ times in the shell diameter.

Monadenia fidelis celeuthia (note the presence of three dark bands): SW Oregon

Similar species: A variety of small *Monadenia* in the southern Oregon Cascades and throughout much of the range of *M. fidelis* appear similar to this and the following subspecies, providing opportunities for further study of the *Monadenia*.

Distribution: *M. fidelis celeuthia* is known from Jackson Co., Oregon, in the Rogue River watershed.

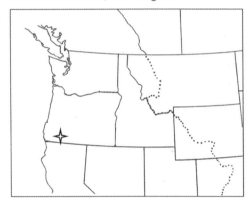

Monadenia fidelis columbiana   Pilsbry, 1939

Description: This variety is a dark form. The shells are approximately the normal shape and dimensions, but the supraperipheral band is wide with a narrow yellow band below and a very narrow light border above. The base is dark brown, and the spire is chestnut brown. However, other specimens that appear to be closely allied to this form have a spire that is nearly the same color as the base except for slight mottling.

Similar Species: The melanistic color variety may be darker with even fainter light bands, but unless there is a significant population, or it is found within the known area for *columbiana*, it may not be that subspecies. Similar dark-appearing specimens occur in other forested areas.

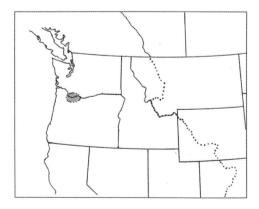

Distribution: This subspecies is most common in the Columbia Gorge in Skamania Co., Washington, and around Mount Hood in Oregon.

Monadenia fidelis flava Hemphill, 1892
Synonym: *Monadenia fidelis beryllica* Chace, 1935

Description: Shells of this variety are full-sized and similar to typical *fidelis* and its variations, except the base is a dark green color and there is a green tint over the spire. Also the umbilicus is narrower and is about half-covered by the columellar lip margin. Specimens from the Oregon State Arthropod Collection (OSAC) appear to average a height/diameter ratio that is higher than normal for *M. fidelis fidelis*.

Note: This subspecies was originally known as *M. f. beryllica*, until Coan and Roth (1987) pointed out that "the only extant type specimen [for *M. f. flava*] . . . is from within the range of *M. f. beryllica*, and is the earliest available name for that subspecies." The loss of other specimens is unfortunate, since it is apparent that Hemphill's intent was to apply this subspecific epithet to the yellow specimens that occur in many populations of *M. fidelis* throughout its range (Henderson 1929a; Pilsbry 1939), and which are today recognized only as a color variation as discussed by Pilsbry.

1

2

3

1. *Monadenia fidelis flava* (*beryllica*): Curry Co., OR (OSAC #SIS05-004); scale bar = 1 cm
2. *Monadenia fidelis flava* (*beryllica*): Curry Co., OR (OSAC #SIS05-008); scale bar = 1 cm
3. *Monadenia fidelis flava* (*beryllica*; atypical): Curry Co., OR

Similar Species: The population from Curry County Oregon, now referred to as *M. fidelis flava*, is distinct in its color and shell shape, although the shape can also be seen occasionally among other *M. fidelis*. Other greenish forms of *M. fidelis* occur. There is a small population with a greenish tint on their shells in the Cispus watershed of Lewis Co., Washington, discovered by Mr. Tom Kogut, and a greenish shell was found on Lopez Island, San Juan Co., Washington, among a normal population of *M. fidelis fidelis*. The color of these is much lighter than that of *M. fidelis flava*, and their shells are shaped more like typical *M. fidelis fidelis*.

Distribution: Pilsbry (1939) reported *M. f. flava* from "five strong colonies" and scattered shells from six other localities, all in Curry Co., Oregon. Although not the same as this subspecies, greenish-colored or tinted *Monadenia* occur in other areas as well. A population of greenish-shelled *M. fidelis* occurs along the Cispus River in Lewis County, Washington (specimens courtesy of Tom Kogut). Many quite small *Monadenia* from the Cascade Mountains from northern to southern Oregon have a faint greenish sheen on their shells, and one *M. fidelis* collected on Lopez Island, San Juan County, Washington, had a translucent light green colored shell. These and other variations in the shells of *Monadenia* indicate opportunities for further study of this interesting family of snails.

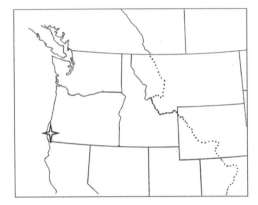

Monadenia fidelis leonina   S. Berry, 1937

Description: This snail is small for the species, 21 to 27 mm wide by about 15 (13 to 21) mm high, with 5⅔ to 6¼ whorls. Umbilicus about one-ninth the shell width. Shell thin,

1. *Monadenia fidelis leonina* (dark form): Siskiyou Co., California
2. *Monadenia fidelis leonina* (light form): Siskiyou Co., California

translucent, depressed to low-conic. Aperture ovate, slightly squared at the columella. Peristome little thickened palatally and narrowly reflected basally. Protoconch sculptured with crossing rows of papillae. Early teleoconch whorls with strongly developed, nearly rib-like growth-wrinkles, becoming less distinct and irregularly spaced on latter whorls where weak spiral striae may also be seen. Color of the shell is light, creamy buff to a reddish or rather dark yellowish-brown, lighter than the basal color of *M. fidelis fidelis*. There is usually a very dark brown to nearly black supraperipheral band, which may or may not be bordered above and below by narrower yellowish or olive-buff bands; there may be a second lighter brown band dorsally above the yellowish band. The animal is brown dorsally, lighter on the sides and head; the sole is lighter buffy brown.

Similar Species: Compare this species with *Monadenia chaceana*. Its reflected columellar lip ascends more abruptly to the umbilical insertion while that of *M. chaceana* angles upward more gently.

Distribution: *Monadenia fidelis leonina* has been found one mile up from the mouth of Beaver Creek, Siskiyou Co., California, and is believed to occur in Jackson and Josephine counties, Oregon.

Monadenia fidelis minor   (W. G. Binney, 1885)

Monadenia fidelis minor: The Dalles, Wasco Co., OR; scale bar = 1 cm

Description: This is a small subspecies, measuring about 20 to 25 mm wide by 12 to 15 mm high with 5½ to 6 whorls. It further varies from the typical form by having one or two faintly traced spiral bands between the supraperipheral band and the suture and rather distinct spiral striae. The animal is grayish-brown with bluish pigment granules.

Similar Species: A similar race is known from the east side of Upper Klamath Lake (Pilsbry 1939), but it is now considered a separate race (Frest and Johannes 1995). Small varieties similar to *M. fidelis minor* are found in various locations through the Oregon Cascades. There is a population of similar-appearing *Monadenia* southeast of Mount Hood, in forested habitat where they are mostly associated with logs and shrubs. Many of the *Monadenia* on

Monadenia fidelis minor and similar variations: (A) *M. fidelis minor*, The Dalles, Wasco Co., OR; (B&C) Lopez Island, San Juan Co., WA; (D&E) Mt. Hood National Forest, Wasco Co., OR

the east slopes of the Cascades in Washington, Chelan, and Yakima counties are also of this size. Most of these populations occur east of the Cascade Range, where annual precipitation is less than in the western forest habitats of typical *M. fidelis*.

Distribution: The type locality for *M. fidelis minor* is The Dalles, Wasco Co., Oregon, where the subspecies is still extant, living among rocks and in and under bluffs around

springs and in riparian areas. The population extends east to about the mouth of the John Day River, Sherman Co., Oregon, up the Deschutes River Valley for about ten miles, and to Klickitat Co., Washington (Frest and Johannes 1995). *Monadenia* fitting the size of *M. fidelis minor* were also found among rock outcroppings on Lopez Island, San Juan Co., Washington; however, typical *M. fidelis* were more commonly associated with the same type of habitat, so those small snails are considered to be merely small individuals among the population of *M. fidelis fidelis*.

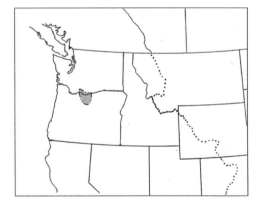

Monadenia fidelis semialba   J. Henderson, 1929

Description: This is a unique form in which the creamy-white peripheral band extends from the black supraperipheral band down under the base to a broad dark brown patch encircling the umbilicus. Colors above the supraperipheral band are as in typical *M. fidelis*. It was found among rocks and grasses on a steep slope about one-quarter-mile long by 250 yards wide (Pilsbry 1939). Because typical *M. fidelis fidelis* were found with it, and Henderson (1929a) said it was known only from the type specimen, its status as a subspecies is questionable. However, Pilsbry implied that others of the *M. f. semialba* pattern were taken from the same area. Therefore, additional surveys of the area and study of any

specimens resembling the description of this snail may be advantageous in determining whether or not it should be listed as a distinct taxon.

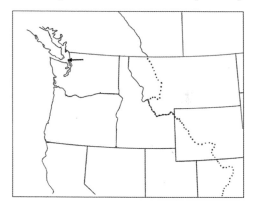

Similar Species: Unique forms or color patterns are not unusual among the *Monadenia*, but no other population resembling this subspecies is known to me.

Distribution: This uncertain taxon is known only from Rosario State Park, Fidalgo Island, Washington.

Monadenia New to the Pacific Northwest Fauna

Other species of *Monadenia* and subspecies occur in California, where the genus is much more speciose than in the northern states. Some of the California taxa could potentially be discovered in the Pacific Northwest by chance discovery of new populations or by accidental or intentional importation.

Dr. Barry Roth proposed a revision of the genus in a report titled "Taxonomy and Classification of *Monadenia*," dated 25 July 2002, to Dr. Paul Hoenlohe, Corvallis Forestry Sciences Laboratory. In his proposal, Dr. Roth listed 3 subgenera, 14 species (including 2 new ones), and 28 subspecies. All of these occur exclusively in California except *Monadenia chaceana*, *M. fidelis* and some of its subspecies, and possibly two new species.

Frest and Johannes (1995) listed two new, undescribed taxa of *Monadenia* from Oregon: *Monadenia (Monadenia) fidelis* n. subsp. 1, Deschutes Sideband, from along the Deschutes River, Wasco and Sherman counties, Oregon, and *Monadenia (Monadenia)* n. sp. 1, Modoc Rim Sideband, from Upper Klamath Lake, Klamath Co., Oregon.

Ms. Nancy Duncan (personal communication) discovered a new species in southwestern Oregon with a bristly shell, perhaps an undescribed species or at least new to the Oregon fauna.

Selected references for the families Helminthoglyptidae and Monadeniidae: Berry 1937, 1940; Coan and Roth 1987; Frest and Johannes 1995, 1996; Kelley et al. 1999; Pilsbry 1939; Roth 1980, 1981, 2001; Roth and Pressley 1986; Roth and Sadeghian 2006; Talmadge 1960.

Family: POLYGYRIDAE

This family is of small- to large-sized snails with well reflected lip margins and often with apertural teeth. Periostracal hairs adorn the shells of many of these species, but in most of the *Cryptomastix* the hairs are lost by adulthood.

Genus: *Allogona*

Allogona are fairly large snails, measuring 18 to 35 mm wide with 5 to 6½ whorls. They have a relatively large body whorl, and most have a moderately elevated spire. The umbilicus is small and usually about half-covered by the reflected basal-columellar lip margin. Color of the periostracum ranges from straw yellow to dark brown. The peristome is well reflected, and apertural dentition is generally lacking, although there may be a basal ridge with a slightly raised cusp toward the columellar end. The genus is found across the Pacific Northwest from northern Oregon into British Columbia. *Allogona townsendiana* inhabits the more humid forests west of the Cascades, while the other species are found in the dryer, though not arid, forests east of the Cascades. Species may be locally common.

Key to the Species of *Allogona*

Pilsbry (1940) referred all *Allogona* west of the Continental Divide to subgenus *Dysmedoma*.

(a) Found east of the Cascade Crest. Color variable . . . .[b/bb]

  (aa) Found west of the Cascade Crest. Color brown and/or buff . . . . [d/dd]

(b) Shell moderately elevated, large, about 27 mm wide, rather solid, somewhat glossy, rich dark brown or less often buff-colored. Umbilicus narrow, contained 10 to 13 times in the shell width. Last whorl with fine, well-spaced, buff-colored ribs. . . . .*Allogona lombardii*

  bb) Shell smaller, 18 to 24 mm wide. Color variable, nearly white, straw-yellow to dark brown. Not especially glossy . . . . [c/cc]

(c) Shell 18 to 24 mm wide (occasionally up to 28 mm), with rather thin, moderately elevated spire, usually cinnamon to dark brown but may be lighter. Peristome reflected but little recurved except basally. Fine, distinct rib-like growth lines and distinct, close and regularly spaced spiral striae. Umbilicus narrow, contained 8 to 11 times in the shell width. Widespread mostly east of the Cascade Crest to the Continental Divide, from British Columbia through the Idaho Panhandle and into northeastern Oregon . . . . *Allogona ptychophora ptychophora*

  (cc) Shell 18 to 23 mm wide with relatively low, conic spire. Growth-wrinkles relatively weak and spiral striae irregular. Peristome well-reflected and recurved at its edge. Snake River Canyon, along Washington-Idaho border and Lapwai Creek watershed, Nez Perce and Lewis counties, Idaho. . . . . *Allogona ptychophora solida*

(d) Relatively large shells 26 to 35 mm wide. With low to moderately elevated spire, the earlier whorls are brown with fine, rib-like growth lines. The last three-fourths whorl with malleations, brown within the indentations and buff on the surrounding ridges. Southern British Columbia, western Washington, and northwestern Oregon . . . .*Allogona townsendiana*

  (dd) Shell 24 to 30 mm wide. Spire slightly higher than that of typical *townsendiana* and not as glossy. Umbilicus narrow, contained 8 to 9 times in the shell width. Color tan or buff to rather dark brown, usually with a buff

colored area near the end of the last whorl extending to the aperture. Malleations on the last whorl are generally lacking or indistinct if present. South Puget Sound to Pacific Co., Washington.
. . . . *Allogona townsendiana frustrationis*

Allogona lombardii   A. G. Smith, 1943   Selway Forestsnail

Allogona lombardii: Allison Cr., Idaho Co., ID (MNHP collection)

Description: Shells of 5¾ to 6⅛ whorls are about 27 mm wide by 18 to 19 mm high. The spire is moderately elevated, convexly-conic. The umbilicus is contained 10 to 13 times in the diameter. It is nearly uniform, expanding very slightly to near its end; then it straightens, opening a little in the last half-whorl. The shell is a uniform rich dark brown except for the fine ribs, which are light buff on the last 1½ to 2 whorls. Occasionally shells occur that are all light buff. Other than the distinct, fine, usually well-spaced ribs, the sculpture includes prominent spiral striae, both dorsally and ventrally, beginning faintly near the end of the second whorl, showing distinctly on the last three whorls, and extending through the outer surface of the reflected peristome to the edge of the aperture. The body whorl deflects abruptly but very shortly for about the last two millimeters. The peristome is well reflected, white anteriorly and buff posteriorly. There may be a slight callus ridge or small denticle near the inner end of the basal aperture.

Similar Species: *Allogona ptychophora* is smaller and has a much broader range, which includes that of *A. lombardii*. *Allogona townsendiana* is about the same width but has a lower spire and does not occur within the same range as *A. lombardii*.

Distribution: Known from Idaho Co., Idaho, *Allogona lombardii* may be found in moist forests and riparian areas, hiding under shrubs and other vegetation and in the open during wet periods.

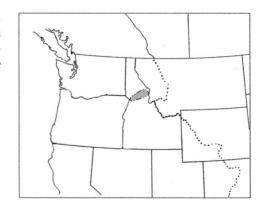

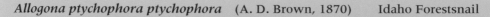

Allogona ptychophora ptychophora   (A. D. Brown, 1870)   Idaho Forestsnail

1. *Allogona ptychophora*: Smackout Creek, Stevens Co., WA
2. *Allogona ptychophora*: (A) Pend Oreille Co., WA; (B) Chelan Co., WA; (C) Stevens Co., WA

Description: Shells of 5¼ to 6¼ whorls measure 18 to 24 mm wide by 12 to 18 mm high. The shell is rather thin; the spire is moderately elevated, conic or slightly convexly so. The narrow umbilicus is contained 8 to 11 times in the diameter, increasing very slowly until the last whorl, in which it doubles its width. The shell is usually cinnamon to dark brown, but colors may vary, and individuals of from nearly white, grayish, buff, straw yellow, to dark brown may be found. The sculpture is of coarse growth-wrinkles and very fine, rib-like structures. Spiral striae, beginning faintly with the third or fourth whorl, are distinct and prominent dorsally, laterally, and ventrally on the latter whorls, and they extend through the outer surface of the peristome to the aperture. The aperture is roundly-lunate, or is often rather auriculate from a lobe in the basal-columellar angle formed by a basal callus ridge or slight denticle. The peristome is well-reflected; it is white anteriorly and buff posteriorly. The body whorl is deflected abruptly but shortly, just before the aperture.

Similar Species: *Allogona townsendiana* may be found with *A. ptychophora* in the Columbia Gorge, but it is larger and generally has irregular malleated sculpturing rather than the rib-like growth-wrinkles of *A. ptychophora*. *A. townsendiana* form *frustrationis* appears more like *A. ptychophora* than *A. townsendiana*, but it is much larger than typical *ptychophora*. *A. lombardii* is limited to the area of the southwestern Idaho Panhandle. It is larger than *A. ptychophora*, and

(A) *Allogona ptychophora*:
     Pend Oreille Co., WA;

(B) *Allogona ptychophora solida*:
     Nez Perce Co., ID

it has a more solid shell and stronger, more buff-colored ribs. *A. ptychophora solida* is known from the Snake River Canyon and eastward into Idaho along Lapwai Creek. See its description below. *Cryptomastix devia* usually has a distinct parietal denticle and is seldom found outside of the lower to mid-elevation forested areas of the foothills of western Washington and northwestern Oregon.

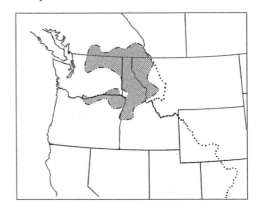

Distribution: *Allogona ptychophora ptychophora* can be found from the eastern Cascades and the Columbia Gorge, east to the Continental Divide in Washington, Idaho, Montana, and British Columbia, and south into Oregon in Umatilla and Union counties. It occurs in moist or open forest and riparian areas, hiding under shrubs and other vegetation, and in the open during wet periods.

Allogona ptychophora solida   (Vanatta, 1924)

Allogona ptychophora solida: E of Culdesac, Nez Perce Co., ID; scale bar = 1 cm

Description: Shells of about 5½ whorls are 18 to 22.5 mm wide by 11 to 13 mm high. This subspecies differs from the typical form by its lower spire (low, conic), weaker rib-like growth lines and irregular spiral striae. Its color range is generally from light buff to cinnamon-brown. The aperture is relatively small; the peristome is relatively broad and may have a distinctly recurved edge.

Similar Species: As do the *Allogona,* two species of *Cryptomastix* may lack a parietal tooth. *C. hendersoni* is generally less than 17 mm wide. It has a distinct constriction before the peristome, and often a small parietal tooth, especially in the eastern part of its range. *C. populi*, also normally less than 17 mm wide, has a smooth, glossy shell, a rather wide elliptical aperture, and a relatively narrow peristome without a constriction behind it.

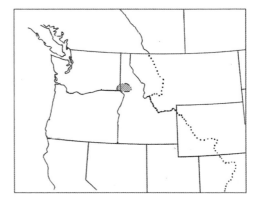

Distribution: This subspecies occurs in the Snake River Canyon, Asotin Co., Washington, and Nez Perce Co., Idaho. It has also been found along Lapwai Creek in Nez Perce Co., and into Lewis County, Idaho. It is found in rock slides and rocky riparian areas.

Allogona townsendiana   (I. Lea, 1838)   Oregon Forestsnail

Allogona townsendiana: Rock Creek, Lewis Co., WA; scale bar = 1 cm

Description: Shells of about 5 to 6½ whorls measure 26 to 35 mm wide by 16 to 25 mm high. The spire is relatively low and convexly-conic. The umbilicus is about one-sixth the shell width, narrow and uniform internally but about doubling its width in the last whorl. The early whorls are brown and sculpted with fine, irregular but distinct growth-wrinkles, less rib-like than in our other *Allogona* species. The last whorl is sculpted with rather dense malleations as well as growth-wrinkles that are somewhat rib-like. The elevated ridges or edges around the malleations are a glossy buff color, which is co-dominant with the brown even though the indented areas are the same brown color as the spire. Fine regular spiral striae also adorn the latter whorls, extending faintly onto the posterior of the reflected peristome. The peristome is well-reflected, white anteriorly and buff posteriorly. The aperture is widely ovately-lunate, flattened basally by a slight callus ridge inside.

Vanatta (1924) described *A. townsendiana* "form *brunnea*," in which the shell is very dark, from near Kelso, Cowlitz Co., Washington. Because typical forms occur in the same populations, and dark forms have been seen from other areas, Pilsbry (1940) did not consider these to be racially distinct.

Similar Species: *A. townsendiana townsendiana* is distinguished from *A. ptychophora*, which might rarely be found within its range, by its larger size and malleations on the last whorl rather than fine rib-like sculpturing. *A. townsendiana frustrationis* (see below), except for its larger size, is more likely to be confused with *A. ptychophora*. *Cryptomastix devia*, which is likely to be found within the same range as *Allogona townsendiana*, is smaller and normally has a parietal tooth.

Distribution: *A. townsendiana* occurs in southwestern British Columbia, through the lowland forests of western Washington, into the western Cascades, the Willamette Valley and Coast Range of northwestern Oregon, and through the western Columbia Gorge. It inhabits lush forested areas and riparian zones, hiding under vegetation and woody debris.

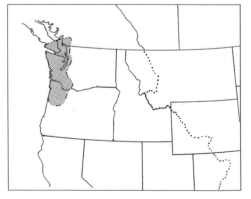

Allogona townsendiana frustrationis   Pilsbry, 1940

Allogona townsendiana frustrationis: Thurston Co., WA; scale bar = 1 cm

Description: As described by Pilsbry (1940): With about 6 whorls the shells measure approximately 24 to 29 mm wide by 18 mm high. The shell is thinner and smoother than typical *A. townsendiana*, mostly cinnamon-brown but sometimes lighter, with cinnamon-brown streaks or occasionally darker. The periostracum is often lost. Ribbing is irregular, often lacking on parts of the last whorl. "Malleation is sparse or wanting, but impressed spiral lines are about as in typical *townsendiana*."

Some specimens may be darker brown. The outer surface of the reflected peristome and the constriction behind it is usually buff-colored; the apertural side of the peristome is white.

Similar Species: Although only designated a form by Pilsbry, this snail may be found in local populations of distinctly appearing specimens, which would seem to warrant subspecies status. It is a frustrating species to identify, as it is the size of *A. townsendiana* and occurs within the range of that species, but it is more like *A. ptychophora* in appearance, and may be the source of reports of that species west of the Cascades. *A. ptychophora* is smaller and generally occurs east of the Cascade Crest. Typical *A. townsendiana* has an irregular buff-colored pattern on the last whorl that surrounds malleations in the shell, the indented portions of which are darker brown as are the earlier whorls. *Cryptomastix devia* is somewhat smaller, and its aperture contains a parietal denticle.

Distribution: *Allogona townsendiana frustra-tionis* occurs in southwestern Washington, with populations occurring in (at least) Pacific, Thurston, and Lewis counties. It inhabits forest floor litter and vegetation, coming out during cool, wet periods.

Genus: *Cryptomastix*

Cryptomastix are medium to moderately large snails, sub-globose to low-conic, narrowly umbilicate to imperforate. The aperture is lunate or auriculate, often widely so, and often tri-lobed by the intrusion of the characteristic apertural denticles. The peristome is distinctly reflected and often recurved or revolute. Dentition consists of a parietal tooth, a basal cusp, usually at the outer end of a basal ridge or fold on the inside of the basal lip margin, and one inside the palatal or the outer lip margin. The teeth, however, may be reduced to a mere callus or lamina, or may be absent, so any combination from zero to three may occur. The young of many *Cryptomastix* are hirsute, but few species retain the hairs in adulthood. Hair-scars or associated papillae can sometimes be found on the periphery or on the penultimate whorl in front of the aperture. The shells may be sculptured with spiral striae, although these are often faint or indistinct.

Pilsbry (1940) named *Cryptomastix* a subgenus of the larger and more widespread genus *Triodopsis*. Since elevated to genus status (Emberton, 1991), *Cryptomastix* are endemic to the Pacific Northwest.

Species may be distinguished by combinations of shell characteristics, many of which show extreme intraspecific variation, and forms intermediate between species or subspecies are common. *Cryptomastix mullani* is so polymorphic that the variation within that group alone led Pilsbry (1940) to accept nine subspecies with a range of shell characteristics intergrading between many of the other species.

Key characteristics are given as normal for the species or subspecies, often with exceptions in parentheses. The general characteristics for the species should be representative of the majority of a population with an occasional specimen displaying the variations.

Key to the Species of *Cryptomastix*

(a)   Shells small, diameter 6 to 8 mm, perforate or imperforate and with scattered hairs often retained into adulthood. There are 5 to 5½ closely coiled whorls, the last conspicuously constricted behind the aperture. The peristome reflected but not recurved; the aperture with a long parietal tooth. Western Oregon and Washington to southwestern British Columbia . . . . *Cryptomastix germana*

    (aa)   Shells larger, diameter greater than 8 mm. Umbilicus narrow, open or partially to nearly (rarely wholly) covered by the reflected basal-columellar lip margin . . . .[b/bb]

(b)   Shell large, diameter 20 to 24 mm, height 12 to 16 mm. With a small but prominent white parietal tooth and usually a very low, blunt basal cusp or lamina. Umbilicus partly covered. Western Washington into northwestern Oregon and Vancouver Island, British Columbia. Rarely on the eastern slope of the Cascades. . . . . *Cryptomastix devia*

    (bb)   Shells smaller, diameter 8 to 18 mm. Peristome reflected and recurved, or revolute. Primarily found east of the Cascade Crest, but some may occur in the Columbia Gorge . . . . [c/cc]

(c)   Apertural teeth prominent, all three distinct and relatively large, covering nearly half of the cross-sectional area of the aperture. Last whorl constricted behind the lip . . . . [d/dd]

    (cc)   Apertural teeth not all prominent, although they may or may not all be distinct. Either they are relatively small (covering less than one-fourth of the apertural cross section) or one or more of the teeth is/are reduced to a callus or lamina, or is/are absent. Constriction behind the apertural lip is variable . . . . [e/ee]

(d)   Whorls 5¼ to 6, closely coiled. Diameter 10 to 13 mm, height 5 to 8 mm. Shell light grayish-brown or buff. Peristome reflected and recurved a little at the edge. Umbilicus little covered . . . . *Cryptomastix sanburni*

  (dd)   Whorls about 4¾, less closely coiled and usually more depressed. Diameter 10 to 11 mm, height 5 to 6 mm. Color a medium amber-brown. Peristome revolute. Umbilicus one-third to one-half covered. . . . .*Cryptomastix magnidentata*

(e)   Dentition generally lacking (but C. *hendersoni* may have a small parietal tooth). . [f/ff]

  (ee)   One to three denticles normally present; however, some may be represented by raised calluses, ridges, or folds (laminae) inside of the peristome . . . . [g/gg]

(f)   Shell relatively large, diameter to about 17 mm with 5½ whorls. Umbilicus contained about 6 to 7 times in the diameter. Whorls increasing slowly to the last one, which increases rapidly until, in its last quarter turn, it is more than twice as wide as the adjacent penultimate whorl. The last whorl descends rather abruptly to the aperture but is not significantly constricted behind the peristome. The peristome is narrowly reflected and strongly recurved or revolute. . . . . *Cryptomastix populi*

  (ff)   Shell 14 to 16 mm wide, by 8 to 9 mm high. Umbilicus about one-tenth the width of the shell. Aperture widely lunate and well constricted behind the lip. Peristome is narrowly reflected and very narrowly recurved at the outer edge, or it may be nearly revolute basally. Normally lacks dentition, but similar shells within its range have a small parietal cusp. . . . . . . *Cryptomastix hendersoni*

(g)   Shell small, diameter 8 to 9 mm with about 4¾ whorls. Whorls increase in size slowly, the last descending in front and constricted behind the peristome. The peristome is reflected, white and a little recurved around the edge. Apertural teeth are small but all three are distinct . . . . *Cryptomastix harfordiana*

  (gg)   Shells with 5 to 6 whorls range from 10 to 18.5 mm wide. The spire is low to moderately-low and convexly-conic. Peristome normally revolute. Shell characteristics vary (see subspecies), but typically the last whorl descends little in front, and is well-constricted behind the lip. Tooth patterns vary by subspecies, but typically the parietal tooth is short and triangular; the basal tooth is a long ridge or lamina; and the palatal or outer tooth is moderately developed or absent. . . . . *Cryptomastix mullani* and ssp.

Cryptomastix devia   (Gould, 1846)   Puget Oregonian

Cryptomastix devia: Kraus Ridge, Lewis Co., WA; scale bar = 1 cm

Description: The shell is large for the genus. With about 6 whorls it measures 20 to 24 mm wide by 12.5 to 15 mm high. The color is medium to dark brown. The spire is moderately elevated; the periphery is rounded. The lip is well-reflected and slightly recurved at the edge; the peristome is white to tan. There is a small to medium parietal tooth and a basal callus with a small cusp inside the median half of the basal lip margin. The shell is sculptured with coarse, sometimes faint, spiral striae. Very short hooked bristles can sometimes be seen on immature shells but are absent from adults.

Similar Species: This species is significantly larger than any other *Cryptomastix*. However, it may be confused with *Allogona townsendiana* form *frustrationis*, which occurs within the same range but is larger and lacks a parietal tooth.

Distribution: *Cryptomastix devia* occurs from Vancouver Island, British Columbia, through the Puget Trough of Washington (to elevations of about 2000 ft) and into the Oregon Coast Range, south to about McMinville, Yamhill Co. There is a single record from the eastern Cascades near Cle Elum, Kittitas Co., Washington.

Cryptomastix germana   (Gould, 1851)   Pygmy Oregonian

Cryptomastix germana: Kraus Ridge, Lewis Co., WA; scale bar = 2 mm

Description: This is the smallest *Cryptomastix*. With 5 to 5½ whorls it measures only 6.5 to 8.5 mm wide by 4.3 to 5.5 mm high. The shell is depressed-globose, thin and tightly coiled. The umbilicus is perforate or imperforate. The last whorl contracts before the aperture, which is narrowly but distinctly reflected. A long, white parietal tooth angles across the aperture. Fairly long, well-spaced or scattered, usually curved hairs are retained into adulthood.

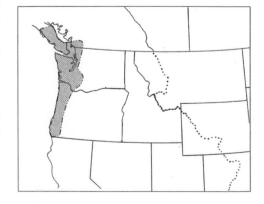

Similar Species: See below.

Distribution: The species occurs from British Columbia, south through western Washington and Oregon.

Subspecies of *Cryptomastix germana*

Two subspecies are described as follows, but there is a lot of variation in the size, crest, constriction, height of parietal tooth, bristles, and umbilicus across the species range. While the typical subspecies is described for the southern part of the species range, and the subspecies *vancouverinsulae* for the northern part of the range, there does not appear to be a clear distinction geographically, and the differences described, although distinctive, do not appear consistent and may simply be variations in shell characteristics.

Cryptomastix germana germana   (Gould, 1851)

Description: With 5 to 5½ whorls it measures approximately 8.3 mm wide by 5.5 mm high. It is imperforate. The last whorl has a prominent crest, then a deep constriction before the moderately reflected peristome. The hairs are moderately long and moderately spaced, and they are slightly to moderately curved. The basal-columellar curve of the aperture is rather abruptly angled.

Similar Species: See below.

Distribution: This subspecies is recorded from southwestern Washington to southwestern Oregon.

Cryptomastix germana vancouverinsulae   (Pilsbry & Cooke, 1922)

Description: Somewhat smaller than the typical form, with the same number of whorls it measures 6.5 to 7.7 mm wide by 4.3 to 5.3 mm high. It is perforate. The crest is not prominent, and the constriction is not as deep. The hairs are more scattered and may be lost by adulthood. The basal-columellar curve of the aperture is more rounded.

Similar Species: See below.

Distribution: This is the northern subspecies, occurring from British Columbia, including Vancouver Island, through western Washington into northwestern Oregon.

Note on Similar Species

In general, the shell of *C. germana* resembles that of the *Vespericola* more than other *Cryptomastix*. Shells of *Vespericola* are usually hirsute, while most other *Cryptomastix* lack hairs as adults (*C. mullani tuckeri* has short, flat scales). However, all *Vespericola* from within the same range are larger than *Cryptomastix germana*. *Vespericola sierranus* from northern California is near the same size but seldom has a parietal tooth, and its periostracal hairs are fine and short. *Trilobopsis loricata* is slightly smaller than *Cryptomastix germana* and is readily distinguished by its three small denticles and crescentic, scale-like periostracal sculpture.

Cryptomastix harfordiana   (W. G. Binney, 1886)   Salmon Oregonian

Description: *C. harfordiana* is a small *Cryptomastix*, with a low-rounded spire. With 4¾ whorls the shells measure about 9 mm wide. The umbilicus is about one-sixth the shell diameter. The whorl is slightly constricted behind the aperture. The peristome is white, well-reflected and slightly recurved at the edge. The aperture contains three prominent denticles.

Similar Species: *C. harfordiana* is smaller than any other *Cryptomastix* from within its range. Pilsbry (1940) discussed and included pictures of shells of a form intermediate between *C. clappi* and *C. harfordiana*, from Idaho Co., Idaho, in which the size and shape of the shell resembles *clappi*, but the dentition is like that of *harfordiana*.

Distribution: Salmon River, Idaho; exact site not described but Pilsbry (1940) believed it to be somewhere north of Lucile, Idaho Co.

Cryptomastix hendersoni   (Pilsbry, 1928)   Columbia Oregonian
Synonym: *Cryptomastix mullani hendersoni*

1

Description: *C. hendersoni* is a medium-sized snail, with 5⅓ to 5½ whorls measuring 14.5 to 16.2 mm wide by 8.2 to 9 mm high. The spire is low to moderately elevated. The umbilicus is narrow, contained 9½ to 10 times in the shell width, with the reflected peristome covering about one-fourth of it. The periphery is rounded. The aperture is widely lunate and generally lacks dentition, although some specimens, especially from the eastern part of their range, have a small parietal tooth. The last whorl descends only little before the aperture. It has a rather strong constriction before the white, reflected, and narrowly recurved peristome. Shells are sculptured with weak or very faint spiral striae. There are no apparent traces of hairs.

Similar Species: It has been questioned whether denticulate specimens from the eastern part of the

1. *Cryptomastix hendersoni* (dentition either lacking or with only a small parietal tooth): Rowland Cr., Klickitat Co., WA; scale bar = 1 cm

2. *Cryptomastix hendersoni* (oblique view showing constriction prior to apertural lip): Rowland Cr., Klickitat Co., WA

2

species range might be a separate, undescribed species. These are found with the typical variety at the west end of their range as well and are treated here as forms of the same species, recognizing the need for further study. *C. populi* is the only other *Cryptomastix* that typically lacks dentition; however, the shape of its shell is dissimilar to that of *C. hendersoni*.

C. populi has a very low spire and relatively large elliptical aperture without a constriction of the last whorl. Other edentate shells may be aberrant forms of other species.

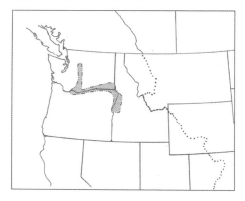

Distribution: *C. hendersoni* occurs from The Dalles, Wasco Co., Oregon, east along the Columbia and Snake Rivers and the Washington-Oregon border, to Adams and Washington counties, Idaho, and north from the Columbia along the Yakima River through Yakima County, Washington.

Cryptomastix magnidentata (Pilsbry, 1940)   Mission Creek Oregonian
Synonym: *Cryptomastix mullani magnidentata*

Cryptomastix magnidentata (dark brown; revolute lip; umbilicus about half covered): Asotin Co., WA; scale bar = 1 cm

Description: As implied by its specific epithet, this snail is striking for its large apertural teeth (but see also *C. sanburni*). It is fairly small, measuring 10.8 mm wide by 5.7 high with about 4¾ whorls. The spire is quite low and slightly rounded; the whorls increase slowly but regularly. The shell is a rather dark amber-brown color. The umbilicus is about half-covered by the reflected columellar lip margin. The last whorl is constricted behind the strongly revolute peristome. Sculpture is of microscopic spiral striae and some scattered hair-scars.

Similar Species: The shell of *C. sanburni* is similar in size, shape, and dentition, but is light buff colored and more tightly coiled, shells of the same width having one-half to a full additional whorl. *C. m. latilabris* is somewhat larger, has a broader reflected lip margin, and generally less-well-developed lip teeth.

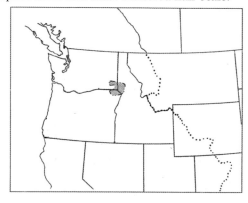

Distribution: *Cryptomastix magnidentata* has been found in Nez Perce Co., Idaho, along the Snake River in Asotin Co., Washington, and in Joseph Canyon, Wallowa Co., Oregon.

Cryptomastix mullani Complex   Coeur d'Alene Oregonian

Cryptomastix mullani is a composite taxon, with enough variation that Pilsbry (1940) accepted nine subspecies. Two of these subspecies, *C. hendersoni* and *C. magnidentata*, have since been elevated to full species. The variation within the species that led to so many sub-taxa is often not clearly defined within a population, and many intergrades can be found. Although not diagnostic by itself, *C. mullani* tend to have strongly revolute peristomes.

Described here by their distinguishing shell characteristics are all of the subspecies described by Pilsbry (1940). However, there is so much variation within the species that a continuum exists between these subspecies, making them difficult to separate. Several specimens may need to be examined from a population to discern to which subspecies they best conform.

Key to the Subspecies of *Cryptomastix mullani*

(a)   Apertural teeth 3, but the outer (or palatal) and/or basal lip teeth may be
       in the form of a low ridge or lamina . . . .[b/bb]

   (aa)   Normally one or two apertural teeth, including a small to well-developed
           parietal (sometimes absent in *C. m. tuckeri*), and either a basal or outer
           cusp or lamina. . . . . [d/dd]

(b)   Shell 12 to 15 mm wide. Umbilicus about one-half to nearly wholly covered by
       the reflected basal-columellar lip margin. Parietal tooth short, triangular; outer
       lip tooth moderately developed (sometimes absent); basal tooth a long ridge or
       lamina with a low cusp on its outer end. Shell globose-depressed. .*C. mullani mullani*

   (bb)   Umbilicus open to about one-half covered by the reflected lip margin.
           Spire lower than that of *C. mullani mullani* . . . . [c/cc]

(c)   Reflected apertural lip very broad. Parietal tooth usually rather large; basal
       tooth a low ridge; outer lip tooth small but usually distinct (sometimes absent).
       Spire quite low . . . . *C. mullani latilabris*

   (cc)   With or without a parietal tooth. Palatal as well as basal armature in
           the form of a low lamina, with or without low cusps. Periostracal
           hairs scale-like and retained as adults. . . . .*C. mullani tuckeri*

(d)   A variable shell, mostly with a short, triangular parietal tooth and low basal
       ridge or lamina. Umbilicus nearly or completely covered by the reflected
       basal-columellar lip margin (occasionally more open). Shell low to
       moderately elevated, convexly-conic. Peristome revolute . . . . . . *C. mullani hemphilli*

   (dd)   Parietal tooth small to well-developed or sometimes absent. If absent,
           then with sparse, short, scale-like hairs. Umbilicus open to about half-
           covered by the basal-columellar lip margin. Shell more depressed . . . . . [e/ee]

(e)   Parietal tooth triangular, small to well-developed or absent. Ridge-like lamina
       inside of upper and lower lip margins. Periostracum sparsely covered with
       short, flat, scale-like hairs. Lip light brown, expanded rather thinly at the
       outer-basal curve. . . . .*C. mullani tuckeri*

   (ee)   Parietal tooth small to well-developed. Basal lamina present but no
           upper lip tooth. Hairs usually absent from adult shells, but if present
           they are minute and closely spaced . . . . [f/ff]

(f)   Basal lip margin curves rather sharply upward to meet the columella, forming a sub-
       angular basal-columellar curve. Umbilicus wide, slightly to partly covered. Parietal

tooth small to well-developed. Shell diameter 11 to 18 mm wide by 7 to 10 mm high. Peristome variable, wide to rather narrow. Eastern Washington and northeastern Oregon through northern Idaho into western Montana. . . . .*C. mullani olneyae*

(ff)   Basal lip margin curves gradually into the columella, forming a gently rounded curve. Parietal tooth small and the basal ridge indistinct, although there may be a small cusp about mid-basally. Width 11 to 15 mm, height 6 to 7 mm. . . . . [g/gg]

(g)   Umbilicus contained about 6½ times in the diameter. Northern Idaho and western Montana. . . . . *C. mullani blandi*

(gg)   Umbilicus wider, contained about 4½ times in the diameter. Idaho and Clearwater counties, Idaho . . . . *C. mullani clappi*

C. mullani blandi   Hemphill, 1892

Description: With a very low conic spire, the shell is 13 to 13.5 mm wide by 6.3 to 6.5 mm high. The umbilicus is between one-sixth and one-seventh the shell diameter and is about half-covered by the reflected lip margin. The periphery is rounded or slightly widest above the midpoint. The aperture is widely auriculate and contains a small parietal denticle and a very small mid-basal cusp. Sculpture of oblique rows of microscopic papillae on the base may or may not be present.

Similar Species: *C. m. blandi* differs from the typical subspecies by its very low spire and only two small denticles, the basal being very minute. It is most similar to *C. m. clappi*, from which it differs by having a narrower, partly covered umbilicus and slightly lower spire.

Distribution: *C. mullani blandi* has been reported from the vicinity of Post Falls, Kootenai Co., Idaho, and Ravalli Co., Montana.

C. mullani clappi   Hemphill, 1897

Description: The shell is similar to that of *C. m. blandi* but the spire is a little more elevated and conical. It has about 5¼ whorls and measures 11 to 15 mm wide. The umbilicus is contained about 4½ times in the width of the shell and is open, being little if at all covered by the reflected lip margin. The periphery is rounded or slightly wider above its midpoint. The aperture is widely auriculate and contains a small parietal denticle and a very small basal cusp.

Similar Species: *C. m. clappi* is most similar to *C. m. blandi*; it differs in its slightly more elevated spire, its larger, more open umbilicus, and its distribution.

Distribution: *C. m. clappi* is known from the Salmon River Mountains, White Bird, Lucile, and Slate Creek, Idaho Co., and near Orofino, Clearwater Co., Idaho.

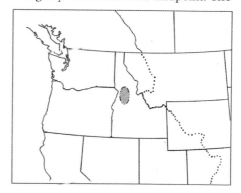

C. mullani hemphilli   W. G. Binney, 1886

1. *Cryptomastix mullani hemphilli*: Cataldo Mission, Kootenai Co., ID; scale bar = 1 cm
2. *Cryptomastix mullani hemphilli*: Swakane watershed, Chelan Co., WA; scale bar = 1 cm

Description: This medium-sized snail has a low to moderately elevated spire of a rather rich brown color (cinnamon or darker). With 5½ to 5⅔ whorls it measures 12.8 to 17 mm wide by 7 to 9.7 mm high. It is narrowly umbilicate, but the umbilicus is partly to completely covered by the reflected lip margin, so it may appear rimate or imperforate. The peristome is white, broad, thickened, and strongly revolute or recurved at the edge. The aperture contains a small to moderate parietal tooth and a basal ridge (lamella) with a low cusp at its outer end. Sculpture is of shallow spiral striae, and in some specimens sparse hair-scars may be seen.

Similar Species: *C. m. hemphilli* is unique in its variable umbilicus. It differs from the typical *C. mullani* by generally lacking an outer or palatal denticle; from *olneyae* by its less tightly revolute peristome and generally more covered umbilicus; and from *clappi* and *blandi* by its more elevated spire and strong basal lamella.

Distribution: *C. m. hemphilli* occurs in Northern Idaho around Lake Coeur d'Alene, Kootenai, Benewah, Shoshone, Clearwater, and Idaho counties. In Montana it occurs in Sanders and Missoula counties. In Washington an apparent disjunct population of this subspecies occurs in Chelan Co.

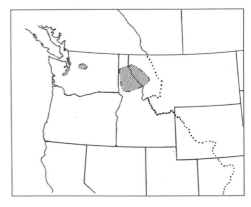

C. mullani latilabris   Pilsbry, 1940

1. *Cryptomastix mullani latilabris*: Slate Cr., Idaho Co., ID; scale bar = 1 cm
2. *Cryptomastix mullani latilabris*: John Day Cr., Idaho Co., ID; scale bar = 1 cm

Description: This medium-sized snail, about 11.4 to 15.5 mm wide by about 7.2 mm high, has a very low, roundly-conic spire. It has a narrow umbilicus, open but up to half covered by the reflected lip margin. The peristome is very broad and white. The aperture is generally tridentate with a strong, triangular parietal, but the teeth are variable, from fairly strong to barely visible, if at all. Hair-scars may or may not be seen.

Similar Species: *C. mullani latilabris* is similar to *C. m. olneyae*, but it has a lower spire, the peristome is wider, and the parietal tooth is usually more prominent. *C. magnidentata* and *C. sanburni* are somewhat smaller, and their dentition, especially the lip teeth, is more consistently larger. Also, the range of *latilabris* is farther south than these latter two species.

Distribution: *C. mullani latilabris* occurs along Slate and John Day Creeks and the Salmon River, Idaho Co., Idaho.

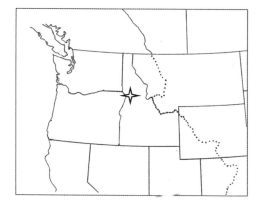

C. mullani mullani   (Bland & J. G. Cooper, 1861)   Coeur d'Alene Oregonian

Cryptomastix mullani mullani: Beauty Cr., Kootenai Co., ID; scale bar = 1 cm

Description: The low, rounded spire of 5½ to 6 whorls measures 13 to 15 mm wide by 7.9 to 8.7 mm high. The umbilicus is mostly covered by the reflected lip margin. The aperture is somewhat auriculate but normally with a gap in the outer-basal curve between the two lip teeth. The last whorl is constricted before the white, reflected, and recurved peristome. The aperture is tridentate; the three denticles are distinct if not always prominent. Look for the distinct cusp in the outer lip margin. The parietal may be fairly large to quite small. Shell sculpture consists of irregular growth striae and weak spiral striae. Hair-scars may be seen in front of the aperture.

Similar Species: The other *Cryptomastix* from east of the Cascade Range differ as follows: *C. harfordiana* is small; *C. hendersoni* and *C. populi* are normally without dentition (*hendersoni* may have a small parietal); *C. magnidentata* and *C. sanburni* are smaller, more tightly coiled, and have much larger denticles. Among the subspecies of *C. mullani, C. mullani mullani* is the one with three distinct cusps. *C. mullani latilabris* may or may not have similar dentition, but it differs by its wider, thicker peristome.

Distribution: Typical *C. mullani* is found in Kootenai and Shoshone counties in northern Idaho and in adjacent western Montana.

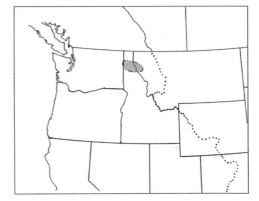

C. mullani olneyae   (Pilsbry, 1891)

Description: This medium-sized snail has a low conic spire with about 5½ whorls. The shell is 11 to 19 mm wide by 7.4 to 11.6 mm high. It has a narrow, open umbilicus (wider than that of typical *mullani*) that is less than half-covered by the reflected lip margin. The periphery is rounded. The aperture is widely lunate and constricted before the peristome, which is white and strongly revolute. Dentition includes a small parietal and a basal ridge with a low rounded cusp. The basal-columellar curve of the apertural lip is rather abrupt, forming an angle. Spiral striae are weak or not apparent, and hair-scars are few if any.

Cryptomastix mullani olneyae: O'Hara Campground, Idaho Co., ID; scale bar = 1 cm

Similar Species: *C. m. olneyae* differs from typical *C. m. mullani* by its lack of an outer or palatal denticle and its basal columellar lip margin covering less than half of the umbilicus.

C. m. blandi and *C. m. clappi* are similar but have a narrower peristome and a very minute basal cusp. The basal-columellar angle is more gradually rounded in these latter two subspecies.

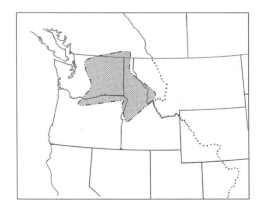

Distribution: *C. mullani olneyae* occurs in eastern Washington into adjacent northeastern Oregon, northern Idaho, western Montana, and southeastern British Columbia.

C. mullani tuckeri   Pilsbry & Henderson, 1930

Description: The low, roundly-conic spire of 5 to 5½ whorls measures 12.5 to 13 mm wide by 6.5 to 7 mm high. It is rather tightly coiled, the whorls increasing gradually. The open umbilicus is partly covered by the reflected lip margin. The aperture is wide, the peristome rather widely reflected but narrowed at the outer-basal curve. There may or may not be a small, triangular parietal tooth. The palatal and basal denticles are in the form of ridges or lamella, sometimes with low cusps. There may be sculpturing of irregular spiral striae, but especially unique are the short, flat scale-like hairs on the shell surface.

Similar Species: This subspecies stands alone among *Cryptomastix* because of its scale-like hairs even as an adult. The hairs of *C. germana*, which are also retained into adulthood, are long and bristle-like, and the shells of that species are smaller and more globose than others of the genus.

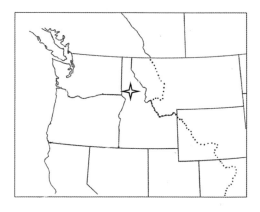

Distribution: *C. mullani tuckeri* occurs along the Clearwater River in Idaho, near the mouth of Fourth of July Creek, and near Orofino, Clearwater County.

Cryptomastix populi (Vanatta, 1924)   Cottonwood Oregonian

Cryptomastix populi: 1 mile S of Asotin, Asotin Co., WA; scale bar = 1 cm

Description: This is a medium-sized snail with a very low, convex-conic spire. With 5½ whorls it measures about 17 mm wide by 8.8 mm high. The umbilicus is open, less than one-half covered by the reflected lip margin and nearly one-sixth the width of the shell. The periphery is well rounded. The whorls increase slowly in size until the last, which is much larger than the penultimate. There is a constriction just before the off-white, narrowly reflected and revolute peristome. Apertural dentition is lacking. Shell sculpture is of fine growth-wrinkles and spiral striae, usually weak at the periphery but strong above and below. No hairs or hair-scars are apparent.

Similar Species: Although other species occasionally produce an edentate individual, *C. hendersoni* is the only other *Cryptomastix* that is described as normally lacking dentition. The form of the *C. hendersoni* shell is similar to other *Cryptomastix* of similar size with elevated spires, while *C. populi* is somewhat unique in appearance with its relatively flat, rapidly enlarging body whorl, and relatively wide, open aperture. *Allogona ptychophora solida*, which also lacks dentition, may also be found with *Cryptomastix populi*. The *Allogona*, however, has a more capacious body whorl and a much wider, flatter peristome that partially covers the umbilicus.

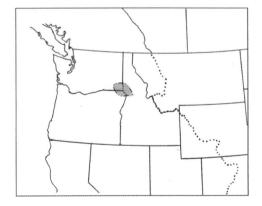

Distribution: *Cryptomastix populi* is found along the Snake River in Whitman and Asotin counties, Washington, and in Cottonwood Canyon, Nez Perce Co., Idaho. It is found in rock slides and brushy draws.

Cryptomastix sanburni (W. G. Binney, 1886)   Kingston Oregonian

Description: This fairly small snail with 5¼ to 6 whorls measures 10 to 12.2 mm wide and 5.7 to 7.8 mm high. The spire is low and slightly rounded; the whorls are rather tightly coiled and increase slowly in size. The shell is a light buff color. The umbilicus is narrow and about one-half covered by the reflected columellar lip margin. The last whorl is constricted behind a white, widely reflected, and somewhat recurved peristome. The aperture contains three quite large denticles similar to those of *C. magnidentata*. There are no apparent hairs.

Cryptomastix sanburni (tan or buff colored; reflected lip; open umbilicus): Snake River Canyon, Whitman Co., WA; scale bar = 1 cm

Similar Species: The shell of *C. magnidentata* is similar in size, shape, and dentition, but that species is a rich brown color and is more loosely coiled. Shells of about the same width have one-half to one less whorl than those of *C. sanburni*, and its peristome is tightly revolute.

Distribution: *C. sanburni* is known from Shoshone, Coeur d'Alene, Kootenai, and Bonner counties, Idaho, and it can also be found in the Snake River Canyon, in Whitman and Asotin counties, Washington.

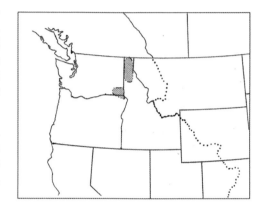

Genus: *Hochbergellus*

Hochbergellus hirsutus Roth & W. B. Miller, 1992   Sisters Hesperian

Description: Shell tan, 13.7 to 17.2 mm wide by 9 to 12.5 mm high in about 5¾ to 6⅝ whorls. The narrow umbilicus is partly covered by the reflected basal lip margin. The reflected peristome is somewhat concave and is white to pinkish-tan. It is narrowly recurved at the edge, most strongly so basally. There is usually a small white parietal tooth, and there may be a thickened basal ridge with a slight cusp toward its inner end. There are 2–3 short periostracal hairs per square millimeter.

Similar Species: *Hochbergellus* resembles *Vespericola*, and there are no specific external characteristics by which to distinguish the two genera. Characteristics in the above description are extracted from Roth and Miller (1992), from which the complete description can be read, including that of the anatomy.

Distribution: The type locality of *Hochbergellus hirsutus* is in Curry Co., Oregon.

Genus: *Trilobopsis*

Key to the Species of *Trilobopsis*

Several species of *Trilobopsis* occur in California, but the range of only one is known to extend into the Pacific Northwest. A second species is included here because it has been found within 12 miles of the Oregon-California border.

(a)   Spire low-conic or slightly convex but distinctly elevated. Narrowly umbilicate. Apertural teeth relatively small . . . . *Trilobopsis loricata nortensis*

   (aa)   Shell discoidal, spire nearly flat. Umbilicus wider. Apertural teeth prominent . . . . *Trilobopsis tehamana*

Trilobopsis loricata   (Gould, 1846)   Scaly Chaparral

Trilobopsis loricata (right inset shows close-up of sculpture on dorsal surface of shell): North Umpqua Ranger District, Douglas Co., OR; scale bar = 1 mm

These small snails measure 6.4 to 6.9 mm wide by 3.6 to 4.1 mm high in 4¾ to 5¼ whorls. The spire is rather low, convexly-conic; the periphery is sub-angular or somewhat shouldered, being widest above the midpoint. It is narrowly umbilicate or perforate. The peristome is narrowly reflected following a constriction, and there are three small teeth—a parietal, a basal, and an outer. The shell has a unique sculpture of short scale-like crescents arranged collabrally.

Trilobopsis loricata nortensis   (S. S. Berry, 1933)

Description: Besides the typical species, Pilsbry (1940) described five additional subspecies, all but one of which is confined to California. *T. loricata nortensis*, which occurs in our study area, differs from the typical species "in its being smaller, with a thinner lip, [and] reduced apertural dentition. The sculpture is altogether finer, more even, and less crude" (Pilsbry 1940). He also suggested: "It is probably a distinct species." Our specimens from the North Umpqua drainage, Oregon, are neatly sculptured little shells, with fairly well-developed dentition, which agrees with Pilsbry's description of specimens from Oregon.

Similar Species: The general appearance of *T. loricata*, the only member of this genus known to be found as far north as Oregon, is somewhat like that of *Cryptomastix germana*. That species, however, is a little larger, more globose, lacks the dentition on the apertural lip, and has rather long periostracal hairs rather than the scale-like sculpturing of *Trilobopsis loricata*.

Distribution: *T. loricata* occurs from as far south as Fresno and Mariposa counties, California. *Trilobopsis loricata nortensis* is known from Del Norte Co. in northern California, north to the North Umpqua River drainage, Douglas Co., Oregon.

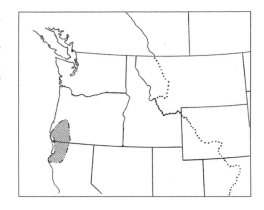

Trilobopsis tehamana (Pilsbry, 1928)   Tehama Chaparral

Description: The shell is small and discoidal; with 5⅓ to 5½ whorls it measures 7 mm wide by 3.2 mm high. The spire is nearly flat; the whorls tightly coiled, the last strongly constricted behind the reflected peristome. The aperture is small, with three well-developed denticles: a long, high parietal; a fairly large outer tooth; and a smaller one near the outer end of the basal aperture. The umbilicus is funnelform and rather large, about one-fifth the shell width. The shell is cinnamon or darker brown, the surface with a matte texture from fine, closely spaced radial wrinkles and weak spiral striae.

Similar Species: Similar in appearance to *Trilobopsis roperi*, which is also found in Shasta Co., California, but in the Sacramento River watershed; it is not described in this work. *T. roperi* is less tightly coiled and has a slightly larger aperture and umbilicus (contained 4½ times in the shell width). The peristome of *T. roperi* is flared more gradually outward, while that of *T. tehamana* is abruptly reflected.

Distribution: *Trilobopsis tehamana* is known from Tehama, Shasta, Butte, and Siskiyou counties, California. It inhabits rocky areas, under limestone talus, and nearby woody debris and leaf litter.

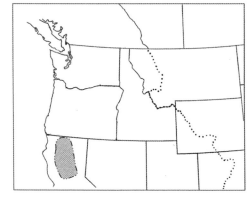

Genus: *Vespericola*

Shells of *Vespericola* are of a distinctive and fairly uniform shape, although their size and other minute details may differ. The body whorl is rather capacious, and the spire is moderately elevated and generally conic. There is usually a distinct constriction just before the reflected lip margin. The northern species are usually without dentition, but some of those from southwestern Washington and farther south have a parietal tooth. The shell is thin and fragile; the umbilicus is perforate to very narrowly umbilicate. Most are adorned with periostracal hairs (seldom lacking) of varying size and density.

General shell characteristics are quite similar, making identification from shells alone difficult or impossible. In a series of papers from the 1990s, Roth and Miller showed that dissection is necessary for positive identification of the species. Location of the collection site can be helpful, but the taxa are probably not all described, and known ranges of the described taxa may yet be expanded. In keeping with the intent of this work, the following keys and species accounts are based on shell characteristics and location as described by Pilsbry (1940) and, to some extent, modified in later works. A similar genus, *Hochbergellus* Roth and Miller, 1992, is indistinguishable from *Vespericola* by external characteristics.

Vespericola is endemic through the Pacific states and provinces, from southern Alaska to west-central California, and has been reported to the east through northern Idaho and into Montana.

Key to the Species of *Vespericola*

(a) Periostracal hairs lacking, or only occasional hairs remaining on adult shells. Surface somewhat glossy. Spire low for the genus (compares to *V. c. depressus*). Umbilicus quite narrow, little covered by the reflected peristome. Western Washington, northwestern Oregon, and the Columbia Gorge.
. . . . *Vespericola columbianus columbianus*

   (aa) Adult shells with periostracal hairs . . . .[b/bb]

(b) Shell small (8 to 9 mm wide), thin. Spire moderately elevated and rounded. Slightly angled at the periphery. Very narrowly open umbilicate. Normally edentate but rarely with a small parietal tooth. Northern California into southern Oregon. . . . .*Vespericola sierranus*

   (bb) Shell larger, adults normally greater than 12 mm wide . . . . [c/cc]

(c) Reflected apertural lip, rather broad; umbilicus quite narrow or perforate, and half or more covered by the reflected columellar lip margin . . . . [d/dd]

   (cc) Reflected lip margin narrower; umbilicus more open, one-fourth or less covered by the columellar lip margin. Relatively large, 12 to 18 mm wide.
. . . .[g/gg]

(d) Columellar lip hooked around the umbilicus leaving an angular perforation or narrow chink . . . .[e/ee]

   (dd) Columellar lip margin reflected over, but not hooked around the umbilicus. Shell thin but relatively solid; spire relatively high; body whorl rather capacious. Umbilicus quite narrow and normally at least half-covered by the reflected lip margin. Hairs dense and rigid. Rarely denticulate. Western Washington to Salem, Oregon, northern Idaho, and Western Montana. . . . . *Vespericola columbianus latilabrum*

(e) Relatively large, about 18 to 20 mm wide with 6 to 6¾ whorls. Parietal tooth strongly developed, white, and curved. Edge of basal lip margin sinuous when viewed from below . . . . *Vespericola euthales*

    (ee) Smaller, about 13 to 16 mm wide with 5⅓ to 6 whorls. Parietal tooth usually weak or lacking, seldom well developed. Edge of basal lip margin straight when viewed from below . . . .[f/ff]

(f) Spire conical, moderately elevated. Umbilicus pilose within. Body whorl deep; periphery widest just above the middle of the whorl. Fine short bristles, 12 to 34 per square mm . . . . *Vespericola eritrichius*

    (ff) Variable, with a fairly low to moderately elevated conical to convexly-conic spire. Body whorl deep; periphery broadly rounded, slightly shouldered. Peristome fairly wide, thickened inside and slightly recurved. Bristles very fine, dense, and short, 24 to 70 per square mm. West-central California to southwestern Oregon.. . . . .*Vespericola megasoma*

(g) Shell, thin, spire relatively low, dull, light brown. Umbilicus small, open (a little larger than that of *V. c. columbianus*), little covered by the reflected lip margin. Hairs very short, dense, evenly spaced, weaker basally. No dentition. The Dalles and Columbia Gorge . . . . *Vespericola columbianus depressus*

    (gg) Shell thin, spire moderately high. Umbilicus very narrow to perforate, about one-fourth covered by the reflected lip margin. Hairs dense, rather short. Occasionally denticulate, especially in the more southern parts of its range. See discussion under species accounts. . . . .*Vespericola columbianus* ssp.

Vespericola columbianus Complex

Pilsbry (1940) listed 5 subspecies of this taxon. Records of *V. c. orius* (Berry, 1933) are confined to California, so that subspecies is not described in this work. Roth and Miller (1993) redescribed *V. c. pilosa* (Henderson, 1928) and *V. c. orius* from California specimens, and elevated them to species status. That work raised taxonomic questions about the Washington and Oregon specimens previously referred to *V. c. pilosa*. In separating it from *V. columbianus*, they found distinct differences, in the internal anatomy and some shell characteristics, between *V. pilosus* from the type locality around San Francisco and *V. columbianus* from the western Columbia River Valley and northward.

    V. c. columbianus, depressus, and *latilabrum* are subspecies of the Pacific Northwest, as is *V. columbianus* ssp. that used to be considered a part of *V. c. pilosus* (see below). Characteristics of the three subspecies are: *columbianus* is mostly without periostracal hairs; the other two have hairs, but *depressus* has a lower spire than the typical form; and, *latilabrum* has a higher spire than the other two, a wider lip than other *Vespericola* from within its range, and an umbilicus more covered by the reflected peristome. *V. columbianus* ssp. (previously the northern *V. c. pilosus*) differs from *latilabrum* by its narrower peristome and its less-covered umbilicus.

Vespericola columbianus columbianus   (Lea, 1838)   Northwest Hesperian

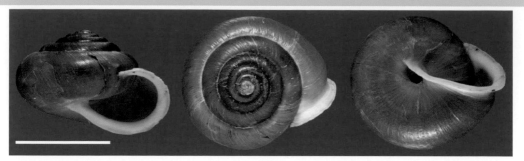

Vespericola columbianus columbianus (bristles lacking or very sparse; umbilicus narrow and little covered): Burdoin Mountain, Klickitat Co., WA; scale bar = 1 cm

Description: The thin shell, typically shaped for the genus, has a moderately elevated spire. In about 6 whorls it measures 13 to 16.5 mm wide by 9.3 to 11 mm high; height/diameter ratio (H/D) = 0.69. It is light yellowish to light brown and very slightly shouldered to the last half or two-thirds whorl. The umbilicus is perforate to narrowly umbilicate, partly covered by the reflected lip margin. The aperture is widely ovate or somewhat auriculate. The body whorl is a little constricted before the peristome, which is white and slightly thickened within. Dentition is lacking. Hair-scars are inconspicuous, the hairs normally being lost from adults, but occasionally some scattered bristles remain. The 1½ whorls of the protoconch are sculptured with curved radial wrinkles and faint granules.

Similar Species: *Cryptomastix hendersoni* is found near the same area and generally lacks dentition. However, the general *Vespericola* shell shape and auriculate aperture are distinct, and the umbilicus is narrower than those of *Cryptomastix*, which has a more narrowly reflected and recurved peristome.

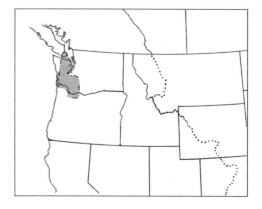

Distribution: *V. c. columbianus* is a snail of forested habitats of western Washington, the Columbia Gorge, and into northwestern Oregon. It is found mostly on and under logs, bark, and other woody debris.

Vespericola columbianus depressus   (Pilsbry & Henderson, 1936)

Description: The thin shell is similar to that of other *Vespericola* but more depressed, the spire being less elevated. With 5½ to 6 whorls it measures 14 to 15 mm wide by 8.7 to 9 mm high; H/D = 0.61. It is dull light brown or tan. The aperture is rather elliptically auriculate, being slightly flattened basally. The lip is reflected following a constriction of the last whorl. It covers the narrow umbilicus very little. There is no apertural dentition. Sculpture is of short, fine, closely spaced hairs aligned in forwardly descending rows, weaker basally.

Similar Species: The spire is more like that of *V. c. columbianus* but with rows of very short and/or fine hairs.

Vespericola columbianus depressus (spire lower; dense, short bristles; small umbilicus less than half covered): E of The Dalles, Wasco Co., OR; scale bar = 1 cm

Distribution: Type locality is The Dalles, Wasco Co., Oregon. *V. c. depressus* is a species of forest habitats.

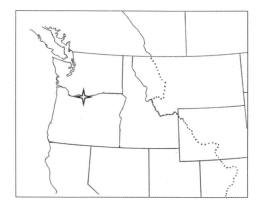

Vespericola columbianus latilabrum   (Gould, 1846?)

Description: This snail has a relatively solid shell with a moderate to fairly high spire. With 5¾ whorls it measures 12.5 to 18 mm wide by 8 to 13 mm high; H/D = 0.69. The

1. *Vespericola columbianus latilabrum* (spire higher; rather dense bristles; widely reflected lip nearly covers but does not curve around umbilicus): Discovery Bay, Jefferson Co., WA; scale bar = 1 cm

2. *Vespericola columbianus latilabrum* (with broken spire): Tillamook Co., OR; scale bar = 1 cm

umbilicus is quite narrow and half or more covered by the reflected lip margin. The last whorl is well-constricted before the widely reflected peristome. The auriculate aperture generally lacks dentition, but Pilsbry (1940) included illustrations of dentate specimens (with a small parietal and/or distinct basal cusp) from Oswego, Oregon. The hairs are dense and rigid, 7 to 19 per square mm. The animal is slender, light rusty colored with a pinkish or purplish tint, and covered with rows of elliptical tubercles. The bristles on the shells of these animals are often entwined with spiderwebs, soil, and other bits of debris, which appears to camouflage them.

Similar Species: *V. columbianus* ssp. differs in having the umbilicus less than one-fourth covered by the reflected peristome. The peristome of that subspecies is usually narrower than that of *latilabrum*, but that characteristic appears variable in both subspecies.

Distribution: *V. c. latilabrum* occurs from Salem, through northwestern Oregon and western Washington; also reports from the Coeur d'Alene Mountains, Idaho, and the Deer Lodge Valley, Montana (Pilsbry, 1940). A snail of forested habitats, it can be found in hardwood or conifer leaf litter and under logs and other woody debris. It is often found on the underside of Douglas-fir bark, where scraping marks from foraging can be seen along with droppings. These or other *Vespericola* species are often the most abundant snails found in the western Washington and Oregon conifer forests.

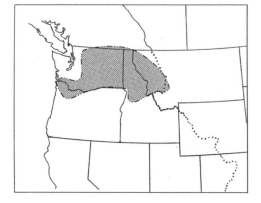

Vespericola columbianus ssp.
Synonym: Previously known as *V. columbianus pilosa* (Henderson, 1928)

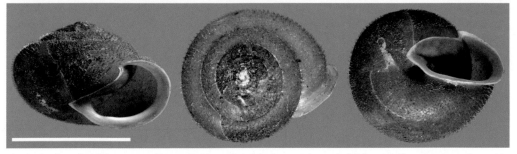

Vespericola columbianus ssp. (formerly *V. c. pilosa*) (spire moderately high; umbilicus narrow, about ¼ covered; dense, short hairs; occasionally denticulate in southern part of range): southwest WA; scale bar = 1 cm

Description: The shell is thin, light brown, with a moderately elevated (high for the genus), conic spire. With 6 whorls it measures 12.7 to 15 mm wide by 9 to 10.3 mm high. It is very narrowly umbilicate or perforate, and the umbilicus is about one-fourth covered by the reflected lip margin (Pilsbry1940). Normally edentate, but occasionally with a long, low, parietal tooth. The peristome is relatively narrow to moderately broad. Sculpturing is of closely-spaced, short hairs in forwardly descending rows. Embryonic whorls are sculptured with fine granulations.

Similar Species: Previously known as *V. columbianus pilosa*, Roth and Miller (1993) elevated the west-central California populations to full species, pointing out distinct differences between those and specimens from the Columbia River Valley northward. As well as anatomical differences, these researchers found the shells of the southern populations differing by being "broadly rounded; the shell of *V. columbianus* is usually weakly subangular." They also found that the shells of *V. pilosa* have about 19 to 30 hairs per square mm, while *V. columbianus* they examined had only 7 to 19 hairs per square mm.

These new findings leave us with the unanswered question of where to place the northern subspecies previously called *V. columbianus pilosa*. For the time being, it is apparently a subspecies of *V. columbianus*. It seems to differ from *V. c. latilabrum* in shell characteristics. Its umbilicus is about one-fourth or less covered by the columellar lip margin, while the umbilicus of *V. c. latilabrum* is over three-fourths covered. More study of the varieties of *Vespericola columbianus* is needed to sort out the northern species and subspecies.

Distribution: This taxon is apparently found from Alaska, through the Cascades and westward and south possibly into northern California.

Vespericola eritrichius (Berry, 1939)   Velvet Hesperian
Synonym: Pilsbry (1940) believed it to be synonymous with *Vespericola megasoma*. Roth and Miller (1995) elevated it to full species status.

Description: A rather small *Vespericola*, the spire is moderately elevated, conic or slightly convexly-conic. With 5⅓ to 6 whorls it measures about 12 to 16 mm wide by 9.5 to 11 mm high. The shell is reddish-brown or yellowish-brown; the last whorl lighter on the break into the constriction behind the peristome. The aperture is auriculate with a cream-colored, reflected peristome, the columellar lip of which hooks around the umbilicus, nearly covering it completely. There is usually a small, white lamella on the parietal callus, absent from some specimens. There are short, closely-spaced bristles on the shell (12 to 34 per square mm).

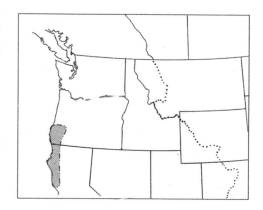

Similar Species: *V. eritrichius* is most like *Vespericola euthales* and *V. megasoma*; it is smaller than *V. euthales*, and lacks the strong parietal tooth of that species. Compared to *V. megasoma*, *V. eritrichius* has a higher, more conic spire than *V. megasoma* and a slightly more open umbilicus.

Distribution: *V. eritrichius* occurs from Mendocino Co., California, to southwestern Oregon. It is found in riparian zones among dense vegetation and under logs and leaf litter in mixed forest sites.

Vespericola euthales   (Berry, 1939)
Synonym: *Vespericola megasoma euthales*

Description: Fairly large for the genus, with 6 to 6¾ whorls its shell measures 18 to 20 mm wide by 15 to 17 mm high (Berry also listed a smaller form that measured 13 by 11 mm with about 6 whorls). The shell differs from that of *V. eritrichius* by its strongly developed parietal tooth, a wider, more sinuous peristome, and a much greater size (Berry).

Similar Species: *V. euthales*, *V. eritrichius*, and *V. magasoma* all have a small angular perforate umbilicus with the basal peristome curving around it. *V. euthales* is generally larger than the other two and has a much stronger parietal tooth. The peristome of the other *Vespericola* of the PNW does not distinctly hook around the umbilicus as in these species.

Distribution: *Vespericola euthales* occurs in southwestern Oregon, south to Humboldt Co., California. Originally found in redwood forest.

Vespericola megasoma   (Pilsbry, 1928)   Redwood Hesperian

Description: This snail has a rather capacious body whorl and a spire that varies from fairly low to moderately elevated; the body whorl is slightly shouldered. With about 5½ to 6 whorls, it measures 6.9 to 10.3 mm high by 11.3 to 15.1 mm wide, excluding the

1. *Vespericola megasoma* (dense, fine, short, bristles; peristome hooked around and over the small, angled umbilicus; often denticulate): Jackson Co., OR (OSAC #PGC07-066a); scale bar = 1 cm
2. *Vespericola megasoma*: Curry Co., OR (OSAC #ROS05-005); scale bar = 1 cm

reflected peristome. The umbilicus is an angular perforation more than half covered by the reflected lip margin, which wraps around it leaving an angular chink. The last whorl is constricted behind the lip; the aperture is widely auriculate. The peristome is fairly wide, thickened within, and slightly recurved at the edge. The inner edge of the basal margin is straight when viewed from below, and there is a small parietal tooth, which varies from well-developed to weak or absent. The periostracum is brown, or occasionally yellowish; the peristome is white. Dense, very fine, short bristles or papillose hair-scars cover the shell.

Similar Species: Look-alike species include *Hochbergellus hirsutus* and other *Vespericola* that may be found within the same areas (see their descriptions). Especially compare *V. eritrichius* and *V. euthales*, which have been considered synonymous to and a variety of *V. megasoma*, respectively. The range of *V. megasoma* in California may overlap that of *V. sasquatch*, which is about the same size but with an even lower, more convex spire, a lighter-colored periostracum, and usually with a low straight parietal lamella on a granulose parietal callus. Although the columellar lip of *V. sasquatch* is reflected to cover most of the umbilicus, it does not hook around it like that of *V. megasoma*.

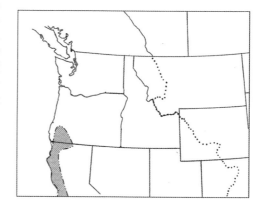

Distribution: *V. megasoma* occurs in western California from Marin to Del Norte and Siskiyou counties, and in Curry and possibly Josephine and Douglas counties, Oregon.

Vespericola sierranus (Berry, 1921)   Siskiyou Hesperian

Vespericola sierranus (small [8-9 mm wide] with dense, close, short bristles; sometimes with small parietal tooth): Jackson Co., OR (OSAC #PGC07-066b); scale bar = 1 cm

Description: The shell is small and thin with a moderately elevated, rounded spire that is slightly angled at the periphery. With 5¼ to 5¾ whorls, it measures 8.4 to 9 mm wide by 5.2 to 5.8 mm high. The umbilicus is very narrowly umbilicate, and open (contained 11 to 14 times in the shell diameter). The last whorl is deeply constricted before the peristome, which is fairly narrowly reflected, thickened, and light brown in color. The aperture is auriculate and normally edentate, but rarely with a small, narrow, white parietal tooth. The protoconch is smooth, without wrinkles or granulations; the teleoconch is covered with minute hairs.

Similar Species: Smaller than other *Vespericola*, and with a more convex spire, this species may more closely resemble *Trilobopsis loricata* or *Cryptomastix germana*. *Vespericola sierranus*

rarely has a parietal tooth and lacks lip teeth. The shell sculpture of these three species also differs. *Trilobopsis loricata* has crescent-shaped scales. *Cryptomastix germana* has relatively long hairs, and while the ranges of *C. germana* and *Vespericola sierranus* may approach, they are not known to overlap.

Distribution: *V. sierranus* occurs in southern Oregon and northern California. It may be found in marshy areas under woody debris.

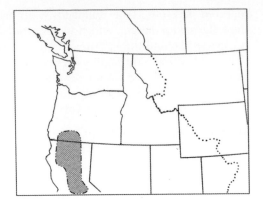

Selected references for the family Polygyridae: Pilsbry 1940; Roth and Miller 1992, 1993, 1995, 2000; Roth and Sadeghian 2006; Smith 1943.

Family: THYSANOPHORIDAE
Genus: *Microphysula*

Small white snails (about 4 mm diameter) with thin, nearly flat, glossy, transparent, rather tightly coiled shells with open umbilicus, narrowly lunate apertures, and microscopic spiral striae. They are often found under alder or aspen stands, or in rock slides in moist conditions.

Key to the Species of *Microphysula*

(a) Diameter 3.6 to 4.4 mm wide by 1.7 to 2.2 mm high with 4½ to 5 whorls. The basal lip margin not sinuous. Cascades Range westward. . . . . *Microphysula cookei*

    (aa) Diameter 4.0 to 4.8 mm wide by about 2.5 mm high with 5½ to 5¾ whorls, with a sinuosity of the basal lip margin when viewed from below. Cascades Range in Washington, Oregon, British Columbia, southward and eastward. . . . .*Microphysula ingersolli*

Microphysula cookei   (Pilsbry, 1922)   Vancouver Snail

Description: A small, white, tightly coiled snail, measuring 3.6 to 4.4 mm wide by 1.7 to 2.2 mm high with 4½ to 5 whorls. The transparent to translucent shell has a nearly flat spire (slightly rounded) and an open umbilicus of moderate size (about one-quarter the shell width). The whorls increase regularly but fairly slowly. The aperture is oblique and narrowly lunate. The periphery of the last whorl is rounded. The basal lip margin is not perceptibly sinuous when viewed from below. The latter whorls are sculptured with distinct microscopic spiral striae, at least dorsally.

Similar Species: *Microphysula ingersolli* is very similar but a little larger. It has a sinuous basal lip margin and is not from the same region. Juvenile *Megomphix* species would have a wider aperture, thicker shell, and fewer whorls when of comparable size.

Distribution: *Microphysula cookei* inhabits western Washington and western British Columbia including Vancouver Island, to southern Alaska, although the species needs to be confirmed for specimens from the western Washington Cascades. It is found in litter under stands of alder and possibly other hardwood trees.

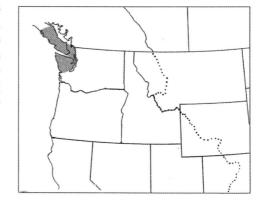

Microphysula ingersolli   (Bland, 1875)   Spruce Snail

Microphysula ingersolli: Pass Cr. Pass, Pend Oreille Co., WA; scale bar = 1 mm

Description: *M. ingersolli* is similar to *M. cookei* but a little larger, measuring 4.0 to 4.8 mm wide by 2.5 mm high with 5½ to 5¾ whorls. The shell is whitish, transparent or translucent. The spire is usually nearly flat but may vary to low and rounded. The umbilicus is open (about one-quarter the shell width or smaller) and increases regularly. The aperture is usually narrowly lunate and less oblique than that of *M. cookei* but may vary in this characteristic to be more similar to it. The basal margin is slightly sinuous when viewed from below. Spiral striae on the latter whorls are very faint. The animal is white.

Similar Species: *M. cookei* is a little smaller. Its basal lip margin is not apparently sinuous, and it is not from the same region. Juvenile *Megomphix* would have a wider aperture, thicker shell, and fewer whorls when of a size comparable to *Microphysula*.

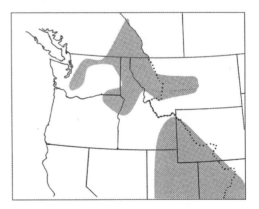

Distribution: *M. ingersolli* inhabits the inter-mountain and Rocky Mountain regions from southwestern British Columbia and adjacent Alberta, through eastern Washington, northeastern Oregon, Idaho, western Montana, northwestern Wyoming, and south to Kansas, Arizona, and New Mexico.

Selected references for the family Thysanophoridae: Forsyth 2004; Frest and Johannes 1995; Henderson 1929a; Pilsbry 1940.

Superfamily: PUNCTOIDEA

The superfamily Punctoidea includes the Oreohelicidae and also the families Punctidae, Charopidae, Discidae, and Helicodiscidae; in past works such as Pilsbry (1948), Burch (1962), and Kerney, Cameron, and Riley (1994) these last four were considered to be included in the family Endodontidae.

Family: OREOHELICIDAE
Genus: *Oreohelix* Pilsbry, 1904

The family Oreohelicidae in North America contains two genera, *Oreohelix* and *Radiocentrum*. The latter contains taxa of the southwestern United States and Mexico and does not occur in the Pacific Northwest. *Oreohelix* appears to have originated in the Rocky Mountains, and a variety of species can be found within that region from Canada to Mexico. The species are variable in size and sculpturing, but they generally have a low to moderately elevated spire and a fairly large open umbilicus. Many have a pair of lateral bands, one just below the periphery and the second somewhat above it, and some species have ribs or other sculpturing on the shells. The genus is usually readily separable from others, but our two species of *Anguispira* (family Discidae) might be confused with some oreohelices because of the size, shell shape, and banding of *A. kochi* and the ribbing of *A. nimapuna*. Morphologically, the oreohelices are Holopoda,[1] in which the pedal furrows are not apparent from the lateral view. If the whole snail is available, this characteristic can be useful in separating members of Holopoda families from those of Aulacopoda families (e.g., Discidae) in which the pedal furrows are visible.

A greater challenge, however, is in separating similar-appearing *Oreohelix*. There are many species, and intraspecific variations may be great. For example, individuals from arid sites are often much smaller than those from more moist or mesic locations. Therefore, snails on the mesic, concave, lower slope may be the same species as a much smaller specimen found on a more arid, convex, mid- or upper slope. In other instances they may be different species. If dissection is done, variations in genitalia are useful in distinguishing between some species (e.g., *O. strigosa* and *O. subrudis*).

The *Oreohelix* are diverse and include many described and yet-to-be-described species and subspecies. As with other lesser-known groups of organisms, the nomenclature for *Oreohelix* is still in flux, and there may be disagreement as to the proper relationships among species. For the most part, this work follows the nomenclature in Bouchet and Rocroi (2005) for family groups and Turgeon et al. (1998) for genus and species, and relies on other works (see references) and personal experience for subspecific taxonomy. A number of species or subspecies are known from just outside of the defined study area, and the ranges of some cross the boundary. In order to minimize confusion over closely appearing species, and recognizing that additional populations are likely to be discovered, several species have been included in this work that are not known from our study area but that occur in close proximity to its borders. There are many such *Oreohelix* species that occur in northern Utah. Several undescribed species or varieties are also recognized, and informal descriptions of some of those are included at the end of the species accounts for this genus. Other undescribed species are also known, and there are no doubt others yet to be found.

The following keys to species and subspecies are based primarily on shell characteristics and species ranges. Intraspecific variation and similarities among many species make

---

1 Superfamily of Pilsbry (1946); Division of Burch (1962); Infraorder of Boss (1982) and Burch and Pearce (1990).

identification of some difficult without dissection, and molecular analysis will often prove valuable in determining relationships. However, it is assumed that few of our readers will have the knowledge, equipment, or desire to perform these laboratory procedures on their specimens, so localities in which the species are known to occur will be a step forward in determining which species might be expected. For some groups, notes on the anatomy are included in the species accounts, but if dissection is desired, the reader is referred to the referenced literature.

Discussion of subspecies is generally included under species accounts for the typical species, but if distinctly different they may also be included separately in the key to the species. Subheadings to some more obvious characteristics are provided within the keys for the purpose of streamlining the identification process. For example, if a shell known to be an *Oreohelix* is ribbed but lacks a distinct peripheral carina (ridge or keel), one can skip forward in the key to that subheading.

Key to the Species of *Oreohelix*

(a) Sculptured with spiral lirae (raised ridges) other than just at the periphery. (See also *O. tenuistriata* under Ribbed Shells With Peripheral Carina). . . . . *Oreohelix haydeni*

    (aa) Sculpture of other than spiral lirae except sometimes with peripheral carina. . . . . [b/bb]

(b) Sculpture of radial ribs, with or without peripheral carina. . . . . [c/cc]

    (bb) Without regular radial ribs; sometimes with coarse growth-wrinkles and sometimes with peripheral carina . . . . [d/dd]

(c) Ribbed shells with peripheral carina. . . . . [j/jj]

    (cc) Ribbed shells with very weak or no peripheral carina . . . . [n/nn]

(d) Periphery with a distinct carina, but without radial ribs or spiral lirae. . . . . [e/ee]

    (dd) Periphery without prominent raised keel or carina, usually angled in shells of immature snails, often onto the last whorl, but rounded at the aperture of mature animals . . . . [p/pp]

Shells Without Ribs but Periphery With a Distinct Carina

(e) Small, less than 13 mm wide. Last whorl descending very little to the aperture; palatal insertion just below the peripheral carina. . . . . [f/ff]

    ee) Larger, greater than 13 mm wide . . . . [g/gg]

(f) Shell low to moderately elevated, convexly-conic, width about 9 to 10 mm in 4½ whorls. The sutures well impressed, umbilicus contained 4½ to 5 times in the shell width. Lateral bands, when present, light and confined mostly to the last whorl. Growth-wrinkles coarse and rather sharp, and there are fine beaded lirate lines and irregular spiral striae basally. Juab Co., Utah. . . . . *Oreohelix eurekensis*

    (ff) Shell low to moderately elevated, convexly-conic, the sutures well impressed. Width 9 to 12 mm in 4½ to 5 whorls. Umbilicus contained 4⅓ times in the shell width. Rather coarse growth-wrinkles and weak, irregular spiral striae followed by rows of scale-like granules. Powell Co., Montana . . . . *Oreohelix carinifera*

(g)   Salmon River drainage, Idaho into Nevada. Last whorl descending slowly so
      the aperture is below the periphery, but the palatal and columellar insertions
      are separated by a parietal callus. . . . . [h/hh]

   (gg)   East of Continental Divide in Gallatin watershed, Montana. Sutures not
          impressed, generally being filled by the keel of the previous whorl.
          Last whorl descending rather abruptly to the aperture, palatal and
          columellar insertions often joined . . . . . *Oreohelix yavapai* and subspecies [i/ii]

(h)   Sutures fairly well impressed. No abrupt descent of the last whorl near the
      aperture. Medium-sized, 16.7 mm wide by 10.6 mm high in 5½ whorls.
      Shell thin with a low spire. Umbilicus about one-fifth the shell width.
      Peripheral carina strong and serrated. Sculpture of coarse, regular growth-
      wrinkles and weak spiral striae. Custer Co., Idaho, and into Nevada.
      . . . . *Oreohelix hemphilli*

   (hh)   Sutures shallow. Spire rounded, moderately elevated to high domed.
          Last whorl descending abruptly shortly before the aperture.
          Fairly large, 18 to 20 mm wide by 10 to 14 mm high. Periphery
          angled in immature shells, then keeled near the aperture.
          Idaho Co., Idaho.. . . . . *Oreohelix strigosa goniogyra*

(i)   Medium to large, 14 to 19 mm wide in 4 to 5¼ whorls, with low conic spire.
      Keel may or may not extend to the aperture. Lateral bands distinct to
      nonexistent. Widely-spaced, granulose spiral lines most distinct below
      the periphery. . . . . *O. yavapai extremitatis*

   (ii)   A little larger, 18 to 23 mm wide in about 5 ½ whorls, with a low
          rounded spire. Very narrow brown lateral bands. Periphery keeled
          to carinate to within one-quarter whorl of the aperture . . . . *O. yavapai mariae*

Ribbed Shells With Peripheral Carina

(j)   Adults greater than 18 mm wide. Periphery strongly angled and carinate. . . . [k/kk]

   (jj)   Adults less than 18 mm wide. Spire moderately elevated to rather high.
          Ribs strong or very fine. . . . . [l/ll]

(k)   Large, 21 to 28 mm wide, spire low-conic. Umbilicus about one-quarter
      the shell width. Sutures very little impressed, being filled by the carina of
      the previous whorl. Peripheral bands faint or lacking. Mission Mountains,
      Montana. . . . . *Oreohelix elrodi*

   (kk)   Large, 20 mm wide by 8.4 mm high. Spire very low, nearly flat.
          Umbilicus about one-third the shell width. Sutures impressed but
          shallow. Peripheral bands lacking. Idaho County, Idaho . . . .*Oreohelix hammeri*

(l)   Ribs very fine, more like closely spaced, sharp, wavy radial threads. Often also
      with low spiral lirae. Shell with well-elevated spire, rather small (10 to 13 mm
      wide by 6.5 to 9 mm high with 4¾ to 5¼ whorls). Umbilicus narrow, one-sixth
      to one-fifth the shell width. Bannock County, Idaho. . . . . *Oreohelix tenuistriata*

   (ll)   Ribs strong. . . . . [m/mm]

(m)   Medium-sized and moderately to fairly high spire, 11 to 16 mm wide by 7
      to 10 mm high. Umbilicus contained 4½ to 5+ times in the shell width.
      With strong ribs and peripheral carina. Snake River Canyon, Asotin Co.,
      Washington, to Idaho Co., Idaho . . . . *Oreohelix idahoensis baileyi*

(mm) Small, 8 to 11 mm wide by 4.5 to 7 mm high with 4½ to 5¼ whorls. Spire low-conic, light horn or pinkish-brown color, with hint of typical bands, faint or lacking in some. Umbilicus ¼ to ⅓ the shell width. Periphery angled, somewhat carinate; with strong whitish ribs and strong, coarse spiral striae. Salmon River watershed, vicinity of Lucile, Idaho County, Idaho . . . .*Oreohelix waltoni*

Ribbed Shells, Not Carinate (see also Undescribed Species)

(n) Medium-sized with fairly high spire, 12 to 14 mm wide by 10.5 to 11.5 mm high. Umbilicus about one-sixth the shell width. Strong ribs, but very weak if any carina, and lacking lateral bands. Periphery angled on immature shells but with round aperture in adults. Salmon River watershed, vicinity of Lucile, Idaho County, Idaho . . . .*Oreohelix idahoensis*

    (nn) Ribs weaker, more closely spaced. Lateral bands variable, distinct to lacking . . . . [o/oo]

(o) Shells variable, small to large measuring 8 to 21 mm wide in 4½ to 5½ whorls, and with low, convexly-conic to high and sometimes domed spire. Ribs usually rather fine and closely spaced, but may be weak to well developed. With or without lateral bands of variable width and/or intensity. Northern Utah. . . . . *Oreohelix peripherica*

    (oo) Medium to fairly large size with thin shell and moderately elevated spire, about 13 to 18 mm wide by 7 to 12 mm high. Peripheral carina very weak; ribs dense and moderately strong. Yellowish-brown with typical peripheral bands and sometimes supernumerary basal bands. Little Salmon River, Adams or Idaho County, Idaho . . . . *Oreohelix intersum*

Without Ribs, Carina, or Distinct Keel at or near the Aperture (See also Undescribed Species)

(p) Glossy, streaked with pinkish-brown and white, and lacking the two typical peripheral bands. Rather large, 15 to 22 mm wide by 11 to 16 mm high in about 5½ whorls. Umbilicus about one-fifth the shell width. Solid with a rather high convexly-conic spire. Coarse growth-wrinkles, and faint or indistinct spiral striae. Periphery usually a little angled in front of the aperture, which is generally round or appearing a little oblique. Columbia Gorge, Wasco and Sherman counties, Oregon. . . . . *Oreohelix variabilis*

    (pp) With typical spiral bands, usually one just below the periphery and one just distal to the mid-dorsal area of the whorls. . . . . [q/qq]

(q) Small, 11 mm wide or less. Montana and Wyoming . . . . [r/rr]

    (qq) Larger, greater than or equal to 11 mm wide . . . . [s/ss]

(r) Small, 7 to 10 mm wide by 3 to 5 mm high with 4 to 4½ whorls. Umbilicus contained about 5 times in the shell width. Fine growth-wrinkles; no spiral striae. Mission Range, Montana . . . .*Oreohelix alpina*

    (rr) Small, 9 to 11 mm wide in 4¾ to 5 whorls. Spire high, globose, closely coiled; the last whorl close under the penultimate. Umbilicus narrow, contained 6 to 8½ times in the shell width. With strong growth-wrinkles and spiral striae. Adjacent to our study area, on the east slope of the Rocky Mountains near Billings, Montana, and south into Wyoming . . . .*Oreohelix pygmaea*

(s) Medium to large, 11 to 24 mm wide. Spire low to moderately elevated; umbilicus contained 3 to 4½ times in the shell width . . . .[t/tt]

  (ss) Somewhat larger, 15 to 27 mm wide with medium to high spire (sometimes dome-shaped). Umbilicus contained 4 to 8 times in the shell width. . . . . [w/ww]

(t) Coarse growth-wrinkles, no spiral striae. Small to fairly large, 11 to 18 mm wide by 6 to 11 mm high. Low-rounded spire; umbilicus ¼ to ⅓ the shell width. Bitterroot Mountains, Missoula County, Montana . . . . *Oreohelix amariradix*

  (tt) Very weak to fine growth-wrinkles. . . . . [u/uu]

(u) Spire very low-conic, nearly flat. Shell large 18 to 22 mm wide; umbilicus about one-quarter the shell width. Shell color whitish with the typical two brown to sometimes purplish peripheral bands. Growth-wrinkles quite weak; spiral striae fine and irregular. Talus and rocky slopes in the Grand Coulee and along the Columbia River and tributaries, Grant, Okanogan, and Chelan counties, Washington . . . .*Oreohelix junii*

  (uu) Spire low to moderately elevated, conic or a little convex. Umbilicus contained 3 to 4½ times in the shell width. Last whorl descends abruptly to the aperture; parietal callus narrow or apertural margins may join. . .[v/vv]

(v) Large, 18 to 24 mm wide by 9 to 13 mm high with 4¾ to 6 whorls. Low to moderately elevated conic spire, umbilicus about one-third the shell width. Last whorl descending sharply and often deeply to the aperture. Apertural margins approach, leaving only a narrow parietal callus between the insertions, or are sometimes joined. Spire light brown, lighter to whitish or pinkish on latter whorls, with dorsal flammules, and with typical cinnamon-brown peripheral bands. Growth-wrinkles very weak; spiral striae irregular. Under rocks along the Salmon River near Lucile, Idaho County, Idaho . . . . *Oreohelix jugalis*

  (vv) Medium-sized, 12.5 to 15+ mm wide by 7.5 to 8 mm high, with up to 5½ whorls. Low, slightly roundly-conic spire; umbilicus ¼ to ⅓ the shell width. Fine growth-wrinkles and irregular spiral striae. Shell brownish on the spire, grayish basally. Two darker brown bands, the wider one below the periphery, the narrower one above, with a light peripheral area between. Salmon River near White Bird, Idaho County, Idaho.
   . . . . *Oreohelix vortex*

(w) Shell fairly large with low to moderately elevated spire, brown with typically placed, darker, narrow, sharply-edged lateral bands. Northern Utah.
   . . . . *Oreohelix howardi*

  (ww) Height and width variable (15 to 27 mm wide) but generally fairly large. Brownish on the early whorls, becoming lighter on the latter ones, to grayish or whitish or light brown on the last whorl. With two typical *Oreohelix* spiral bands. Sometimes with supernumerary bands of varying number and width, most often below the periphery. Eastern Cascades through the Rocky Mountains . . . . [x/xx]

(x) Variable in size, shape and height of spire. Fairly to quite large, generally 17 to 27 mm wide by 12 to 16 mm high with about 5 whorls. Spire low, conic to moderately elevated and rounded or sometimes domed. Grayish-white to brownish. Umbilicus contained 4 to 5½ times in the shell width. Weak to moderate growth-wrinkles and variable spiral striae. Eastern Cascades, Washington, to the Rocky Mountains in western Montana. (See also species accounts for *O. strigosa* subspecies below) . . . . *Oreohelix strigosa*

(xx)   Medium to large, 15 to 26 mm wide by 11 to 20 mm high, with 5 to 6
        whorls. Spire generally high, convex-conic to rounded or somewhat
        domed. Umbilicus generally narrow, ⅛ to ⅓ the shell width, varying by
        subspecies. Generally buff or lighter colored. With fine sharp growth-
        wrinkles and fine, close spiral striae. Rocky Mountains, rarely as far
        west as Pend Oreille County, Washington. (See also species accounts
        for *O. subrudis* subspecies) . . . . *Oreohelix subrudis*

Oreohelix Species Accounts

Oreohelix alpina (Elrod, 1901)   Alpine Mountainsnail

Oreohelix alpina: East Saint Mary's Peak, Mission Mountains, MT; scale bar = 2 mm (MNHP collection)

Description: Very small for an *Oreohelix*, with 4 to 4½ whorls it measures 7 to 10 mm
wide by 3 to 5 mm high. The shell diameter is 4½ to 5½ times the width of the umbilicus.
The shell is thin and has a moderately elevated conic spire with rounded whorls and well-
impressed sutures. Pilsbry (1939) described its color as "matt brown" or "brownish-gray"
with a light-colored base. The two dark bands do not appear sharply defined but fade
outward along their edges. As with most of the *Oreohelix*, the immature shell has an angled
periphery, but at maturity the periphery is rounded near the aperture. The last whorl
descends rather sharply as it nears the aperture. Sculpturing is absent from the shell except
for fine growth wrinkling.

Similar Species: Elrod (1901) originally called the species *Pyrimidula strigosa* var. *alpina*.
Oreohelix strigosa is a variable species with seven subspecies accepted by Pilsbry (1939),
who doubted its affinity to *O. alpina*. Its small size should separate *O. alpina* from the other
Oreohelix within its range.

Distribution: The type locality is East St.
Mary's Peak at 8500 ft in the Mission Range,
Montana. *O. alpina* is also known from these
other Montana locations: McDonald Peak,
Swan Range, Lake Co.; Scapegoat Wilderness
and Prairie Reef, Lewis and Clark Co.; and
White River Pass, Powell Co. All sites are in
limestone talus in open but shaded areas,
near or above timberline (Hendricks, personal communication).

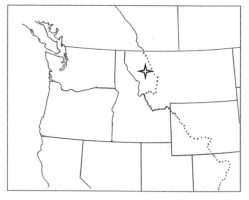

Oreohelix amariradix Pilsbry, 1934   Bitterroot Mountainsnail

Oreohelix amariradix: Lolo Creek, MT; scale bar = 1 cm (MNHP collection)

Description: This medium-sized snail measures about 10.8 to 18 mm wide by 6.4 to 11 mm high in 4¾ to 5½ whorls. The umbilicus is about one-third to one-fourth the width of the shell. It has a low to moderately elevated, rounded spire with the sutures not deeply impressed. The periphery is angled to within about one-third whorl of the aperture. The protoconch is nearly smooth, followed by coarse growth wrinkling on the teleoconch. No regular spiral striae are apparent. There are two reddish bands, one above and one below the periphery.

Similar Species: *Oreohelix jugalis* and *O. vortex* also have rather low spires and are quite similar to *O. amariradix*. Other than its range being well separated from the other two species, *O. amariradix* has a sharper peripheral angle. As with the other two, the body whorl rounds out before the aperture.

Distribution: *O. amariradix* is known from the Bitterroot Mountains above Lolo Creek southwest of Missoula, Missoula Co., Montana. It occurs in small talus patches or slides scattered among grass and shrubland on south-facing open Ponderosa pine/bunchgrass slopes, with scattered serviceberry and ninebark, at about 3500 ft elevation.

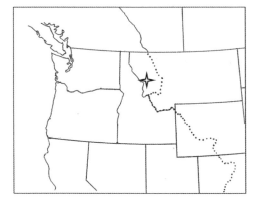

Oreohelix carinifera   Pilsbry, 1912   Keeled Mountainsnail

Oreohelix carinifera: Upper Grouse Gulch, MT; scale bar = 1 cm (MNHP collection)

Description: This rather small *Oreohelix* measures 9 to 12 mm wide by 5 to 7 mm high with 4½ to 5 whorls. The width is about 4⅓ times that of the umbilicus. The spire is low to moderately elevated. The whorls are rounded but become concave at the outer edge, where they extend into a peripheral carina. The shell is a light tan color, the base lighter to whitish and the apex may be darker or reddish-brown, with faint darker brown bands. The sculpturing is primarily of coarse growth-wrinkles, but there may be some faint, irregular spiral striae. Immature shells are adorned with spiral rows of small cuticular scales below the periphery, the remains of which are apparent in some adults, usually as indistinct granules.

Similar Species: *O. hemphilli* is known from east-central Idaho and east-central Nevada and is larger than *O. carinifera*. The embryonic growth-wrinkles of *O. hemphilli* are sharper and show weak spiral striae. Other species differ in their radular dentition (Pilsbry 1939).

Distribution: *O. carinifera* occurs in Montana, near Garrison, Powell County. It is found mostly under shrubs on open slopes. Also found in the Garnet Range and Sapphire Mountains, Granite Co., and near Ravena, Missoula Co., Montana (Hendricks, personal communication). It is found among limestone outcrops and shrubs on open, south slopes with scattered Douglas-fir, junipers, and sagebrush, at elevations between 4100 and 4900 ft.

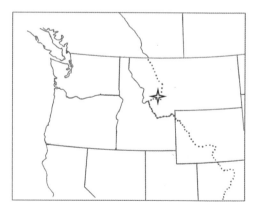

Oreohelix elrodi (Pilsbry, 1900) Carinate Mountainsnail

1. *Oreohelix elrodi*: Montana (MNHP collection)
2. *Oreohelix elrodi*: Montana; scale bar = 1 cm (MNHP collection)

Description: The shell is moderately large and ribbed, with a strongly angular and carinate periphery. With 5 to 5 ½ whorls, it measures 21 to 28 mm wide by 8.8 to 13.3 mm high. The umbilicus is about one-quarter the shell width. The spire is low conic, the whorls flattened along the angle of the spire, with the sutures little impressed, being occupied by the peripheral carina. The ribs are rather sharp on their leading edges, and spiral striae may be seen in the interspaces. The protoconch of about 2½ whorls begins smooth then becomes irregularly finely ribbed. The periostracum is light red or reddish-brown in immature specimens but is mostly lost by adulthood. The two brown peripheral bands may be seen on the last half of the body whorl in some specimens.

Similar Species: Unique-appearing, *Oreohelix elrodi* should not be confused with any other snail within its range.

Distribution: *O. elrodi* occurs in Montana on the shores of McDonald Cirque and between 3000 and 7500 ft in the Mission Mountains. It occurs in coarse talus on south slopes with sparse Ponderosa pine, Douglas-fir and other conifers, quaking aspen, mountain ash, and serviceberry.

Hendricks (personal communication) has records from three locations in the Swan Range, Lake Co., and the Scapegoat Plateau, Lewis and Clark Co., from which site the size and habitat may suggest a distinct new taxon.

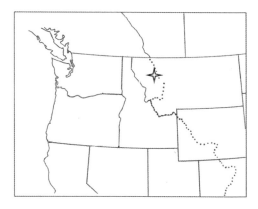

Oreohelix eurekensis   J. Henderson & Daniels, 1916   Eureka Mountainsnail

Description: This small *Oreohelix* with 4⅓ to 4½ whorls measures about 9 to 10 mm wide by 5 to 5.5 mm high. The umbilicus is contained 4½ to 5 times in the shell width. The spire is low to moderately elevated, the sutures well impressed. The spire is light brown, transitioning to grayish-white on the body whorl. The two bands, above and below the periphery, are light-colored. The whorls are sublenticular-shaped, rounded dorsally, then slope flatly or concavely to the carinate periphery. The protoconch of about 2 whorls is very weakly striate, with spiral lines beginning on the second whorl. The teleoconch is roughened, with coarse irregular growth-wrinkles. There are tiny spiral beaded lines with very indistinct striae in between and in the umbilicus, these being most prominent below the periphery and much less distinct above.

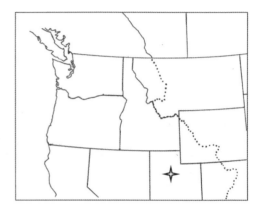

Similar Species: *O. alpina*, of similar size, is relatively smooth, with distinct colored bands, and it lacks the peripheral carina.

Distribution: *Oreohelix eurekensis* is found adjacent to the PNW, "Near Eureka, Juab and Uinta counties, Utah, on the north side of Godiva Mountain, under shrubs, etc." (Henderson and Daniels 1916).

Oreohelix hammeri   Fairbanks, 1984   Seven Devils Mountainsnail

Description: The shell is strongly depressed; the spire is nearly flat, and the sutures are moderately impressed. The body whorl is strongly keeled; the whorls and aperture are lenticular in cross-section. The palatal insertion is below the periphery, but the descent to the aperture is slight and not abrupt. The shell is gray-brown and lacks peripheral bands. With 5¼ whorls the shell is about 20 mm wide by 8.4 mm high. The umbilicus is about one-third the shell diameter. The protoconch is of 2¼ whorls, with faint spiral striae and radial ridges increasing with the size of the shell. The teleoconch is strongly ribbed and retains the fine spiral striae.

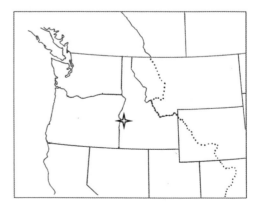

Similar Species: *O. elrodi* is most similar, but its spire is more elevated and its umbilicus narrower. Other ribbed *Oreohelix* in the Pacific Northwest (*O. intersum, O. waltoni, O. idahoensis*) are generally smaller with higher spires and weak peripheral carina, if any.

Distribution: *Oreohelix hammeri* occurs in the Seven Devils Mountains, Idaho Co., Idaho. It was found among rock outcrops, loose rock, and talus on a southerly aspect with scattered low vegetation.

Oreohelix haydeni and Subspecies

Oreohelix haydeni (Gabb, 1869) Lyrate Mountainsnail

Description: This rather large, spirally lirate species has a low to moderately elevated, conic spire of an ash-gray color, lacking the typical *Oreohelix* peripheral bands. The typical form with 5 to 6 whorls measures about 17 to 23 mm wide by 11 to 17 mm high. The width of the umbilicus is about one-fifth the shell diameter. The 8 to 13 spiral lirae are strongly developed both dorsally and ventrally; most are located ventrally. The periphery is angled and supports a carina formed by the stronger peripheral lira. Growth-wrinkles are coarse and irregular. This typical form, from northern Utah, does not occur within our study area, but other subspecies are found from the southern Idaho Panhandle east into Montana and southeastern Idaho.

Similar Species: *Oreohelix tenuistriata* may also have thin, cuticular spiral lirae and may eventually be found to be a subspecies of *O. haydeni* (Pilsbry, 1939). The smaller size and cuticular lirae of *O. tenuistriata* and its fine, closely spaced riblets distinguish it from the *haydeni* group. There are seven subspecies of *O. haydeni* that occur within or near to our study area. For descriptions, see the following key to these subspecies.

Distribution: Typical *O. haydeni* is found in the Weber Canyon area in the Wasatch Mountains of Utah.

Key to the Subspecies of Oreohelix haydeni

The following key includes seven subspecies of *Oreohelix haydeni*, including the typical form for comparison. Five of these subspecies occur within our study area, two exclusively, and two are not known to occur here. One of these subspecies, alien to us, is found near the Idaho border, so it could potentially be found within our area. The other is the typical *O. haydeni*. The subspecies are keyed here, but individual species accounts are not provided for them. Habitat of *O. haydeni* is generally in limestone outcrops and talus on open slopes of Ponderosa pine forest or open grassland with scattered shrubs.

(a) Southeastern Idaho or northern Utah. . . . .[b/bb]

   (aa) Southern Idaho Panhandle or western Montana . . . .[e/ee]

(b) Spiral lirae strong, regularly spaced. Width 17 to 23 mm. Lacking the peripheral bands typical of many *Oreohelix*. . . . . [c/cc]

   (bb) Spiral lirae weak or low, or if rather strong then irregularly spaced. Shell 13 to 20 mm wide . . . . [d/dd]

(c) Spire conic, low to moderately elevated. Umbilicus one-fifth the shell width. Peripheral lira forming a somewhat prominent carina. Color ash-gray. Wasatch Mountains of Utah. . . . . *O. haydeni haydeni*

   (cc) Spire moderately high, obtusely conic. Umbilicus narrower, about one-seventh the shell width. Peripheral lira less prominent. Dorsal and ventral lirae strong and with fine spiral threads in the interspaces between. Color light pinkish to white with darker apex. Cache Co., Utah, near the Idaho border . . . . *O. haydeni corrugata*

Oreohelix haydeni oquirrhensis: MT; scale bar = 1 cm (MNHP collection)

(d)   Spire roundly conic to conic. Umbilicus contained 5½ to 6 times in the shell width. Lirae widely spaced and very low. Color very light to white; spire a little darker. Peripheral bands mostly faint or lacking. Bear Lake, Idaho County, and into Utah . . . .*O. haydeni hybrida*

     (dd)   Spire low, roundly conic to higher and globose. Umbilicus contained about 5 to 6 times in the shell width. Peripheral carina prominent and irregularly spaced, weak to rather strong spiral lirae. Color light cream to buff with brownish spire, and sometimes with two indistinct peripheral bands. Vicinity of Missoula, Montana, and northern Utah. . . . . *O. haydeni oquirrhensis*

(e)   Vicinity of Salmon River, Idaho County, Idaho. . . . .[f/ff]

     (ee)   Western Montana. Peripheral carina apparent. . . . .[g/gg]

(f)   Spire rounded, low to moderately elevated. Width 18 to 25 mm; umbilicus about one-fourth the shell width. Peripheral carina strong, as are the spiral lirae. Color light buff to near white. Peripheral bands variable, from none to two. John Day and Wet Creeks, Idaho County, Idaho.. . . . . *O. haydeni hesperia*

     (ff)   Spire moderately elevated, the whorls terraced, and with a blunt, rounded apex. Small, 13 to 16 mm wide. Umbilicus contained 4 to 5½ times in the shell width. Periphery angular to carinate but aperture nearly circular. Thin ribs form a grid pattern with the spiral lirae. Color light pinkish with zero to two peripheral bands. NE of Lucile, Idaho Co., Idaho . . . . *O. haydeni perplexa*

(g)   Spire low, roundly-conic (more depressed than *O. haydeni haydeni*). Small, width about 15.5 mm. Umbilicus deep and wider than typical form. Carina prominent but not sharp. Dorsal lirae weak or faint, more prominent basally. Color dirty white with two brown peripheral bands. Montana . . . .*O. haydeni bruneri*

     (gg)   Spire low, roundly conic to higher and globose. Umbilicus contained about 5 to 6 times in the shell width. Peripheral carina prominent and irregularly spaced, weak to rather strong spiral lirae. Color light cream to buff with brownish spire, and sometimes with two indistinct peripheral bands. Vicinity of Missoula, Montana, and northern Utah . . . . *O. haydeni oquirrhensis*

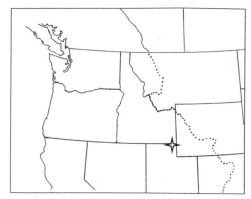

Distribution: *Oreohelix haydeni corrugata*

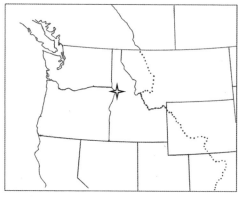

Distribution: *Oreohelix haydeni hesperia*

Distribution: *Oreohelix haydeni hybrida*

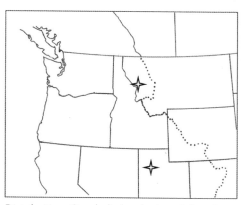

Distribution: *Oreohelix haydeni oquirrhensis*

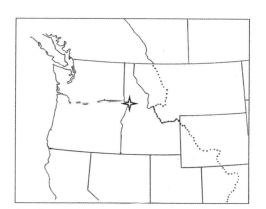

Distribution: *Oreohelix haydeni perplexa*

Oreohelix hemphilli (Newcomb, 1869)   Whitepine Mountainsnail

Description: A medium-sized snail, measuring 16.7 mm wide by 10.6 mm high in 5½ whorls. The umbilicus is about one-fifth the width of the shell. The shell is thin, rather low spired, light brown dorsally, whitish basally, and with irregular whitish or grayish flammules or streaks on the latter whorls. The whorls are strongly rounded dorsally, steeply inclined dorso-laterally, and flared outward concavely just before the peripheral carina. The carina is strong in front of the aperture and finely serrated, but weakens in the last half-whorl. The basal curve is rather broadly rounded, the aperture being a little wider than high. The 2½ whorl protoconch is finely, radially striate. The latter whorls have rather coarse and regularly spaced growth striae and weak spiral striae, which are most prominent basally and irregular dorsally.

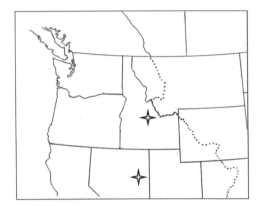

Similar Species: The similar *O. carinifera* is smaller and from western Montana.

Distribution: *O. hemphilli* is a species of high-elevation rocky areas, from 8,000 ft in White Pine Co., Nevada, and 10,000 to 11,000 ft in the Lost River Mountains, Needle Park, Idaho.

Oreohelix howardi Jones, 1944   Mill Creek Mountainsnail

Description: Jones (1944), in a single paragraph in the document cited, describes this species as "readily distinguished from *depressa* (Cockerell) by markings on the nuclear whorls. Fresh specimens also have a very characteristic color pattern, characterized by the narrowness of the two sharply-margined, chocolate-brown color bands set on a deep brown background. The ground color and, to a lesser extent, the color bands fade when exposed to light."

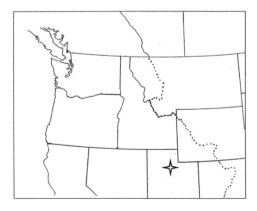

Similar Species: *O. strigosa depressa* is the species with which Jones compared *O. howardi*. He did not elaborate on the distinguishing markings on the protoconch, but *O. strigosa depressa* is generally lighter colored than *O. strigosa strigosa*, being a light yellowish-brown, which should contrast with the "deep brown background" of *O. howardi*.

Distribution: Jones found *O. howardi*. in Mill Creek Canyon at Salt Lake City, Utah.

Oreohelix idahoensis   (Newcomb, 1866)   Costate Mountainsnail

Oreohelix idahoensis idahoensis: Lucile, Idaho Co., ID; scale bar = 1 cm (TJF/Deixis collection)

Description: This is a rather small, ribbed *Oreohelix* with relatively high conic spire. It is a light tan or buff color with a pinkish tint, darker near the apex. With 5 to 5½ whorls it measures 12 to 13.7 mm wide by 10.5 to 11.3 mm high. The umbilicus is about one-sixth the shell diameter or smaller. The 2 whorls of protoconch are finely, radially striate. The latter whorls are sculpted with strong white ribs, rarely with fine spiral striae showing in the interspaces. The whorls are generally rounded, with well-impressed sutures and occasionally with a very weak peripheral keel. Immature shells are angled at the periphery and often show a weak carina between the ribs.

Similar Species: Other ribbed snails from the Snake and Salmon River area of western Idaho include: *O. hammeri, O. idahoensis baileyi,* and *O. waltoni,* all with a lower spire and prominent carina; and *O. intersum* with a lower spire, thin shell, and finer, more closely spaced ribs. Also see the section on undescribed Oreohelix following the species accounts for this family.

Distribution: *O. idahoensis* occurs in the Salmon River drainage in the vicinity of Lucile, Idaho County, Idaho. It occurs in limestone outcrops and talus at low to mid-elevations, among sagebrush, mixed shrubs, and grasses. Four shells, but no live animals, were collected from near the Snake River in Wallowa Co., Oregon in April, 2010.

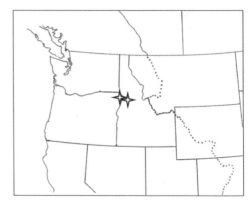

O. idahoensis baileyi   Bartch, 1916

Oreohelix idahoensis baileyi: Lime Hill, Asotin Co., WA; scale bar = 1 cm (TJF/Deixis collection)

Description: Although variable, the shell is similar to typical *O. idahoensis*, but with a lower spire and a strong peripheral carina. Measurements recorded (Pilsbry 1939) range from 11 to 15.7 mm wide by 7 to 9.6 mm high. Height/diameter ratios range from 56.9 to 84.3. Those from Asotin Co., Washington, were smaller and relatively higher, but fall within the above range of measurements.

Similar Species: This subspecies is distinguished from typical *O. idahoensis* by its lower spire and prominent peripheral carina. See comments on other ribbed *Oreohelix* under *O. idahoensis*.

Distribution: *O. idahoensis baileyi* is known from the Seven Devils Mountains and the Snake River Canyon below the mouth of the Salmon River, Idaho County, Idaho, and Lime Point, Asotin County, Washington. It is usually associated with limestone outcrops and talus at mid-elevations in arid land. At Lime Hill, shells were abundantly scattered over the steep, grassy, northeasterly slope.

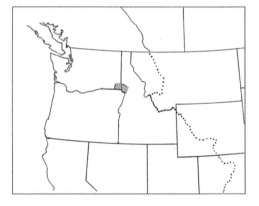

Oreohelix intersum (Hemphill, 1890)   Deep Slide Mountainsnail
Synonym: *O. jugalis intersum* (Hemphill) Pilsbry, 1934

Description: The shell is rather thin, with closely spaced ribs and a low to moderately elevated roundly conic spire. It is yellowish-brown dorsally and with darker flammules, and lighter below the periphery. The two peripheral bands are darker brown. Some shells have narrow lines or bands on the base, as seen in some *O. strigosa*. With 5⅓ to 6 whorls,

the shell measures 12.6 to 18.2 mm wide by 7.3 to 11.5 mm high. The periphery is weakly carinate, and the shell is sculptured with rather dense, moderately strong ribs.

Similar Species: See similar species under *O. idahoensis*.

Distribution: *O. intersum* is known from the Little Salmon River in rock piles below a bluff in Adams or Idaho County, Idaho. It is found in open basalt and schist talus at low elevations on grassland with scattered clumps of shrubs.

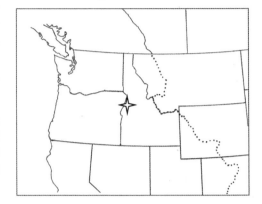

Oreohelix jugalis (Hemphill, 1890)   Boulder Pile Mountainsnail

Description: *O. jugalis* is relatively large for the genus. With 4¾ to 6 whorls it measures 18 to 24 mm wide by 9.3 to 13.2 mm high. Its umbilicus may be nearly one-third the shell diameter. The low to moderately high conic spire is light brown, but the latter whorls are whitish, with pinkish to light brown or gray streaks or dorsal flammules. There is a cinnamon-brown band at or just below the periphery and another about mid-dorsally on the whorls. The angled periphery of immature shells may or may not extend to the last whorl, which descends rather abruptly and often deeply to the aperture. The apertural insertions are quite close together, leaving only a narrow parietal callus, or they are sometimes joined. Sculpturing is of rather fine growth-wrinkles and, in most specimens, very weak, irregular spiral striae, which are most prominent dorsally and laterally.

Similar Species: Pilsbry (1939) says, "*Oreohelix jugalis* is characterized by the very wide umbilicus, rather capacious within, the closely approaching upper and columellar margins of the peristome, which is occasionally continuous, and the weak or obsolete spiral sculpture." It is similar to *O. amariradix* and *O. vortex*, which Pilsbry considered a subspecies of *jugalis*. *O. jugalis* is somewhat larger than either of these and has a relatively larger umbilicus and a more distinctly conic spire. *O. amariradix* occurs in the Bitterroot Range of Montana, but *O. vortex* is found near the range of *O. jugalis*, the type localities being only about 24 kilometers (15 miles) apart.

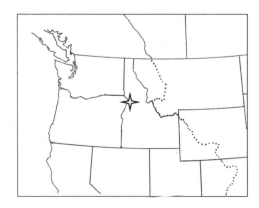

Distribution: *O. jugalis* is found under rocks on the banks of the Salmon River near Lucile, and north of Riggins, Idaho County, Idaho.

Oreohelix junii Pilsbry, 1934   Grand Coulee Mountainsnail

Oreohelix junii: Okanogan Co., WA; scale bar = 1 cm

Description: A fairly large *Oreohelix* with a low conic spire. With 5¼ to 5⅓ whorls it measures 18 to 27 mm wide by 10 to 13 mm high. The shell is usually whitish, with the typical two brown peripheral bands of varying widths and shades. The periostracum is brown dorsally, but it has often been sloughed off. Spiral striae are faint and irregular, or lacking on some specimens. The whorls are somewhat flattened top and bottom, more so dorsally. The aperture is oval, wider than high, with a simple lip, slightly expanded basally and at the umbilical margin. The last whorl descends to below the periphery in the last one-eighth turn or less, sometimes quite abruptly; the aperture expands rapidly in that same distance. The apertural insertions are rather close together, leaving only a relatively narrow parietal callus. The last whorl is weakly shouldered or angular in front of the aperture to the last one-half to one-fourth turn.

Similar Species: The shell of *Oreohelix jugalis* is sometimes similar, but that species is not found within or near the same areas as *O. junii*. The range of *O. strigosa* overlaps that of *O. junii*, and some of the low-spired *O. strigosa* might appear similar to it. In general, the body whorl of *O. strigosa* will be fairly well-rounded before descending to the aperture, while the whorl of *O. junii* is more lenticular in the same area.

Distribution: *O. junii* is known from Grant, Chelan, Okanogan, and probably Douglas counties, Washington. It occurs in talus slopes in the Grand Coulee, Grant Co., north to the Columbia River, then westward and southward along both sides of the Columbia River at least to Rocky Reach Dam, north of Wenatchee, and some distance up the Entiat River, Chelan Co. Its range extends north along the Methow River in Okanogan County at least as far as Twisp, and probably also in Moses Coulee, Douglas County, and other gorges in that vicinity. It is usually found in talus and under shrubs at the bases of talus slopes and rock outcrops. Its range tends to follow the courses of the larger rivers and coulees.

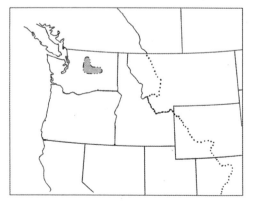

Oreohelix peripherica   (Ancey, 1881)   Deseret Mountainsnail

Description: This snail "varies widely in size, degree of elevation, development of ribs and being plain or banded" (Pilsbry 1939). With 4½ to 5½ whorls it may measure 8 to 21 mm wide. The shell is moderately elevated to high, convexly-conic or somewhat globose, or less often low-conic. It may be light brownish or buff-colored to nearly white, and may have a darker or pinkish spire. The umbilicus is quite narrow but expands abruptly in the last half-whorl. Peripheral bands vary from lacking to one or two. When present they may be faint to distinct, narrow to very wide, and sometimes with a whitish band between the darker ones. They usually have fine to well-developed, closely spaced ribs. In some colonies a few have a low baso-columellar tooth or callus, similar to that of *O. strigosa buttoni*.

Similar Species: Compared to other ribbed *Oreohelix*, the ribs of *O. peripherica* are generally finer but still distinct. The other species, with coarser ribs, are not known from the same region. The most similar-appearing shell is an undescribed *Oreohelix* from a ridgetop north of Entiat, Chelan County, Washington (see Hoder's Mountainsnail under Undescribed Oreohelix).

Distribution: *Oreohelix peripherica* is a species of northern Utah.

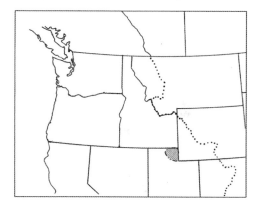

Oreohelix pygmaea   Pilsbry, 1913   Pygmy Mountainsnail

Description: *O. pygmaea* has a small, tightly coiled, high-spired, rather globose shell with an obtusely conic apex. The last whorl is closely coiled under the penultimate. With 4¾ to 5 whorls, it measures 9.7 to 11 mm wide by 7 to 9 mm high. With the tightly coiled body whorl, the umbilicus is small, being contained 6⅓ to 8½ times in the shell diameter. The spire is light brown to buff colored with whitish streaks, a pattern that is reversed on the last whorl. There is a dark brown band below the periphery and usually additional narrow, lighter-colored bands on the spire and the base. There are relatively strong, irregular growth-wrinkles or weak ribs, and spiral striations. Whorls are well rounded and the sutures deeply impressed, causing the spire to appear terraced in some specimens.

Similar Species: The small size and tightly coiled, rather globose shape, separate *O. pygmaea* from most other *Oreohelix* species. *O. strigosa berryi* is very similar in size and appearance but with much weaker growth-wrinkles and spiral striae, and it has a relatively larger umbilicus.

Distribution: A peripheral species from outside of the defined area of this study, *O. pygmaea* is known from southeast of Billings, Montana, southward into Wyoming.

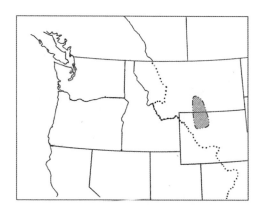

Oreohelix strigosa and Subspecies

Oreohelix strigosa and its subspecies make up a large group of taxa covering a broad range from the Rocky Mountains west to the east slopes of the Cascades. Subspecies and forms, as recognized by Pilsbry (1939) are listed and discussed below, but even within the "subspecies" there are significant variations in shell characteristics, so that the groupings might be further separated or lumped together.

Chelan County, Washington, is recognized as an area of high botanical endemism. It also has a high variety of snail species, including several undescribed forms of *Oreohelix*. That area contains what is considered to be the type locality for *O. strigosa*, but other species, forms, or varieties also occur there, including *O. junii* (the Grand Coulee Mountainsnail), *Oreohelix* n. sp. 1 (the Chelan Mountainsnail listed by Frest and Johannes, 1993), and others (see sections: Undescribed Species and Varieties of *Oreohelix*). Snails of other genera, far removed from their known ranges, also occur in that area; it remains unanswered whether or not they are disjunct populations of known species, or undescribed species or subspecies. Notable are snails appearing similar to *Anguispira nimapuna* and *Cryptomastix mullani hemphilli*, as well as other *Oreohelix*. Recently discovered undescribed species are discussed following these *Oreohelix* species accounts.

1. *Oreohelix strigosa*: Hoodoo Canyon, Ferry Co., WA; scale bar = 1 cm
2. *Oreohelix strigosa*: Lapwai Cr., Lewis Co., ID
3. *Oreohelix strigosa*: Slate Cr., Idaho Co., ID

Oreohelix strigosa strigosa   Gould, 1846   Rocky Mountainsnail
Synonyms: *O. cooperi* Binney; *O. strigosa* var. *parma*, var. *subcarinata*, var. *bicolor*, var. *lactea*, and var. *picta* (all Hemphill); *O. strigosa canadica* Berry; and *O. strigosa delicata* Pilsbry

Description: This snail is variable in size, height of spire, and markings. Generally, with about 5 whorls it measures 17 to 27 mm wide by 12.5 to 16 mm high. The shell diameter is 4 to 5½ times the width of the umbilicus. The spire is low to moderately elevated, conic or slightly rounded. Gould (1846) described the type specimen as "ashy gray, somewhat mottled with dusky, or altogether rusty brown above, with, usually, a single, faint, re-volving band on the middle of each whorl, and often with numerous bands, unequal in size and distance, beneath." The shell is usually calcareous, whitish or gray, with a thin rather light to medium brown periostracum. Banding is variable, but there is typically a subperipheral and a supraperipheral band and sometimes additional narrow, faint or distinct supernumerary bands. The protoconch of about 2½ whorls is sharply angled at the periphery. The angle continues, with diminishing acuteness, to about the last half-whorl in adult shells. Weak impressed spiral lines may begin on the latter whorls of the embryonic shell and continue through the teleoconch as faint or distinct spiral striae. Radial growth wrinkling is weak to moderate.

Similar Species: Genital anatomy of the *O. strigosa* group is characterized by the penis being ridged (plicate) inside for less than half of its length. In the *O. subrudis* and *O. yavapai* groups the internal ridges (plicae) extend for more than half the length of the penis. Similar species are found in the Columbia River Breaks and eastern Cascades foothills of central Washington. Many of these have rows or bands of cuticular scales on the post-embryonic whorls, usually disappearing by adulthood. See however, the Chelan Mountainsnail dis-covered and recorded as *Oreohelix* n. sp. 1, by Dr. Terrence Frest and Mr. Edward Johannes (1993); and informally described in Frest and Johannes (1995); and Kelley et al. (1999).

Distribution: *Oreohelix* from western Montana and eastern Washington between the Continental Divide and the eastern foothills of the Cascade Mountains, north into British Columbia, Canada, south through the Idaho Panhandle to the Salmon river drainage, Idaho County, Idaho, and into Umatilla County, Oregon, have been designated *Oreohelix*
strigosa strigosa as it is currently understood. Several subspecies, forms and varieties of *O. strigosa* are recognized within that range, and other species of *Oreohelix* also occur in the same region. Other subspecies of *O. strigosa* are known from east of the Rocky Mountains in Alberta and Montana and south into Arizona and New Mexico. Habitat of this species is generally in open forested areas or sometimes riparian areas. It may be found in forest floor litter, under shrubs, or in rock talus.

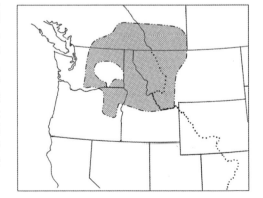

Subspecies of *Oreohelix strigosa*

There are many intraspecific variations as indicated by the number of subspecies and forms that have been described. Pilsbry (1939) and others have pointed out that some of these variations can likely be attributed to environmental conditions. The differences in size appear to vary with moisture, geology, and edaphic conditions between lower and upper slopes and aspects. In northeastern Washington, a distinct variation in the aperture and shell shape can be seen between *O. strigosa* found in the forest floor litter and those living in talus. Those snails found under the litter have a slightly extended and deflected palatal lip, while those among the talus have an apertural lip formed on a flat plane and are often found in aestivation with their apertures sealed to the rocks with dried mucus. Keys to the subspecies of *Oreohelix strigosa* recognized by Pilsbry (1939) follow.

Key to the Subspecies of *Oreohelix strigosa*

(a) Range west of the Continental Divide in Montana and the Idaho Panhandle, south to Adams Co., into the eastern Cascade Range in Washington and the Blue Mountains of northeastern Oregon . . . .[b/bb]

    (aa) Range southeastern Idaho, northwestern Wyoming, northern Utah, and east of the Continental Divide in Montana . . . . [d/dd]

(b) Occurs scattered throughout this described range. Shape, color, and banding variable. Generally with dark brown subperipheral and mid-dorsal bands. Weak to moderately developed growth-wrinkles and weak spiral striae . .*O. s. strigosa*

    (bb) Range more limited, southeastern Washington, northeastern Oregon, and Idaho County, Idaho. Width of shell 20 mm or less, growth-wrinkles and spiral striae well developed, at least dorsally . . . . [c/cc]

(c) Shell thin with prominent spiral striae. Peripheral bands often indistinct. Northeastern Oregon. . . . . *O. s. delicata*

    (cc) Periphery angled to the keeled aperture. Growth-wrinkles and spiral striae fine but well developed, at least dorsally . . . .*O. s. goniogyra*

(d) Small, 7 to 13 mm wide with 4⅓ whorls. Medium to fairly dark brown with whitish radial streaks and patches on the latter whorls. Subperipheral band dark brown, the mid-dorsal band usually lighter. Supernumerary bands present or lacking. Northwestern Yellowstone to central Montana . . . .*O. s. berryi*

    (dd) Small to fairly large, 12 to 23 mm wide (rarely smaller or larger) with 5 to 5½ whorls . . . .[e/ee]

(e) Peripheral bands usually lacking, present in a few. Shell usually of uniform color, whitish with light chocolate-brown tint. Aperture round, thickened inside, insertions sometimes joined. Sometimes with a basal tooth near the columella. Boxelder County, Utah . . . .*O. s. buttoni*

    (ee) Normally with peripheral bands. Aperture not thickened inside and without a tooth. Space between bands lighter than the shell. . . . .[f/ff]

(f) Shell solid, 18 to 24 mm wide (with extremes of 10 to 29 mm wide). Umbilicus one-quarter to one-fifth the shell width. Spire low to moderately elevated, generally more depressed than typical *O. strigosa*. Color variable but often light yellowish-brown with chocolate or darker bands. Northern Utah, southeastern Idaho, western Wyoming, and east slope of the Rockies in Montana. . . . . *O. s. depressa* (ff)   Shell relatively thin and somewhat translucent. Spire low to moderately elevated or sometimes globose. Umbilicus narrow, one-sixth to one-seventh the shell width. Color reddish-brown, lighter on the base, with chestnut-brown or darker bands. Southeastern Idaho and northern Utah . . . *O. s. fragilis*

O. strigosa berryi   Pilsbry, 1915

Description: This small *Oreohelix* measures 7 to 13 mm wide with about 4⅓ whorls and is about two-thirds as high as wide. The umbilicus is contained about 5½ times in the shell width. The spire is moderately elevated and convexly-conic or rounded. The color is medium to fairly dark brown with radial whitish streaks and patches on the latter whorls. There is a dark brown subperipheral band and usually another, lighter band mid-dorsally on the whorls. Multiple narrow and/or faint basal bands may also be present. Growth wrinkling is apparent, and there are weak, irregular spiral striae. The protoconch of about two whorls is sculpted with fine radial striations and with very fine spiral striae.

Similar Species: *O. pygmaea* is very similar in size and appearance but with distinctly stronger growth-wrinkles and spiral striae, and a smaller umbilicus.

Distribution: *O. strigosa berryi* occurs in central Montana from northern Broadwater to southern Fergus and western Musselshell counties, and in southern Montana in Park and Carbon counties, and in northwestern Yellowstone National Park, Wyoming (Hendricks personal communication). It is included in this work as a potential peripheral species, as the southwestern corner of its range is near the Idaho border.

O. strigosa buttoni   (Hemphill, 1890)

Description: The shell has a moderately elevated, convexly-conic spire; with 5 whorls it measures 12 to 23 mm wide. They are whitish or light chocolate colored; most lack the typical darker bands. The whorls and aperture are round, the lip insertions close together, sometimes joined. The lip is unreflected but thickened inside and may be darkened. Some specimens have a distinct denticle in the inner basal lip. In others the denticle may be merely a slight callus tubercle, but it is lacking altogether in most.

Similar Species: This is a rather unique *Oreohelix*, with its uniform color, thickened peristome, apertural insertions joined or nearly so, and some specimens with a basal denticle. While not all specimens show all of these characteristics, enough would be expected within a population to identify the subspecies.

Distribution: *O. strigosa buttoni* was found in Boxelder County near Logan, Utah. Not known from within our defined study area, it occurs in northwestern Utah, not far removed from southeastern Idaho, so it is potentially a peripheral species to our area.

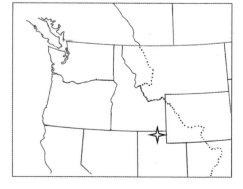

O. strigosa delicata Pilsbry, 1934   Blue Mountains Mountainsnail

Oreohelix strigosa delicata: South Fork Walla Walla River, Umatilla County, OR; scale bar = 1 cm (Xerces Society collection)

Description: With 5 to 5½ whorls it measures 12 to 20 mm wide. The spire is mostly moderately elevated, rounded or roundly-conic with variable colored markings and banding. The dark peripheral bands may be distinct or quite weak to non-existent. The shell is thin, and it may be distinguished by a granular appearance caused by prominent spiral striae, both dorsally and basally, cutting across rather coarse growth lines.

Similar Species: Pilsbry (1934), in naming this subspecies, questioned its distinguishing characteristics as ecological, although, "as distinct in appearance as many named subspecies." In his 1939 monograph, he listed it as a form of *O. strigosa* rather than a subspecies.

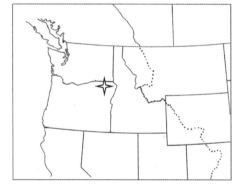

Distribution: *O. strigosa delicata* was found in the Walla Walla River Canyon above the forks, 11 km (7 miles) southeast of Milton-Freewater, Umatilla Co., Oregon, to 6.4 km (4 miles) up the South Fork and 3.2 to 4.8 km (2 to 3 miles) up the North Fork. It was found on south facing lava slides (Pilsbry 1934, 1939).

O. strigosa depressa Cockerell, 1890
Synonym: *Oreohelix depressa* Pilsbry, 1939

Description: This medium-sized snail has a low to moderately elevated spire. Most measure 18 to 24 mm with extremes from 10 to 29 mm wide. The umbilicus is about one-fifth the shell diameter. The periostracum is generally lighter than that of typical *O. strigosa*, being light yellowish-brown, sometimes with a pinkish tint. The subperipheral and supraperipheral bands are chocolate or darker brown with a lighter space between them and are usually rather narrow. Basal bands are most often absent. The last whorl descends a little to the aperture, which is nearly round except for the moderately deflected palatal lip margin and the narrow parietal attachment. The periphery of the last whorl is usually round, but the angled periphery of the immature shell may show in front of the aperture. Sculpture is of fine growth-wrinkles and sometimes weak spiral striae. The protoconch is of about 2⅓ whorls, with radial striae beginning on the second, and sometimes also with the beginnings of weak spiral striae.

Similar Species: *O. s. depressa* differs from typical *O. strigosa* by its generally more depressed spire, usually narrower peripheral bands, and lack of basal supernumerary bands. The ranges of the two subspecies are well separated, *depressa* being known only from southeastern Idaho within this study area. However, other *Oreohelix* occur within the ranges of each.

Distribution: The subspecies is generally found east and south of typical *O. strigosa*; from the east slope of the Rocky Mountains near Billings, Montana, south through western Wyoming, southeastern Idaho, and Utah, to northern Arizona and New Mexico.

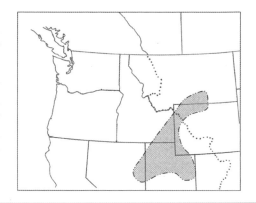

O. strigosa fragilis   (Hemphill, 1890)

Description: This medium-sized snail with a thin, somewhat translucent shell has a low to moderately elevated convexly-conic to rather globose spire. Its shell is a reddish-brown, lighter basally, and lighter yet between the typical chestnut or darker peripheral and supraperipheral bands. With about 5 whorls they measure 17 to 22 mm wide, with a rather

narrow umbilicus about one-sixth to one-seventh the shell diameter. Reported height/diameter ratios range from 0.54 to 0.68, varying somewhat between populations.

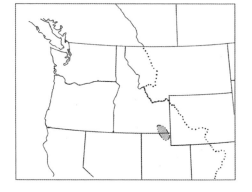

Similar Species: Some *O. s. fragilis* are similar to *O. s. depressa*, but differ by their rather translucent shells.

Distribution: *O. strigosa fragilis* is found in Franklin Co., in southeastern Idaho, and adjacent Cache Co., Utah.

O. strigosa goniogyra   Pilsbry, 1933

Description: A unique-appearing snail, the spire is low, rounded to rather high domed. The angled periphery is exaggerated and extends, as a keel, to the aperture. It measures 18 to 20 mm wide by 10 to 14 mm high, the height being 55% to 77% of the diameter. Color varies from grayish-white to rather dark brown and may be nearly uniform or with a variety of bands, streaks and/or flammules. Radial and spiral striations are usually fine and well-developed, at least dorsally.

Similar Species: Its distinct peripheral keel extending to the aperture in full adults, along with its variable shape, color, and markings, should distinguish this snail from other *Oreohelix*. Its very limited range will also be helpful.

Distribution: Known from Race Creek, two miles north of Riggins, Idaho Co., Idaho, *O. strigosa goniogyra* may be found on rock outcrops within ponderosa pine forests with partially to fully closed canopies.

Oreohelix subrudis and Subspecies

Oreohelix subrudis (Reeve, 1854)   Subalpine Mountainsnail
Synonym: *Oreohelix cooperi* Binney, 1869

Description: This medium to fairly large *Oreohelix* has a rather high, slightly convex-conic to rounded or somewhat domed spire. Varying greatly in size and color, it measures 15 to 26 mm wide by 11 to 20 mm high with a height/diameter ratio of about 0.78 (0.67 to 0.99) in the northern Rocky Mountain races. Populations from farther south appear to be somewhat less elevated. The umbilicus also varies in size. Although variable, the color is often buff or lighter with a peripheral and a supraperipheral band and often supernumerary basal bands of varying widths. Shell sculpturing is distinct as fine, sharp growth lines cut across by fine, closely spaced spiral striae. Genital anatomy of the *O. subrudis* group differs from that of *O. strigosa* by its penis being ridged (plicate) inside for more than half of its length. *O. yavapai* shares this characteristic, but in it, the penis is also distinctly swollen on the end toward the orifice while most *O. subrudis* have little if any swelling in that area.

Similar Species: *O. strigosa depressa* has a lower shell, a less rounded last whorl, a larger umbilicus, and it lacks the sharp, fine striation with close spiral lines cutting the striae (not always distinct in *O. subrudis* either). *O. strigosa depressa* seldom has the supernumerary basal bands often seen in *O. subrudis*. *O. yavapai* generally has a much more depressed spire.

Distribution: *O. subrudis* occurs in the Rocky Mountain states into Canada along the British Columbia-Alberta border, south to eastern Arizona and western New Mexico. Its range is mostly east and south of our defined study area, but there are records from Pend Oreille Co., Washington, across the northern Idaho Panhandle and the Rocky Mountains of Montana, and through southeastern Idaho.
The range of *O. subrudis* is primarily east and south of the defined area of this work, but typical *O. subrudis* and *O. subrudis apiarium* occur within our area. Two other subspecies, *O. subrudis limitaris* and *O. subrudis rugosa*, are included because of their near proximity to this area. Although their ranges overlap some, *O. subrudis* generally occurs farther east and at higher elevations than *O. strigosa*. It may be found in open forests and subalpine meadows, in among rocks and under logs, shrubs, other vegetation, and debris.

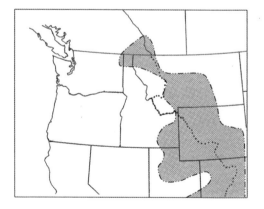

Key to the Subspecies of *Oreohelix subrudis*

(a)   Medium to large with rather high, convexly-conic spire. Height/diameter
       ratio 0.67 to 0.99. Rocky Mountains in northern Montana, west to
       Pend Oreille County, Washington . . . .[b/bb]

   (aa)   Rather large, 24 to 27 mm wide. Spire moderately elevated, obtusely-
           conic. Height/diameter ratio about 0.55. Light-colored with faint or no
           peripheral bands. Umbilicus about one-quarter to one-fifth the shell
           width. Northern Utah to just south of the Idaho border . . . . *O. subrudis rugosa*

(b)   Generally large, 16 to 23 mm wide by 15 to 20 mm high with about 6 whorls. Spire high, somewhat domed, height/diameter ratio 0.8 to 0.99. Rather tightly coiled with a small umbilicus, contained about 6 times or more in the shell width. Growth lines are fine and sharp; spiral striae distinct and closely spaced dorsally and obsolete basally. Color variable; peripheral bands apparent but often faint and with a lighter area between. Glacier National Park through the Mission Range and to Flathead Lake, Montana . . . . *O. subrudis apiarium*

   (bb)   Rather large but spires less elevated and less tightly coiled . . . . [c/cc]

(c)   Generally large, 15 to 26 mm wide. Spire rather high, average height/diameter ratio about 0.78, convexly-conic. Color variable but mostly buff or lighter. With typical darker peripheral bands and often narrow supernumerary bands basally. Umbilicus contained 3 to 5½ times in the shell width. Growth lines fine and sharp, crossed by fine, closely-spaced spiral striae dorsally as well as basally. Rocky Mountains of southeastern British Columbia and southwestern Alberta, and south through Montana, Wyoming, and southeastern Idaho, to New Mexico and Arizona; westward through the northern Idaho Panhandle into Pend Oreille County, Washington . . . . *O. subrudis subrudis*

   (cc)   Generally smaller, 16 to 17 mm wide with relatively high, somewhat globose spire. Not as high as *O. s. apiarium,* or as tightly coiled. Umbilicus narrow, contained 7 to 8 times in the shell width. Rusty-brown with apparent, sometimes faint, peripheral bands. Growth-wrinkles weaker, spiral striae irregular . . . . *O. subrudis limitaris*

Oreohelix subrudis apiarium   Berry, 1919

Description: This variety has a relatively high rounded spire, its beehive shape leading to its specific epithet. With about 6 whorls it measures 16 to 22 mm wide by 15 to nearly 20 mm high, with a height/diameter ratio of 0.8 to 0.99. With a tightly coiled shell, its umbilicus is narrow, being less than one-sixth the shell diameter. Its color is variable, very light cream to buff to chestnut brown, or it may be varying shades of light to dark gray. The darker lateral bands are usually discernible, although often faint, one just below the periphery, the second well above it. The shell is lighter between the bands, and there are sometimes supernumerary basal bands. The dorsal shell sculpture is as in the typical species, but is often indistinguishable on the base. The protoconch is flattened and strongly angled at the periphery; it is sculptured dorsally with fine lines of growth that increase with the size of the shell. These are crossed by very fine, closely spaced, wavy spiral striae that continue onto the base.

Similar Species: *O. subrudis limitaris* from nearby Alberta is very similar. *O. s. apiarium* has a wider umbilicus, a rounder and somewhat thickened aperture, and more distinct sculpturing on the teleoconch. The differences between these two are minor, and they may be the same subspecies, in which case *O. s. limitaris* (Dawson, 1875) would have precedence over *O. s. apiarium* (Berry, 1919).

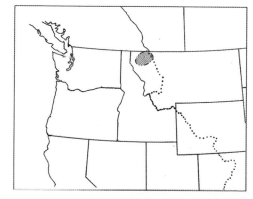

Distribution: *O. subrudis apiarium* is found primarily west of the Continental Divide in the northern Rocky Mountains of Montana and into British Columbia.

Oreohelix subrudis limitaris   (Dawson, 1875)

Description: A medium-sized shell with rather high rounded spire, similar to *O. s. apiarium*, but with about one-half less whorl, a little smaller diameter per whorl, and not quite as domed. In 5½ to 5¾ whorls it measures 16.3 to 17.2 mm wide by 11 to 14.1 mm high. Height/diameter ratio is 0.65 to 0.82, with an average of 0.73. The umbilicus is one-seventh to one-eighth the width of the shell (Berry 1885, cited in Pilsbry 1939) making it narrower than that of *apiarium*. The aperture is simple, a little expanded at the columellar lip. Color is similar to that of *O. s. apiarium*, but the sculpturing on the teleoconch is weaker and irregular.

Similar Species: Similar to *O. s. apiarium*, but the umbilicus is significantly narrower and the aperture appears wider and simpler. See similar species under that subspecies.

Distribution: This subspecies was recorded from Waterton Lake, Alberta, Canada. The area is not within the Pacific Northwest as defined in this paper, but it is immediately adjacent to the boundary.

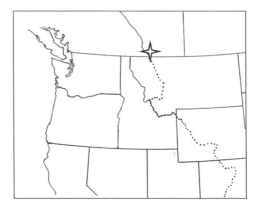

Oreohelix subrudis rugosa   (Hemphill, 1890)

Description: This large *Oreohelix* measures 24 to 27 mm wide by 13 to 15 mm high with about 5 whorls. The conic spire is quite low for *O. subrudis*. The umbilicus is about one-fifth the shell diameter. Its color is a dull yellowish- or grayish-brown with purplish mottling, and it usually lacks the peripheral bands. Shell sculpture is of coarse growth-wrinkles and weak, irregular spiral striae.

Similar Species: Pilsbry (1939) tells us that the shell closely resembles that of *O. depressa* form *carnea*, and dissection of the genitalia is necessary to confirm the species as *O. subrudis*. *O. strigosa depressa* form *carnea* has a higher spire than typical *O. strigosa depressa*, larger caliber whorls, and usually lacks, or has very thin or weak, peripheral bands.

Distribution: *O. subrudis rugosa* occurs in northern Utah, northeast of Great Salt Lake, near the border with southeastern Idaho.

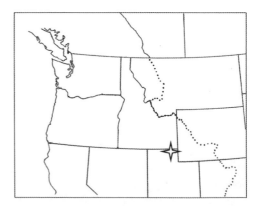

Oreohelix tenuistriata   J. Henderson & Daniels, 1916
Thin-ribbed Mountainsnail

Description: A fairly small snail, measuring about 10.5 to 13 mm wide by 6.5 to 9 mm high in 4¾ to 5⅓ whorls. The umbilicus is narrow, one-fifth to one-eighth the shell diameter. The rather thin shell has a low to moderately elevated conic spire with rounded apex and a simple, thin peristome. The color is light yellowish-brown with two darker brown bands, one above and one just below the periphery. The protoconch is mostly smooth on the first whorl, with faint spiral striae beginning near its end and radial striations beginning on the second whorl. The teleoconch is sculptured with close, sharp, wavy, radial lirae with cuticular edges in young or fresh shells, and well-spaced spiral striae, especially evident basally.

Similar Species: Pilsbry (1939) says, "The structure of the penis shows it to be related to *O. haydeni*." He also states, "It is a small and delicate form of the *haydeni* series, which now seem separable, though possibly further collecting may show the subspecific status more appropriate." The low spiral lines, especially on the base, are similar to the spiral lirae of *O. haydeni*, but it is separable by its small, thin shell, and the very low nature of its spirals.

Distribution: *O. tenuistriata* occurs about 3 km (2 miles) southwest of Lava Hot Springs, Bannock Co., Idaho, in a canyon under *Balsamorhiza* leaves and shrubs growing on limestone rubble.

Oreohelix variabilis   J. Henderson, 1929   Variable Mountainsnail
Synonym: *Oreohelix strigosa variabilis*

Description: The spire is rather high, mostly conic and solid. It is white basally with light mottling of various pinkish or brownish streaks. It appears light caramel-brown dorsally with white streaks along the ridges of the prominent growth-wrinkles, and a white leading edge about 2 mm wide at the aperture. With 5½ whorls it measures 15 to 22 mm wide by 11 to 16 mm high. The periphery is obtusely angled, less apparent in some specimens; the angle normally ends well before the aperture. There are strong growth-wrinkles but generally faint or indistinct spiral striae. The shells lack the typical peripheral bands of *Oreohelix*, but

Oreohelix variabilis: The Dalles, Wasco Co., OR; scale bar = 1 cm

some faint supernumerary bands may be present. The aperture is round and simple, but the palatal lip may be deflected inward a little, making the aperture appear slightly oblique. The parietal attachment is rather narrow, less than one-quarter the circumference of the whorl, and there is very little, if any, descent of the last whorl to the aperture. The protoconch is about 2⅓ whorls. Very fine spiral striae begin on the second half-whorl, and rather coarse, regular radial striae begin on the second whorl. The beginning of the teleoconch is sharply delineated by the change from regular radial striations to very coarse growth wrinkling.

Similar Species: The glossy though coarse sculpturing, the distinction between proto-conch and teleoconch, and the mottled white and pinkish or rich light brown color distinguish *Oreohelix variabilis* within its limited range.

Distribution: *Oreohelix variabilis* occurs in Wasco and Sherman counties, Oregon, and from The Dalles, east along the south side of the Columbia River, and six to ten miles up the Deschutes River. It is associated with shaded basalt talus or outcroppings, most often near springs or small streams.

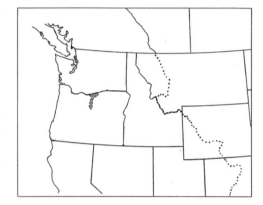

Oreohelix vortex   S. S. Berry, 1932   Whorled Mountainsnail
Synonym: *O. jugalis vortex*   Berry, 1932

Description: A rather small *Oreohelix*, with 5¼ to 5½ whorls it measures 12.6 to 14.2 (rarely to 17) mm wide by 7.6 to 8 mm high. The umbilicus is contained about 3½ times in the shell width. The spire is generally low-conic but slightly rounded. It is of varying shades of brown, and there is a narrow, darker brown band above the periphery and a wider one at or just below it with a light peripheral area between. The shell is grayish below the bands. The angled periphery of immature shells usually extends to in front of the aperture, but the periphery becomes round for approximately the last half-whorl. The last whorl descends fairly abruptly to the aperture. The apertural insertions are quite close together, leaving only a narrow parietal callus. Sculpturing is of rather fine growth-wrinkles and, in most specimens, very weak, irregular, spiral striae. Berry (1932) described a color variation, *O. vortex* form *flammulifer*, which is buff-colored dorsally with brownish streaks and mottling.

Similar Species: *O. vortex* is similar to *O. jugalis*, of which Pilsbry (1939) included it as a subspecies. It differs by being small by comparison. *O. amariradix*, also similar but even smaller, is found in the Bitterroot Range of Montana, across the mountains from *O. jugalis* and *O. vortex*.

Distribution: *O. vortex* is from the Salmon River near White Bird, Idaho Co., Idaho. It is associated with large basalt talus in grassland with scattered patches of shrubs, often on the lower slopes of river valleys.

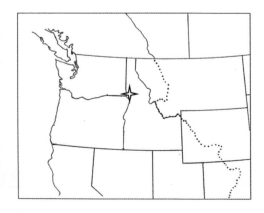

Oreohelix waltoni Solem, 1975   Lava Rock Mountainsnail

Oreohelix waltoni: Upper Butcher Bar, Idaho Co., ID; scale bar = 1 cm (TJF/Deixis collection)

Description: This is a rather small, ribbed snail, with low spire and large umbilicus. Small for *Oreohelix*, with 4⅜ to 5¼ whorls the shell measures 8 to 11 mm wide by 4.5 to 7 mm high. The spire is low, convexly conic; the periphery is angled and slightly carinate. The umbilicus is rather large and funnelform, fitting into the shell diameter about 3 (2.7 to 4.4) times. The protoconch of 2 to 2¼ whorls is sculptured with fine riblets, beginning at about one-half whorl, and very fine spiral striae. At about 1½ whorls the spiral lines become coarse, which leaves rows of blocks cut from the riblets between the striae. The teleoconch is sculptured by coarse ribs and spiral and radial striae that form a faint but distinct lattice pattern on and between the ribs except where the ridges are worn smooth. The typical bands of *Oreohelix* are absent.

Similar Species: *O. idahoensis* and *O. idahoensis baileyi* are larger with higher spires. *Anguispira nimapuna* has a lower spire with finer, sharper ribs, a strongly angled periphery, and a distinct brown periostracum.

Distribution: The known location for *Oreohelix waltoni* is the John Day Creek drainage, Idaho County, Idaho. It is found in open grassland with scattered shrubs and sometimes associated with basalt and schist.

Oreohelix yavapai and Subspecies

Oreohelix yavapai Pilsbry, 1905   Yavapai Mountainsnail

Description: This species is a medium-sized *Oreohelix*, about 17 mm wide by 9 mm high in about 5⅓ whorls. The umbilicus is a little larger than one-quarter the shell diameter. The spire is moderately elevated and generally rounded. The earlier whorls are flattened dorsally, and the sutures are generally filled by the keel of the previous whorls. The color is fairly typical for the genus. The body whorl is whitish and the spire brownish, with two darker brown bands, one just below the periphery, the other well above it. The last whorl is round but with an angled or somewhat keeled periphery that continues to the aperture. The first whorl of the 2⅓ whorl protoconch is smooth. Dense radial striae and rather strong spiral threads begin in the second whorl. The teleoconch begins with sculpturing of fine, sharp growth striae; within a partial whorl these become crossed by irregularly spaced spiral striae. Growth wrinkling on the last whorl is smoother, and the spiral lines are coarser and support rows of granules that are most prominent basally. On shells in the best condition, these granules may also be adorned with cuticular scales.

Similar Species: Compare with the following subspecies, and with *O. tenuistriata*, a smaller snail with a much smaller umbilicus, the range of which is nearer to that of the following subspecies than to the typical variety.

Distribution: The typical *O. yavapai* occurs in Yavapai County in central Arizona. However, Pilsbry (1939) included nine additional subspecies from Montana, Wyoming, Utah, Arizona, and New Mexico. Two of those subspecies are described below, because their known ranges are not far from the defined area of this study, and they might be found within our area of concern.

O. yavapai extremitatis   Pilsbry & Ferriss, 1911

Description: About the same size as the typical species, this race measures 14 to 19 mm wide by 7 to 11 mm high in 4 to 5¼ whorls. The spire is generally lower (low to moderately elevated) and more conical, but the height may measure about the same because the last whorl descends to well below the periphery. The umbilicus is one-quarter to one-fifth the shell width or smaller. The bands are variable from distinct to lacking, and the peripheral keel may or may not extend to the aperture. The widely-spaced granulose spiral lines of *O. yavapai* are normally visible basally. Pilsbry (1939) described one animal from Wyoming as "purplish black above and on the sides, the sole cream colored."

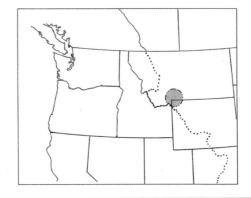

Similar Species: Compare this subspecies with *O. yavapai yavapai*.

Distribution: *O. yavapai extremitatis* is known from Gallatin Co., Montana; Wyoming; and the Grand Canyon, Arizona.

O. yavapai mariae   Bartsch, 1916

Description: Somewhat larger than the typical species, this race measures 18.3 to 22.5 mm wide by about 10 mm high in 5.6 whorls. The spire is low and rounded, and the sutures are as in the typical form. The shell is "flesh" colored, and the brown bands are quite narrow (Bartsch, 1916). The last whorl descends below the periphery. The periphery is keeled and somewhat carinate to within one-quarter whorl of the aperture. Sculpturing is of fine growth-wrinkles and fine spiral striae. The protoconch is little differentiated from the succeeding whorls.

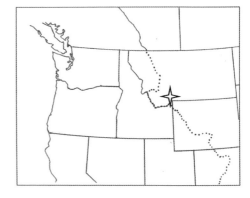

Similar Species: Compare it with *O. yavapai yavapai*.

Distribution: *O. yavapai mariae* is known from Gallatin Canyon, Montana.

Undescribed New Species or Varieties of *Oreohelix*

The following are newly discovered species or varieties of *Oreohelix* that have not yet been described in scientific literature. The number of species in the group of Chelan County, Washington, snails is questionable. The type locality for *Oreohelix strigosa* has been determined to be within this area. *Oreohelix junii* also occurs there, and those familiar with the snails in that area have thought there might be more than one variety of *O. junii*. Following the discovery of the Chelan Mountainsnail (*Oreohelix* new sp. 1) by Frest and Johannes, and its inclusion on the "Survey and Manage" list of federal agencies, surveys for that species led to the discovery of additional species or varieties of *Oreohelix*, as well as other genera, within Chelan County.

Following are descriptions of six recently discovered species from Chelan County and three from other areas. Kat Weaver of the University of La Verne, La Verne, California, is currently analyzing DNA and describing the Chelan County oreohelices.

Key to the Undescribed *Oreohelix*

(a) Species from central Washington, Chelan, Kittitas, or Yakima counties. . . . . . .[b/bb]

    (aa) Species from Idaho County, Idaho. Shells ribbed, with moderately elevated spires. . . . . [h/hh]

(b) Shells with distinct ribs . . . . [c/cc]

    (bb) Shells without distinct ribs . . . . [d/dd]

(c) Spire moderately elevated. With 5 to 5½ whorls it measures 14 to 15.5 mm wide; height about 70% of width. Umbilicus contained about 6 times in the width. Ribs low but distinct. Periphery rounded near the aperture.
. . . . Hoder's Mountainsnail

    (cc) Spire low. With about 5 whorls the shell width is 11.5 to 14 mm; height about 60% to 62% of the width. Umbilicus contained 4½ to 5¾ times in the width. Peripheral bands typical on the body whorl, but faint or obsolete on earlier whorls. Ribs fairly strong; periphery angled or with a very weak carina . . . . Ranne's Mountainsnail

(d) Shell large; spire rather high, somewhat domed. Width 23 to 26 mm; height 65% to 82% of width. Umbilicus contained 5 to 5½ times in the width. Periostracum brown on the spire, lighter on the last whorl and base, but adult shells usually weathered to near white. Peripheral bands typical but faded in some. Often found under quaking aspens . . . . .Aspen Mountainsnail

    (dd) Shells smaller, width generally not exceeding 21 mm . . . . [e/ee]

(e) Shells basically white. Width less than 18 mm; spires moderately elevated. Occurring between the Wenatchee River and Lake Chelan, Chelan County, Washington . . . . [g/gg]

    (ee) Shells whitish, brown or purplish-brown. Width greater than 19 mm; spires low or high . . . . [f/ff]

(f) Shell fairly large, about 21 mm in 5 whorls. Height about 60% of the width. Spire low, rounded, and roughened by fairly coarse growth striae and spiral striae. Umbilicus contained about 5 times in the width. Shell white but often appearing blackish dorsally from soil trapped in the striae. White basally. Bands dark brown, wide and distinct, and with some narrow, faint dorsal supernumerary bands. Periphery angled to subangular at the aperture. . . . . Mad River Mountainsnail (ff)   Shell medium to fairly large, about 20 mm wide in 5 to 5½ whorls. Spire moderately elevated, somewhat domed. Color purplish-brown dorsally, lighter basally. Banding variable, often with multiple supernumerary bands. Yakima River Canyon, Kittitas and Yakima counties. . . . . Yakima Mountainsnail

(g)   Shells 12 to 16 mm wide, with moderately elevated spire. Periostracum light corneous-brown dorsally. Base white. Bands typical, plus often with supernumerary bands. Umbilicus contained 5 to 6 times in the shell width. Early teleoconch whorls with rows of periostracal scales, lost by adulthood. . . . . Entiat Mountainsnail

(gg)   Shells 16 to 18 mm wide in 4½ whorls. Spire moderately elevated. Rows of periostracal scales on the teleoconch retained into adulthood. Occurs on the ridge sloping into the southwest side of Lake Chelan. . . . . Chelan Mountainsnail

(h)   Width about 11 mm with 5 to 5½ whorls. No peripheral bands. Umbilicus contained about 4 times in the width. Protoconch of 2¼ to 2½ whorls with fine regular riblets and fine spiral striae. Lower Salmon River, Idaho Co., Idaho. . . . . Sheep Gulch Mountainsnail

(hh)   Width about 13 mm with 5 to 5¼ whorls. Peripheral bands distinct but narrow, lacking on some shells. Umbilicus contained 6 times in the shell width. Protoconch of 2 to 2⅓ whorls, beginning with low irregular growth-wrinkles, the remainder carinate, and with fine, sharp riblets and spiral threads forming a grid pattern. Pittsburgh Landing, Hells Canyon, Idaho . . . . Pittsburgh Landing Mountainsnail

Oreohelix undescribed sp.   Aspen Mountainsnail

Collected and recognized as unique by employees of the Wenatchee National Forest in June, 1996.

Description: These are large *Oreohelix*, with solid shells and high, rather dome-shaped spires. With 5¾ to 6½ whorls the shells measure 23 to 26 mm wide by 16 to 19 mm high (H/D = 0.65 to 0.82). The umbilicus is narrowly funnelform and contained about 5 to 5½ times in the shell width. The whorls are slightly compressed. The peripheries of some are slightly angled in front of the aperture and slightly wider below the midpoint of the whorl. The last whorl descends a little very shortly before the aperture. The aperture is roundly or ovately lunate; it is slightly wider than high in some, higher than wide in others. The parietal intrusion is narrow. The sutures are shallow or moderately impressed. The apertural lip is simple but a little thickened or narrowly reflected basally, and it is expanded at the columella but normally not enough to cover the umbilicus a significant amount. The shells are mostly weathered and eroded, so color and sculpturing is spotty. Dorsally they are corneous-brown; the last whorl is lighter. The typical dark brown bands above and below the periphery are visible in some shells but faded in others. The protoconch of about 2½ whorls has weak growth wrinkling. The teleoconch has coarse growth wrinkling and shallow, irregular spiral striae that are more coarse and distinct basally, but they are nearly eroded away from some shells.

Similar Species: Pilsbry (1939) provided illustrations of forms or subspecies of *O. strigosa* and *O. subrudis* (not from Washington State) that have high domed spires. However, the

1 and 2. *Oreohelix* undescribed sp. ("Aspen Mountainsnail"): Camas Cr., Chelan Co., WA

aspen mountainsnails are generally found associated with quaking aspen stands in riparian areas or wet meadows, implying a specific ecological niche, thus an ecotype, subspecies, or possible new species.

Distribution: These Aspen mountainsnails were found under quaking aspen stands in the riparian zone of Swakane Creek Beaver Ponds and on an aspen flat in the Camas Creek drainage, both in Chelan County, Washington.

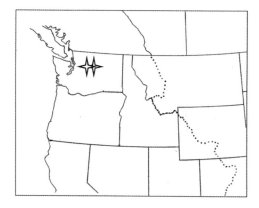

Oreohelix n. sp. 1   Chelan Mountainsnail

Collected and recognized as a new species by Dr. Terrence Frest and Edward Johannes, on slopes above the southwest shore of Lake Chelan, Chelan County, Washington. It was cited as *Oreohelix* n. sp. 1, Chelan mountainsnail, in Frest and Johannes (1993 and 1995) and in Kelley et al. (1999).

Description: With about 4½ whorls the shell is 16 to 18 mm wide with a moderately elevated convexly conic spire. The shell is mostly white or with thin brown periostracum and with the typical peripheral and supraperipheral bands of the banded *Oreohelix*. The periphery and aperture are rounded. The umbilicus is deep, and narrowly funnelform, about one-quarter the shell width. Sculpturing is of low but distinct growth-wrinkles and spiral striae, and there are "periostracal lirations present on both surfaces to adulthood" (Frest and Johannes 1995).

Similar Species: Frest and Johannes (1995) also said, "This land snail may be most closely related to the Yakima mountainsnail (*Oreohelix* n. sp. 2; q.v.); but that species has an angular periphery; is larger, has a smaller umbilicus; and lacks periostracum-fringed lirations." It appears even more similar to the following species, which I have tentatively designated the Entiat mountainsnail, and which occurs adjacent to the limited range of *Oreohelix* n. sp. 1. Outwardly, the two species appear very similar. They differ in that the second species (Entiat mountainsnail) does not retain the "periostracal fringed lirations" into adulthood, although it and some other *Oreohelix* in the vicinity have those scale-like fringes as juveniles.

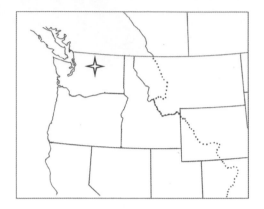

Distribution: The Chelan mountainsnail occurs in schist talus in open Douglas-fir forest on the southwest side of Lake Chelan, Chelan County, Washington.

Oreohelix undescribed sp.   Entiat Mountainsnail

Shells were collected by employees of the Entiat Ranger District, Wenatchee National Forest, following the Tyee wildfire of 1994. Shells were first discovered in burned areas of pinegrass. Live specimens were found by T. Burke in unburned Crum Canyon (9 October, 1997) under forest floor litter and among rocks in open ponderosa pine/Douglas-fir forests. The first live specimen was found on a rock sticking out of 2 to 3 inches of snow from the night before.

Description: The shell of this medium-sized snail with 4⅓ to 5¼ whorls measures 12 to 16 mm wide by 7 to 11 mm high. Its spire is moderately elevated to rather high. The umbilicus

1 and 2. *Oreohelix* undescribed sp. ("Entiat Mountainsnail"): Swakane Cr., Chelan Co., WA

is narrowly funnelform and about one-fifth to one-sixth the shell diameter. Shells are usually weathered and eroded so colors are not always discernible, but they appear to be corneous-brown dorsally and whitish or light corneous on the base. The two typical dark brown bands are present and there may be additional narrow, lighter bands. The aperture is roundly lunate, very slightly oblique, and the parietal intrusion occupies one-eighth or less of its circumference. There is usually a distinct or slight peripheral angle on the penultimate whorl at and in front of the aperture, but the body whorl becomes round before its end. The last approximately one-eighth of the body whorl descends a little to the aperture. Juvenile specimens have rows of periostracal fringes in the form of cuticular scales on the early teleoconch that develops after their birth. The scales do not appear on the protoconch.

Juvenile "Entiat Mountainsnail" showing the periostracal fringes of the juveniles that the Chelan Mountainsnail retains into adulthood; Swakane Cr., Chelan Co., WA

Similar Species: The species is most like *Oreohelix* n. sp. 1 of Frest and Johannes. It differs by not retaining the "periostracal fringed lirations" into adulthood, although it and some other *Oreohelix* in the vicinity have those scale-like fringes as juveniles. Also its juvenile shell is angled at the periphery.

Distribution: The Entiat mountainsnail occurs on the upper slopes of the Columbia River breaks between Lake Chelan and the Wenatchee River, Chelan County, Washington. It can be found under forest floor litter and among rocks in and at the edges of open ponderosa pine/Douglas-fir forests.

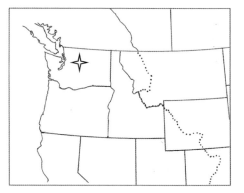

Oreohelix undescribed sp.   Hoder's Mountainsnail

First collected by Judy Murray-Hoder, from Dick Mesa, Chelan County, Washington, in April, 2002.

Description: With 5 to 5½ whorls the shell measures 14 to 15.4 mm wide by 9.8 to 10.8 mm high. The umbilicus, about one-sixth the shell width, is deep and narrow, nearly symmetrical, and the last whorl encroaches slightly on the opening. The aperture is roundly lunate and a little oblique; the parietal intrusion is less than one fourth the circumference of the whorl. There are about 30 major ribs on the last whorl (38 on a sinistral specimen). The ribs are irregular in height, width, and spacing with fine, rather high growth ridges between them, so it is sometimes difficult to distinguish between the individual ribs. The ribs are formed as waves in the shell, so they are visible inside as well as on the outside. The protoconch consists of 2⅓ whorls. Close, fine spiral striae begin early on the protoconch as do well-spaced fine riblets. Coarse ribs begin with the teleoconch. On immature shells, newly-formed ribs are smooth and regularly spaced, rising abruptly from the posterior to a sharp crest, then sloping down anteriorly. Periostracal tongue-shaped scales form rows

1

1. *Oreohelix* undescribed sp. ("Hoder's Mountainsnail"):
   Dick Mesa, Chelan Co., WA; scale bar = 1 cm

2. *Oreohelix* undescribed sp. ("Hoder's Mountainsnail";
   sinistral shell): Dick Mesa, Chelan Co., WA

2

along or near the summits of the ribs and align spirally on very fine raised cuticular edges basally, forming a faint grid pattern. Dorsally, the spiral edges are mostly obsolete, and few of the scales remain on an immature shell of 3½ whorls. The periphery of the immature shell is strongly angled and adorned with a row of the periostracal scales. On the adult shell, the angle of the periphery is slight, but it is distinct on the penultimate whorl. There is no carina. The last one-fourth whorl is below the periphery, and ends on or below the subperipheral band, although there is no abrupt descent before the aperture. The spire is moderately elevated, convexly conic. Its color is light corneous to a corneous gray-brown; the ribs, other raised areas, and much of the base turn white as eroded. The two typical, darker brown bands are narrow but distinct.

Similar Species: Within the limited site where this species occurs, only the Ranne's mountainsnail, another undescribed species that occurs nearby, is similar in appearance. They differ by Ranne's having a much lower spire, a little wider umbilicus, a weak peripheral carina, and lacking the cuticular scales on their immature shells. The most similar-appearing snail may be *Oreohelix peripherica*, an extremely variable snail from northern Utah.

Distribution: Hoder's mountainsnail is known only from Dick Mesa, about three and one-half miles northeast of Entiat, Chelan County, Washington. It is found on or near the ridgetop, in grassland and timber edge, with *Eriogonum* sp. and *Balsamorhiza sagittata*.

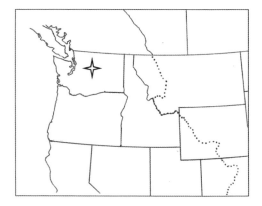

Oreohelix **undescribed sp.**   Mad River Mountainsnail

Collected by T. Burke, from a talus slide under black cottonwood, along the Mad River, Chelan County, Washington, in April, 1996.

Oreohelix undescribed sp. ("Mad River Mountainsnail"): Mad River, Chelan Co., WA; scale bar = 1 cm

Description: A unique-appearing *Oreohelix*, with about 5 whorls the shell measures approximately 21 mm wide by 11.5 to 13.5 mm high. The umbilicus is narrowly funnelform, about one-fifth the shell diameter. The shell is solid with a low rounded spire; the average height/diameter ratio 0.6. The sutures are shallow until the last quarter-whorl, when the body whorl gradually descends below the angular periphery of the penultimate whorl, leaving an overhang at the suture. The aperture is obovate, the whorls somewhat flattened, the periphery angled on the early whorls but only slightly on the last whorl, which becomes round within one-fourth turn of the aperture. The lip is thickened inside. The palatal and outer margins are simple; the basal and columellar are flared.

The protoconch of about 2 whorls is sculpted with very fine riblets and spiral striae, giving it a satiny appearance. The teleoconch is sculpted with very coarse growth-wrinkles and prominent groupings of spiral striae, clustered in regular, closely spaced striae above and below the periphery and below the suture, but also less regular elsewhere, dorsally and basally. The basal area is smooth for about a half whorl in front of the aperture. There are rather irregularly spaced spirals of double striae with slightly raised blocks or granules between them, appearing like railroad tracks. They are most prominent basally, and on one shell were spaced an average of about 0.9 mm apart. The shells are corneous. The periostracum of the first three whorls is brown, followed by a very dirty-appearing ivory-white color, so the shell appears quite dark or diffuse black dorsally. The two typical dark brown bands are rather wide and distinct, and the area between them is white, as is the base of the shell. There may be one or more narrow corneous-gray bands mid-basally.

Similar Species: Within the near vicinity, *Oreohelix junii* is the most similar in appearance. The two species differ in that *O. junii* may be wider, has a flatter more conic spire and a relatively smooth surface with very faint or no spiral striae.

Distribution: The Mad River mountainsnail is known only from near the Mad River in the Entiat Valley, Chelan County, Washington. It is found in talus under black cottonwood or big-leaf maple trees.

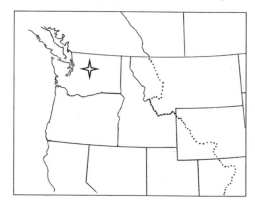

Oreohelix **undescribed sp.**   Ranne's Mountainsnail

First collected by Brigitte Ranne, on Dick Mesa above Entiat, Chelan County, Washington, on October 25, 1999.

Oreohelix undescribed sp. (Ranne's Mountainsnail): Dick Mesa, Chelan Co., WA; scale bar = 1 cm

Description: With about 5 whorls the shell measures 11.75 to 13.75 mm wide by 7.3 to 8.3 mm high. The umbilicus is deep, narrowly funnelform, expanding moderately, contained 4½ to 5¾ times in the shell width. The aperture is roundly lunate, slightly oblique, the parietal intrusion being less than one-fourth of the circumference of the whorl. There are 23 to 32 distinct ribs on the body whorl, formed as waves in the shell, so they are visible inside as well as on the outside of the shell. There is faint growth wrinkling between the ribs, and distinct closely-spaced spiral striation begins early on the protoconch and extends to the aperture. The ribs begin with the teleoconch. Peripheries of the immature shells are strongly angled. The angle is slight on the last whorl, fading to barely discernible at the aperture but retaining a very slight carina, which extends to or nearly to the aperture. The sutures are well impressed. There is no abrupt descent of the last whorl before the aperture; the palatal insertion of the aperture is attached at or just below the periphery of the penultimate whorl. The spire is low to moderately elevated, conic, with well-impressed sutures. It is light brown, corneous, lighter basally and with darker brown bands just below the periphery and mid-dorsally on the last whorl. The dorsal band is faint to obsolete on the earlier whorls.

Similar Species: A short distance from the site where Ranne's mountainsnail occurs is another ribbed *Oreohelix* (Hoder's mountainsnail) that is a little larger and has a distinctly higher spire. Also 10 to 20 miles north, around Lake Chelan, a low ribbed shell similar to *Anguispira nimapuna* has been found, which has a lower, more rounded spire and a larger umbilicus.

Distribution: Ranne's mountainsnail is known only from Dick Mesa about three miles northeast of Entiat, Chelan County, Washington. It is found on a southeasterly aspect near the ridgetop, in grassland with *Eriogonum* and *Balsamorhiza sagittata*.

Deixis Consultants New Species

In records and reports from their surveys and consulting work, Dr. Terrence Frest and Mr. Edward Johannes of Deixis Consultants have documented many additional undescribed taxa of *Oreohelix* from throughout the Pacific Northwest. One of these is *Oreohelix* n. sp. 1, the Chelan mountainsnail, described above because it is found in the same vicinity as the others already discussed. Others include the three discussed below because we have access to shells, and photographs of them are included in this work. For a list of other new species of *Oreohelix* documented by Deixis Consultants, see Frest and Johannes (1995 and 2000).

Oreohelix new sp. 2   Yakima Mountainsnail

This species was informally described as *Oreohelix* n. sp. 2, Yakima mountainsnail, in Frest and Johannes (1995). The following description is of shells identified by Dr. Frest.

Oreohelix new sp. 2 ("Yakima Mountainsnail"): Rattlesnake Cr., Yakima Co., WA; scale bar = 1 cm

Description: A medium-sized snail, with a moderately elevated, convexly conic spire. With 5 to 5½ whorls the shell measures 16 to 20 mm wide by about 10.5 mm high. The whorls are slightly compressed, the last descending slowly in the last one-quarter whorl to below the periphery, which is slightly angled nearly to the aperture. The aperture is roundly-elliptical and slightly auriculate by the intrusion of the penultimate whorl. The color is rather dark brown with very faintly darker brown peripheral and supraperipheral bands, and there may also be supernumerary darker and lighter alternating bands basally. The protoconch is of about 2¼ whorls and is nearly entirely sculpted with growth-wrinkles, which become coarser as the whorls enlarge. Spiral striae begin at about the first half-whorl, are prominent on the protoconch, and continue to the aperture. Growth-wrinkles are quite coarse on the teleoconch.

Similar Species: There is a slight similarity to *O. variabilis*, with its coarse growth-wrinkles and brownish color, but that species normally has a higher spire, even coarser growth-wrinkles, and its shell is more whitish basally and mottled with light brown dorsally.

Distribution: The Yakima mountainsnail occurs in the Yakima River Canyon and along some tributaries in Kittitas and Yakima counties. It is found in dry basalt talus near springs and seeps, in association with shrubs and moist site vegetation.

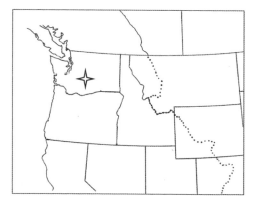

Oreohelix n. sp. 20   Sheep Gulch Mountainsnail

This species was informally described in Frest and Johannes (1995). The following description is of shells provided by Dr. Frest.

Oreohelix new sp. 20 ("Sheep Gulch Mountainsnail"): Lower Salmon River, Idaho Co., ID; scale bar = 1 cm (TJF/Deixis collection)

Description: A small ribbed *Oreohelix*, the shell is white, lacking bands or other markings. With 5 to 5½ whorls it measures up to 12 mm wide by 8.6 mm high. The spire is moderately elevated, convexly-conic, with well-impressed sutures giving it a terraced appearance. The periphery is round but very slightly carinate in some. The aperture is round to oval; the penultimate whorl intrudes only slightly. The umbilicus is deep, narrowly funnelform, about one-quarter to one-third the shell width. The protoconch of about 2¼ whorls is spirally striate on the first half-whorl, at which point small ribs begin. These increase in size with the whorls becoming coarse, widely-spaced ribs. Faint traces of spiral threads and regularly spaced growth lines also appear sporadically on and between the ribs on the teleoconch.

Similar Species: It is most similar to *Oreohelix waltoni*, but that species is smaller, with a little more strongly angled periphery and somewhat finer ribs.

Distribution: The Sheep Gulch mountainsnail occurs near the Lower Salmon River in the vicinity of Lucile, Idaho Co., Idaho. It inhabits limestone outcrops, talus, and schists in open grass and brushland.

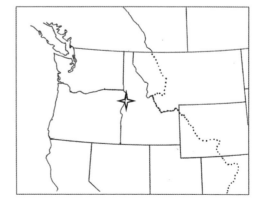

Oreohelix n. sp. 30   Pittsburg Landing Mountainsnail

This species was informally described in Frest and Johannes (1995). The following description is of shells provided by Dr. Frest.

Oreohelix new sp. 30 ("Pittsburg Landing Mountainsnail"): Pittsburg Landing, Hells Canyon, ID; scale bar = 1 cm (TJF/Deixis collection)

Description: A small ribbed *Oreohelix*, the shell color is brownish-horn to white basally, and it has two narrow darker brown bands as typical for *Oreohelix*. With 5 to 5½ whorls it measures up to 13 mm wide by 10 mm high. The spire is moderately elevated to rather high, convexly conic, with moderately impressed sutures. The whorls are round; the aperture is round and only slightly intruded into by the penultimate whorl. The umbilicus is deep, very narrowly funnelform, and about one-sixth the shell width. The protoconch is of about 2½ whorls, the first with widely spaced spiral lirae crossed by fine ribs for about a half whorl, then by radial striae, giving the appearance of rows of braids. Ribs on the teleoconch are weak at first, increasing to moderately coarse, fairly closely spaced on the last whorl. Fairly coarse spiral striae can be seen crossing and between the ribs but become less apparent near the aperture.

Similar Species: The Pittsburg Landing mountainsnail is similar to *O. idahoensis*, but that species has a higher spire, a more acute apex, much stronger ribs, and it lacks the peripheral bands.

Distribution: This snail is known only from the vicinity of Pittsburg Landing, Idaho County, Idaho, inhabiting "a very narrow (<12 inches wide) area along part of a cliff base and scattered talus remnants in the same area" (Frest and Johannes, 1995). The site is surrounded by sagebrush and scattered prairie plants.

Selected references for the family Oreohelicidae: Berry 1932; Fairbanks 1980, 1984; Frest 1999; Frest and Johannes 1993, 1995, 2000; Hanna and Smith 1939; Hemphill 1890; Henderson 1929a; Hendricks 2003; Hendricks et al. 2007; Jones 1944; Kelley et al. 1999; Pilsbry 1933, 1939; Smith 1937; Solem 1975.

Families: PUNCTIDAE, CHAROPIDAE, DISCIDAE, AND HELICODISCIDAE

The following six genera that occur in the Pacific Northwest were previously included in the family Endodontidae; they have since been separated into the following four families:

Family: PUNCTIDAE   Genera: *Paralaoma* and *Punctum*
Family: CHAROPIDAE   Genus: *Radiodiscus*
Family: DISCIDAE   Genera: *Anguispira* and *Discus*
Family: HELICODISCIDAE   Genus: *Helicodiscus*

As a group, these genera are composed mostly of small or minute snails (the exception is the genus *Anguispira*, species of which are medium or rather large). Their shells are generally opaque, fairly solid, without distinctly reflected lip margins, and most lack colored markings; the exceptions are the banded *Anguispira kochi* and the introduced *Discus rotundatus*, the shell of which is marked with reddish flammules.

An aid to separating these, and all of the following families, from some other similar-appearing taxa is their visible pedal furrows, a characteristic of the informal group (division or infraorder of past authors) Aulacopoda. That is, if a specimen includes the whole snail, pedal furrows will be evident, as opposed to a snail of the Holopoda group, in which pedal furrows are not evident. This is useful in separating the aulacopod *Anguispira kochi* from species of the holopod *Oreohelix*, some of which may appear similar.

Key to the Families and Genera of
PUNCTIDAE, CHAROPIDAE, DISCIDAE, and HELICODISCIDAE

When a taxon is identified from these keys, proceed to the family section in which it is found (see the list above).

(a)   Relatively large heliciform shells (normally greater than 17 mm wide by 13 mm high). Whorls smooth, periphery well-rounded, and usually with two reddish-brown spiral bands separated by a yellowish peripheral band. Family: DISCIDAE (in part) . . . . *Anguispira kochi*

    (aa)   Shells minute to medium (less than 15 mm wide), more depressed, with radial ribs or spiral threads or ridges (lirae) . . . .[b/bb]

(b)   Small to medium-sized snails (4 to 15 mm diameter) with solid ribs or spiral lirae . . . . [c/cc]

    (bb)   Minute snails (less than 2.5 mm diameter) with low rounded spires and cuticular riblets (sometimes inconspicuous), or fine solid riblets.
. . . . Family: PUNCTIDAE

(c)   Small discoidal snails (4 to 5 mm diameter) with broad, dish-shaped umbilicus about one-half the shell width. Spiral lirae (raised threads) run parallel to the whorls. Family: HELICODISCIDAE. . . . .*Helicodiscus salmonaceous*

    (cc)   Small to medium-sized snails (5 to 15 mm wide), heliciform with solid radial ribs. Umbilicus not dish-shaped, broad (to one-third the shell width) or narrow but relatively deep . . . . [d/dd]

(d)   Medium-sized snails (11 to 15 mm wide) with low domed spires, angular peripheries, and radial ribs. Idaho County, Idaho, and Chelan County, Washington. Family: DISCIDAE (in part). . . . . *Anguispira nimapuna* (dd)   Small snails (5 to 11 mm wide) with low-domed or nearly flat spires. The outer whorl may be round or shouldered, or sharply angled at the periphery. . . . . [e/ee]

(e)   Shell about 6.5 mm wide with a very low to nearly flat spire, a narrow umbilicus, and slightly shouldered whorls. The first 2 whorls are sculptured with spiral threads and striae, and the remaining whorls have fairly evenly spaced, radial ribs as well as fine spiral striae that continue to the aperture. Family: CHAROPIDAE . . . . *Radiodiscus abietum*

(ee)   Ribbed shells, 5 to 11 mm wide, low to moderately high rounded spire. Last whorl round, shouldered, or angled at the periphery. Family: DISCIDAE (in part) . . . . Genus: *Discus*

Family: PUNCTIDAE

Tiny snails with a brownish or reddish-brown, heliciform shell, open umbilicus, simple lip, and sculpture of fine riblets and microscopic striae.

Key to the Species of the Family PUNCTIDAE

(a)   Shells 1.5 to 2.1 mm wide by 0.9 to 1.3 mm high, with low to moderately elevated spire. Sculpture of rather high cuticular riblets with microscopic radial and spiral striae in the interspaces (also compare with *Planogyra clappi* in the family Valloniidae, which has a nearly flat spire) . . . . *Paralaoma servilis*

(aa)   Shells mostly smaller, 1 to 1.9 mm wide. Sculpture of fine, solid, or low cuticular riblets . . . . [b/bb]

(b)   Spire rounded to dome-shaped, chestnut or reddish-brown. Umbilicus one-quarter to one-fifth the shell width . . . . [c/cc]

(bb)   Spire low to moderately elevated, more conical, light brown. Width 1.1 to 1.5 mm by less than 1 (0.7 to 0.94) mm high. Umbilicus one-quarter to one-third the shell width, funnelform but expanding more rapidly in the last two-thirds whorl. Aperture obliquely lunate, wider than high. . . . .*Punctum minutissimum*

(c)   Shells 1 to 1.4 mm wide by 0.7 to 0.95 mm high, somewhat dome-shaped. Aperture oblique and obovate. Umbilicus contained 4⅓ to 5 times in the shell width, more well-shaped, symmetrical, expanding evenly but more rapidly in the last ¾ whorl. Sculpture of microscopic cuticular riblets with interspersed radial striae . . . . *Punctum randolphi*

(cc)   Shells larger, 1.5 to 1.9 mm wide by greater than 1 (1.05 to 1.25) mm high. Spire rounded, low to moderately elevated. Umbilicus funnelform, expanding evenly. Aperture obliquely lunate, wider than high. . . . . *Punctum californicum*

Genus: *Paralaoma*

Paralaoma servilis (Shuttleworth, 1852)   Pinhead Spot
Synonym: *Punctum conspectum* (Bland, 1865); *Paralaoma caputspinulae* (Reeve, 1852)

Paralaoma servilis servilis: Patos Island, San Juan Co., WA; scale bar = 1 mm

Description: The shell width is about 1.5 to 2.1 mm, height 0.9 to 1.3 mm, in about 4 to 4¼ whorls. It has a moderately elevated, conic spire. The umbilicus is open, about one-fourth or a little more of the shell diameter, and expands rapidly in the last half-whorl. The aperture is broadly lunate and obliquely wider than high. Shell sculpturing consists of rather high cuticular riblets with very fine radial and spiral striae between.
Subspecies: Pilsbry (1919) described a subspecies *Paralaoma servilis alleni* from South Oswego, Oregon. It is said to have a higher spire and more of the larger riblets than the typical species. Pilsbry (1948) gave shell measurements of 2.4 mm wide by 1.5 mm high, somewhat larger than the typical species.

Similar Species: *Planogyra clappi*, of the family Valloniidae, is similar in its size, open umbilicus, and sculpturing. It differs in the spire, which is relatively flat, and the aperture, which is slightly higher than wide.

Distribution: *Paralaoma* is known in North America from Alaska to California, Arizona, and New Mexico, and from western Montana to the Pacific Coast. Populations are scattered throughout the Pacific Northwest, usually in litter of moist forests or riparian zones. It can be found on small woody debris or by searching through forest floor litter; piles of moist dead leaves and live native vegetation such as sword ferns are good places to search.

Genus: *Punctum*

Punctum californicum Pilsbry, 1898   Ribbed Spot

Description: This *Punctum* is a little larger than the following two species. With 3½ to 4⅓ whorls it measures 1.5 to 1.85 mm wide by 1.05 to 1.25 mm high. The umbilicus is one-quarter to one-fifth the width of the shell and expands rapidly in size. The shell is light chestnut-brown, with a moderately elevated, rounded spire, with deep sutures. The width of the whorls increases slowly but regularly. The periphery is rounded; the aperture is roundly lunate but a little wider than high. The fine riblets appear more solid, without apparent periostracal extensions (cuticular riblets).

Similar Species: Other *Punctum* to compare with *P. californicum* within the Pacific Northwest include: *P. randolphi* and *P. minutissimum*. These are both slightly smaller than *P. californicum*; *P. minutissimum* has a lower, more conic spire. The spire of *P. randolphi* is somewhat dome-shaped, and its umbilicus is more well-like and symmetrical than those of the other species.

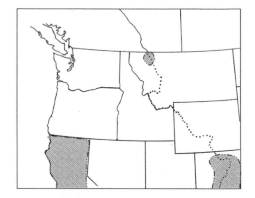

Distribution: *P. californicum* is from California, Arizona, and north through Colorado. It has also been reported from Glacier National Park, Montana, and from eastern South Dakota. Search forest floor litter and small woody debris for these snails.

Punctum minutissimum   (I. Lea, 1841)   Small Spot

Punctum minutissimum: Upper Mill Cr., Umatilla Co., OR; scale bar = 1 mm

Description: This very tiny snail measures 1.1 to 1.51 mm wide by 0.7 to 0.94 mm high with 3¾ to 4¼ whorls. Its umbilicus is about 0.3 to 0.4 mm, or one-quarter to one-third the width of the shell, and it expands more rapidly in the last half-whorl. The shell is very light brown, with a low to moderately elevated, roundly conic spire. The whorls increase slowly but regularly in width, the last about the same width as the penultimate. The periphery is well-rounded, the aperture roundly lunate. The shell is sculptured with uneven, closely spaced riblets and very faint spiral striae.

Similar Species: Other minute, heliciform snails with a deep, narrow umbilicus and brownish shell include *Paralaoma servilis* and *Planogyra clappi*, which are larger than *Punctum minutissimum* and have high cuticular riblets; and *P. randolphi* and *P. californicum*, which have higher, more dome-like spires and narrower umbilici.

Distribution: *P. minutissimum* is generally found in the eastern United States and Canada but has been recorded as far west as Shoshone County, Idaho; Wallowa and Umatilla counties, Oregon; and south-central New Mexico. It inhabits forest floor litter.

Punctum randolphi   (Dall, 1895)   Conical Spot

Punctum randolphi: Patos Island, San Juan Co., WA; scale bar = 1 mm

Description: This minute species measures 1.07 to 1.4 mm wide by 0.73 to 0.95 mm high with 3¾ to 4¼ whorls. Its umbilicus is a little smaller than that of *P. minutissimum*, about 0.25 mm, or one-quarter to one-fifth of the shell width, and expanding less in the last half-whorl. The reddish-brown shells have a moderately high, roundly conic or nearly dome-shaped spire. The whorls increase in width slowly but regularly. The periphery is round; the aperture is obliquely, ovately, or roundly lunate. Shell sculpture is of microscopic, cuticular riblets (often indistinguishable), with even finer radial and spiral striae in the interspaces.

Similar Species: *Punctum minutissimum* has a lower, more conic spire with a wider, more funnelform umbilicus. *P. californicum* is larger, with more solid riblets, and the larger *Paralaoma servilis* has high major riblets. Hatchlings of some of the Pupillidae might require close microscopic examination to avoid mistaking them for a *Punctum* because of their tiny size. Shell shape and width of the umbilicus are clues to their identities.

Distribution: *P. randolphi* is generally found in southwestern British Columbia, through western Washington, into northwestern Oregon, and in the Idaho Panhandle south to Adams County. It can be found by searching through piles of moist, dead leaves or other forest floor litter.

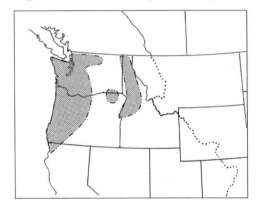

Family: CHAROPIDAE
Genus: *Radiodiscus*

Radiodiscus abietum   H. B. Baker, 1930   Fir Pinwheel

Description: This is a small snail with a discoidal shell. With 5¼ to 5¾ whorls it measures 6.2 to 6.7 mm wide by 3.3 to 3.5 mm high. The umbilicus is narrow, about one-fifth to one-sixth the width of the shell, and symmetrical. The shell is light brown to light or brownish horn-colored with a very low, nearly flat spire. The width of the whorls increases slowly; the body whorl is a little shouldered. The aperture is narrowly lunate; the lip is

1. *Radiodiscus abietum*: Touchet River, Umatilla National Forest, Columbia Co., WA
2. *Radiodiscus abietum*: Idaho Panhandle National Forest, Shoshone Co., ID; scale bar = 2 mm
3. *Radiodiscus abietum*: Idaho Panhandle National Forest, Shoshone Co., ID

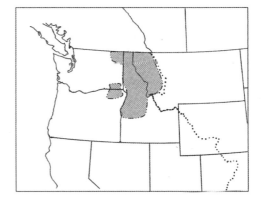

simple and thin. Besides the flattened spire and small umbilicus, the shell sculpturing is unique. The 2 to 2¼ whorl protoconch is sculpted with parallel spiral threads separated by fine striae that continue to the aperture. In addition to the spiral striae, the teleoconch is adorned with low but distinct, fairly evenly spaced ribs.

Similar Species: *Discus* species are ribbed, but they lack the spiral threads on the protoconch and have a roundly-elevated spire, a roundly-lunate aperture, and a relatively wide umbilicus.

Distribution: *Radiodiscus abietum* is endemic to the Pacific Northwest from eastern Washington through the Idaho Panhandle, western Montana, and into northeastern Oregon. A species of conifer forests, it can usually be found on the underside of logs, under bark and other woody debris.

Family: DISCIDAE
Genus: *Anguispira*

Anguispira kochi   (Pfeiffer, 1821)   Banded Tigersnail

The typical *Anguispira kochi* is a species of eastern North America, represented in the west by the subspecies *A. kochi occidentalis,* which does not appear to be readily distinguishable from the typical species; the distinction is retained because, as Pilsbry (1948) pointed out, they are "separated by more than a thousand miles where no related form occurs." A second subspecies, *A. kochi eyerdami,* occurs at the western extreme of the species range.

The shells are globose, brown, solid, and opaque as adults, or a honey-yellow in younger shells. There is a light, supraperipheral band bordered by darker, reddish-brown or sometimes brown bands above and below. The darker color is most intense immediately adjacent to the light band and may be as little as 1 mm wide, or it may extend as a lighter hue over most of the shell.

(A) *Anguispira kochi occidentalis*: Kootenai Co., ID; (B) *Anguispira kochi eyerdami*: The Dalles, Wasco Co., OR. Note that *A. k. eyerdami* is smaller than *A. k. occidentalis* and its bands are less distinct.

Key to the Subspecies of *Anguispira kochi*

(a)   Shell with 6 to 6½ whorls, 19 to 29 mm wide by 14 to 22 mm high. Peripheral bands of narrow, dark brown, or reddish-brown, with a lighter band between. Dorsal and basal colors distinctly or slightly lighter, so the dark bands are apparent. Blue Mountains of Washington and Oregon, Idaho Panhandle, and northwestern Montana . . . .*Anguispira kochi occidentalis*

    (aa)   Shell smaller, 17 to 21 mm wide by 11 to 16 mm high in 5½ to 6 whorls. Peripheral bands separated by a lighter band between but blend into the dorsal and ventral coloration. Yakima Co., Washington, to The Dalles, Wasco Co., Oregon . . . . . *Anguispira kochi eyerdami*

Anguispira kochi occidentalis   (Von Martens, 1882)

Description: This fairly large snail with about 6 to 6½ whorls measures 19.5 to 29 mm wide by 14.5 to 23 mm high. The umbilicus is deep, narrowly funnelform, expanding very slowly until the last three-fourths whorl where it opens a little more rapidly, to about

Anguispira kochi occidentalis: Kootenai County, ID; scale bar = 1 cm

one-fifth the shell diameter. The shell is varying shades of brown to yellowish-brown, the younger shells sometimes lighter, and it has two darker brown or reddish-brown lateral bands, a peripheral and a supraperipheral. A third, lighter band can be seen on the dorsal surface of the whorl in some specimens. The spire is rather high, depressed globose, the body whorl large, and the shell width increasing regularly. The periphery is round; the aperture is roundly lunate. The protoconch, of two whorls, is smooth for the first 1 to 1½ whorls, then is sculpted with rather coarse ribs for the last half-whorl or more. Sculpturing of the teleoconch is of coarse growth-wrinkles and spiral striae.

Similar Species: Shells of some *Oreohelix* species can be difficult to separate from those of *Anguispira kochi*, especially when weathered. However, the distinct rib-like sculpture of the protoconch of *A. kochi* is helpful. Also, with the living animals, *Oreohelix*, being of the Holopoda group, has sides primarily vertical, blending into the angle of the foot without apparent pedal furrows. *Anguispira*, being of the Aulacopoda division, has an apparent pedal furrow, and the lower sides flare outward.

Distribution: *A. kochi occidentalis* is fairly widespread, occurring from southeastern British Columbia and Lincoln Co., Montana, through the north Idaho Panhandle, and scattered in southeastern Washington and into northeastern Oregon. It is a species of moist conifer and mixed forests, and may be found in brushy or hardwood riparian zones along rivers even in non-forested areas.

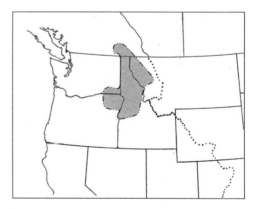

Anguispira kochi eyerdami   Clench & Banks, 1939

Description: This subspecies is smaller than typical *A. kochi*, and it lacks the distinct darker lateral bands. With 5½ to 6 whorls it measures 17 to 21 mm wide by 11 to 16 mm high. The umbilicus is sometimes wider than that of *A. kochi occidentalis*. The shells are medium to dark brown. The dark lateral bands are faint or lacking altogether, blending into the shell color, but there is a lighter yellowish band in the area between where the peripheral and supraperipheral bands of *occidentalis* would be. Aperture, umbilicus, and shell sculpturing are as for *A. kochi occidentalis*.

Anguispira kochi eyerdami: The Dalles, Wasco Co., OR; scale bar = 1 cm

Similar Species: Pilsbry (1948) questioned the validity of this subspecies, in that it "appears to have no special racial characters . . . not occurring in other lots scattered over the area of *occidentalis*." However, a rather large population from just east of The Dalles, Oregon, is quite consistent, with no typical *A. kochi occidentalis* among them, lending credibility to the concept of a distinct subspecies.

Distribution: This subspecies occurs in Yakima County, Washington, south through the Horse Haven Hills and across the Columbia River into Oregon. Photographs are of specimens from near The Dalles, Wasco Co., Oregon. It is found among rocks and brush around springs and small streams and ponds, not necessarily in forested areas.

Anguispira nimapuna   H. B. Baker, 1932   Nimapu Tigersnail

Anguispira nimapuna: Idaho Co., ID; scale bar = 1 cm

Description: This is a small to medium-sized snail, measuring about 12 to 15 mm wide by 4.5 to 7 mm high with 5¾ to 6 whorls. The umbilicus is large, to nearly one-third the shell diameter. The shell is olive-brown to straw-colored; young may have a greenish tint. The spire is low to moderately elevated and rounded, the whorls increasing rather slowly but regularly. The periphery is angular. The aperture is ovate with an angular outer margin. The shell is sculptured with distinct, solid, collabral ribs, with very fine, low secondary ribs and spiral striae (the fine ribs not always apparent).

Similar Species: The *Discus* are mostly smaller with finer ribs. Ribbed oreohelices of similar width (e.g., *Oreohelix idahoensis, O. waltoni*) are more tightly coiled with narrower umbilici, higher spires, and heavier ribs.

Distribution: *Anguispira nimapuna* occurs in the watershed of the South Fork of the Clearwater River and upstream into the Lochsa and Selway drainages, Idaho County, Idaho, and a shell was collected in Wallowa Co., Oregon. There is a disjunct population of this or a similar species, but with weaker ribs, found rarely around Lake Chelan in Chelan Co., Washington. The species is found in forests, often among hardwoods, or in talus or rocky sites.

Genus: *Discus*

Discus in the Pacific Northwest are small heliciform snails, measuring less than 11 mm in width. They have low to moderately elevated spires, large open umbilici, and solid ribs.

Key to the Species of *Discus*

(a)   Shell 5 to 7 mm wide by less than 4 mm high . . . .[b/bb]

    (aa)   Shell 8 or more mm wide by greater than 4 mm high. Periphery angled. Whorls increase slowly. Umbilicus one-third the shell width. . . . . [d/dd]

(b)   Shell of 4½ whorls or less, periphery round . . . . [c/cc]

    (bb)   With 5½ to 6 whorls the shell measures 6 to 7 mm wide by 2.7 to 3 mm high. Periphery somewhat angular. Spire low, rounded, yellowish-brown, with reddish radial stripes or flammules. . . . .*Discus rotundatus*

(c)   With low, rounded spire, umbilicus about one-third the shell width. Well-spaced, solid ribs encircle the whorls . . . . *Discus whitneyi*

    (cc)   With low to moderately elevated, roundly-conic spire. Low, closely spaced ribs weakening on the last whorl and fading below the periphery. Umbilicus about one-quarter the shell width . . . . *Discus shimekii*

(d)   Shell 8 mm wide with 6½ whorls. Spire a broad, high dome. Ribs fade out on the base. Idaho Co., Idaho . . . . *Discus marmorensis*

    (dd)   With 5½ whorls the shell is about 10.5 mm wide by 4.2 mm high. Spire moderately high, roundly conic. Fine distinct ribs weaken on the latter whorls and fade out basally near the end of the last whorl. Lake Co., Montana . . . . *Discus brunsoni*

    Discus brunsoni   S. S. Berry, 1955   Lake Disc

Description: Our largest *Discus*; with about 5½ whorls the shell measures approximately 10.5 mm wide by 4.2 mm high. The large, open umbilicus is about one-third the shell width. The shell is olive brown. The spire is moderately high, convexly-conic (somewhat dome-shaped, similar to *D. marmorensis*), and with an obtuse point to the apex. From the

Discus brunsoni: Montana; scale bar = 1 cm (MNHP collection)

apical view the whorls appear to increase slowly. The periphery is strongly angled, and the whorls are somewhat flattened basally. The aperture is horizontally ovate, the outer lip angled, thin and simple. The protoconch of 2 whorls has a few radial wrinkles, then the teleoconch is sculptured with fine but distinct ribs, which weaken on the outer whorls, nearly disappearing on the base of the body whorl.

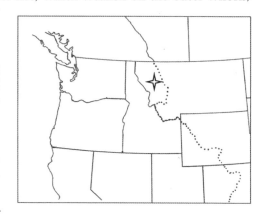

Similar Species: Among the *Discus*, this species most resembles *D. marmorensis*, differing by its larger size, more conic spire, and lack of flammules. *Anguispira nimapuna*, which also has a ribbed shell with an angular periphery, is larger with stronger ribs.

Distribution: *Discus brunsoni* was found in limestone talus at about 3500 ft elevation in the McDonald Lake basin, Lake County, Montana (Berry, 1955).

Discus marmorensis H. B. Baker, 1932   Marbled Disc

Description: This small snail is rather large and unique-appearing for a *Discus*. With 6½ whorls it measures about 8 mm wide by 4.5 mm high. Its umbilicus is large, greater than one-third the shell width. The shell varies from buff to brown with faint chestnut-colored flammules. The spire is domed. The whorls widen very slowly and are flattened basally. The periphery is angular below mid-whorl and may be somewhat carinate. The aperture is a little oblique and lenticular in shape. The shell is sculptured with irregular spiral striations and low, distinct ribs, which fade out on the base.

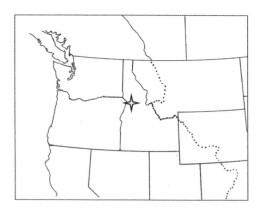

Similar Species: *D. brunsoni* is similar but larger; it has a more conic spire and lacks flammules. The ranges of these two species are well separated.

Distribution: *Discus marmorensis* is found in John Day Creek Canyon and a few other locations on the east side of tributaries to the Lower Salmon River, Idaho Co., Idaho.

Discus rotundatus   (Müller, 1774)   Rotund Disc

Description: This snail measures 6 to 7 mm wide by 2.7 to 3 mm high with 5½ to 7 whorls. Its spire is low and rounded. The umbilicus is wide and deep. The whorls increase slowly but regularly; the periphery is somewhat angled. The aperture is ovately-lunate, wider than high. The shell is yellowish-brown with reddish flammules or radial streaks. Fine, rather closely-spaced ribs encircle the whorls and are nearly as strong basally as dorsally.

Similar Species: Most similar to *Discus whitneyi*, this species differs by its horizontally ovate aperture, angled periphery, and flammulate markings.

Distribution: *Discus rotundatus* is introduced from Europe and known from the eastern United States and Canada. Forsyth (2004) cites records from Vancouver Island, British Columbia; Bellingham, Washington; and California. An introduced species, it is most likely to be found around human habitation in yards and gardens.

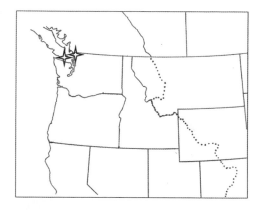

Discus shimekii   (Pilsbry, 1890)   Striate Disc

Discus shimekii: Big Cr., Park Co., MT; scale bar = 1 mm (MNHP collection)

Description: This small snail measures 6 to 6½ mm wide by 3.7 to 4 mm high with about 4½ whorls. The umbilicus is open, about one-quarter the shell width or larger. The spire is moderately elevated and somewhat roundly conic. The whorls increase regularly; the periphery is round, the aperture roundly lunate. The shell is sculptured with low rather closely spaced ribs, weakening on the last whorl and fading out below the periphery, so the base is unribbed.

   Discus shimekii cockerelli Pilsbry, 1948, is a subspecies from the Rocky Mountains. It has a lower spire, a slightly wider umbilicus, and weaker ribbing than typical *D. shimekii*. With 4½ whorls the shell measures 6.9 mm wide by 3.2 high.

Similar Species: *Discus whitneyi* has a lower spire than typical *D. shimekii*, but it is more similar to *D. shimekii cockerelli*. *D. whitneyi* differs in having a wider umbilicus, and it is ribbed basally as well as dorsally.

Distribution: *D. shimekii* is found sporadically from the Yukon watershed, British Columbia, Alberta, and Ontario, Canada. It is widely scattered in western Montana, south through the Rocky Mountain states to Arizona, and west to Klamath Lake, Oregon, and California. Hendricks, Maxell, and Lenard (2006) found less than ten sites known in Montana, only two west of the Continental Divide. It is a species of spruce/fir forests, and occurs in litter, woody debris, and rocks under stands of quaking aspen, cottonwoods, and other hardwoods.

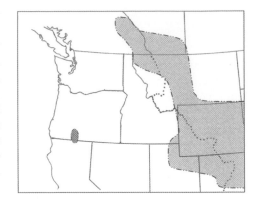

Discus whitneyi   (Newcomb, 1864)   Forest Disc
Synonym: *Discus cronkhitei* (Newcomb, 1865)

Discus whitneyi: Cottonwood Cr., Ferry Co., WA; scale bar = 1 cm

Description: With 3⅔ to 4½ whorls *D. whitneyi* measures 5 to 6.7 mm wide by 2.7 to 3.6 mm high. It has a large funnelform umbilicus, about one-third the shell diameter. The shell is brown; it has a moderately low rounded spire. The whorls increase regularly in size; the periphery is round, and the aperture is obliquely, roundly lunate. Sculpturing is of well-spaced ribs. The animal is translucent and light grayish, with blackish head, neck, and tentacles.

Similar Species: *D. shimekii* differs from *D. whitneyi* primarily by its smooth basal whorls, the ribs not extending onto the base of the shell. Other *Discus* in our area are larger, with angled peripheries. *Radiodiscus abietum* is ribbed and slightly larger, but its light-colored shell has a nearly flat spire and a small umbilicus, among other differences.

Distribution: *Discus whitneyi* can be found in forest habitats in and under woody debris, across Canada and throughout most of the United States. In the Pacific Northwest it occurs in the mountainous areas from the Cascades east, but it is occasionally found west of the Cascade Crest. It occurs on Lopez Island, in San Juan County, Washington; in prairie habitat against rock outcroppings above the sea; and on steep, rocky, forested slopes.

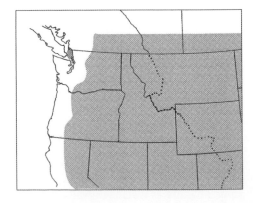

Family: HELICODISCIDAE
Genus: *Helicodiscus*

Helicodiscus salmonaceus Hemphill, 1890   Salmon Coil

1

1 & 2. *Helocodiscus salmonaceus*: Spokane Co., WA; width 3.8 mm

2

Description: This is a very small, unique, discoidal snail. With 5 to 5½ whorls it measures 4.3 to 5 mm wide by 1.3 to 1.8 mm high. The umbilicus is wide, dished, measuring about half the width of the shell. The shell is translucent, whitish or with a horn-colored tint. The spire is flat and discoidal. Individual whorls are of small diameter and increase very slowly, so even though they grow to the outer edge of the previous whorl, the width of the last whorl is little greater than that of the penultimate. The periphery is rounded, the aperture small, narrowly lunate, and only slightly oblique. There are paired denticles on the outer and basal walls inside the whorl. These teeth are often farther back inside the shell, or there may even be more than one pair spaced back around the inside of the whorl. Shell sculpture is of parallel spiral threads (lirae).

Similar Species: No other species native to the Pacific Northwest should reasonably be confused with *H. salmonaceus*.

Distribution: *H. salmonaceus* is found in southeastern Washington, from Spokane south through Asotin County and into Joseph Canyon in Wallowa County, Oregon. It also occurs in the Idaho Panhandle, from Cataldo, Kootenai County, south to Idaho County. It is found in talus and other rocky habitats.

Selected references for the families Punctidae, Charopidae, Discidae, and Helicodiscidae: Berry 1955; Brunson 1956; Forsyth 2004; Frest and Johannes 1995, 1996; Frest and Johannes 2000; Henderson 1924; Henderson 1929a; Hendricks 2003; Hendricks, Maxell, and Lenard 2006; Hendricks et al. 2007; Pilsbry 1948; Roth 1987.

Superfamily: GASTRODONTOIDEA

The Pacific Northwest families included in this superfamily are four of the taxa that were previously combined under the family Zonitidae. In addition the Vitrinidae, now included under the superfamily Limacoidea, was also previously considered one of the genera of Zonitidae.

Directory to the Genera of Superfamily GASTRODONTOIDEA

Family	Subfamily	Genus	Page
EUCONULIDAE	EUCONULINAE	*Euconulus*	242
GASTRODONTIDAE	GASTRODONTINAE	*Striatura*	244
		Zonitoides	245
OXYCHILIDAE	OXYCHILINAE	*Oxychilus*	247
	GODWINIINAE	*Nesovitrea*	250
PRISTILOMATIDAE	VITREINAE	*Hawaiia*	253
		Pristiloma	255
		Ogaridiscus	264
		Vitrea	265

Key to the Genera of GASTRODONTOIDEA

Families EUCONULIDAE, GASTRODONTIDAE, OXYCHILIDAE, and PRISTILOMATIDAE
When a taxon is identified using these keys, proceed to the family section in which it is found (see table above).

(a)   Diameter 1.5 to 1.8 mm in 3 to 3½ whorls. White with greenish tint. Sculpture of low, closely spaced ribs crossed by spiral striae, giving it a ropy or beaded appearance. Aperture broadly ovate; umbilicus broad . . . . *Striatura pugetensis*

  (aa)   Shells larger, mostly greater than 2 mm wide, transparent or translucent and glossy . . . .[b/bb]

(b)   Diameter of adults greater than 7 (8 to 17) mm. Whorls increasing moderately in size. Shell transparent or translucent, often waxy-white basally around the umbilicus. Umbilicus small to medium (about one-sixth the diameter or less). Introduced in North America . . . . Genus: *Oxychilus* (in part)

  (bb)   Generally smaller, diameter of adults 2 to 7 (8) mm. Shells generally uniformly colored, transparent and glossy, although they may have microscopic striations. Umbilicate to imperforate . . . . [c/cc]

(c)   Mostly minute snails, diameter less than 4 mm. Umbilicate to imperforate. If wider than 3.6 mm, the shell imperforate . . . . [d/dd]

  (cc)   Diameter of adult 3.5 mm or larger; shell umbilicate . . . . [f/ff]

(d)   Umbilicus broad, about one-third the shell diameter, which is 2 to 2.5 mm in about 4 whorls. Gray to white; sculpture merely of growth-wrinkles. Aperture quite round or only slightly oblique . . . . *Hawaiia minuscula*

(dd)   Shell imperforate, perforate, narrowly umbilicate or rimate. Mostly
       minute, but some to 4 mm wide. Shell thin and fragile . . . . [e/ee]

(e)   Shell domed, 3 to 3.4 mm wide by 2.4 to 3.5 mm high, nearly as high as wide.
      Brown, translucent shell, perforate to imperforate, with fine, close, radial striae.
      . . . . *Euconulus fulvus*

(ee)   Shell generally with very low conic spire. Regularly spaced, transverse,
       indented lines (sulci) are key characteristics on two species, but fainter
       versions may appear on dorsal shell surfaces of some other species as
       well and can be indicators of this group . . . . [h/hh]

(f)   Whorls narrow, increasing in size regularly, last not greatly wider than
      penultimate. Diameter 5 to 7 (8) mm in 4½ to 4¾ whorls. Umbilicus
      narrow to moderate, one-fifth or more of shell diameter. Aperture
      roundly to ovately lunate . . . . Genus: *Zonitoides*

(ff)   Whorls increasing more rapidly, last noticeably wider than
       penultimate. Aperture broadly, ovately-lunate. . . . . [g/gg]

(g)   Whorls increasing in size regularly to last, which is noticeably wider than
      the penultimate whorl. Diameter 3.5 to 5.5 mm (H/D= 0.55). Microscopic
      indented radial lines may be seen on dorsal shell surface. Umbilicus narrow
      to moderate . . . . Genus: *Nesovitrea*

(gg)   Whorls increase more regularly to the end. Diameter 5.5 to 7 mm in 4
       to 4½ whorls (H/D= 0.4). Whorls somewhat flattened basally. Animal
       dark colored, with a strong garlic odor . . . . *Oxychilus alliarius*

(h)   Shell imperforate. . . . . [i/ii]

(hh)   Shell narrowly umbilicate, perforate or rimate. . . . . [j/jj]

(i)   Spire conic, low to moderately elevated. Shell tightly coiled, the last whorl
      little wider than the penultimate. Aperture narrowly lunate; lip simple.
      Six species ranging from 2 to 3.8 (4) mm wide with 4¾ to 7 whorls
      (imperforate tightcoils) . . . . Genus: *Pristiloma* (in part)

(ii)   Spire low (nearly flat); shell waxy-white. Whorls increase rather rapidly,
       2 to 2.5 mm wide in 3½ to 4 whorls. Aperture broadly lunate. Imperforate
       but may have a rather deep umbilical depression. Western Washington,
       Oregon, and southwestern British Columbia . . . .*Pristiloma johnsoni*

(j)   Diameter 2 to 3.6 mm in 4½ to 5¼ whorls. Height/diameter ratio (H/D) greater
      than 0.5. Spire moderately elevated; perforate to very narrowly umbilicate.
      Aperture narrowly lunate. Two species. Washington, Oregon, Northern
      California, and Western Montana (perforate tightcoils) . . . Genus: *Pristiloma* (in part)

(jj)   Diameter 2 to 3.5 mm. Height/diameter ratio less than or about equal
       to 0.5. Spire very low conic. . . . . [k/kk]

(k)   Diameter 2 to 2.5 mm in about 4 to 4½ whorls. Height less than half
      the diameter. Umbilicus about one-sixth the diameter. Color clear to
      cloudy white . . . . *Vitrea contracta*

(kk)   Diameter 2.7 to 3.5 mm in about 4 to 4½ whorls. Height about half the
       diameter. Shell transparent, clear, perforate or rimately imperforate by
       an expansion of the columellar lip margin. Spire strongly depressed,
       low conic. Umatilla County, Oregon, to Great Salt Lake, Utah, and
       east-central California. . . . . *Ogaridiscus subrupicola*

Family: EUCONULIDAE
Genus: *Euconulus*

Euconulus fulvus fulvus   (Müller, 1774)   Brown Hive

1. *Euconulus fulvus*: Lewis Co., WA; scale bar = 1 mm
2. *Euconulus fulvus*: Patos Island, San Juan Co., WA; scale bar = 1 mm

Description: This is a small snail with a dome-shaped shell. Adults are about 3 to 3.4 mm diameter by 2.4 to 3.5 mm high in about 5½ to 6 whorls. The shell is translucent, brown, and nearly as high as wide. The width increases very slowly, the shell growing downward as much as or more than outward. The umbilicus is very narrowly perforate or imperforate. The aperture is wider than high and oblique, extending from the umbilical depression to just below the periphery of the penultimate whorl. The lip is simple, but expands at the columellar margin. Sculpturing is of fine radial striae, giving the appearance of a fine thread wrapped side-by-side around the whorls, distinct and regular enough to be useful in distinguishing juveniles of this species from similar-appearing young of others. The animal is grayish, with darker gray tentacles and black blotches on the mantle that show through the shell.

Similar Species: The distinct dome-shaped, nearly imperforate shell of *Euconulus* distinguishes it from most other genera. Very small juveniles might be confused with immature pupillids or other juveniles, but the shell sculpturing of *Euconulus* can be used to distinguish this genus. Mature *Pristiloma chersinella* have a rather high, slightly rounded spire, and a very small umbilicus, but the spire of *Euconulus fulvus* is higher and more dome-shaped, and the body whorl is slightly angled at the periphery and blends smoothly into the dome. The spire of *Pristiloma chersinella* is lower, more conic, and the body whorl has a round periphery and is more dominant appearing.

Distribution: *Euconulus fulvus* occurs in most of the Holarctic region and is generally found throughout forested areas of the Pacific Northwest. It is most abundant in riparian areas, moist forests, and brushy sites.

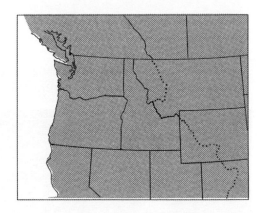

Euconulus fulvus alaskensis   (Pilsbry, 1899)

Description: This subspecies is similar to *E. fulvus fulvus*, but its shell is generally not as high and expands more rapidly, shells of the same diameter having one less whorl. Diameter about 3.25 mm, height 2.6 mm in 4½ whorls. The animal is grayish, with darker gray tentacles and black blotches showing through the shell from the mantle.

Similar Species: Generally, *Euconulus fulvus fulvus* has a more elevated spire and an additional whorl for the same diameter. Some *Pristiloma chersinella* may have a high enough spire to be confused with *Euconulus*, but it has a more distinct umbilicus, a more prominent body whorl attached to the penultimate whorl above the periphery, and lacks the radial thread-like sculpturing.

Distribution: *E. fulvus alaskensis* has been reported from Alaska, British Columbia, and the Rocky Mountain and Pacific states, south to near the Mexican border, but at high elevations in the southern part of its range. As with the typical subspecies, *E. f. alaskensis* is most abundant in riparian areas, moist forests, and open brushy sites within forests.

Family: **GASTRODONTIDAE**
Genus: *Striatura*

Striatura pugetensis (Dall, 1895)   Northwest Striate
Synonym: *Radiodiscus hubrichti* (Branson, 1975)

1. *Striatura pugetensis*: Lewis Co., WA; scale bar = 1 mm
2. *Striatura pugetensis*: Snoqualmie Pass, King Co., WA; scale bar = 1 mm

Description: *Striatura* is a minute snail, measuring 1.5 to 1.85 mm wide by 0.5 to 0.8 mm high in about 3 whorls. It has a whitish shell with a greenish tint, a low spire, and a wide, shallow umbilicus about one-third the width of the shell. The aperture is obliquely ovate. It can be recognized by its shell sculpturing, which is beaded by spiral striae cutting across closely spaced ribs. The protoconch also has spiral threads and striae prior to the ribs beginning.

Similar Species: No other Pacific Northwest species of similar size closely resembles *S. pugetensis*.

Distribution: Found from British Columbia, Canada, to Baja California, Mexico, *Striatura pugetensis* is common in forest floor litter in conifer forests of western Washington and Oregon. It has also been found east of the Cascades in British Columbia, Canada; in northern Idaho; in Glacier National Park, Montana; and Umatilla Co., Oregon; there are also records from the island of Kauai, Hawai'i.

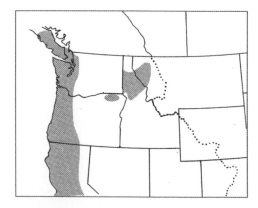

Genus: *Zonitoides*

Zonitoides are small snails, with shells 4 to 8 mm wide in 4½ to 4¾ whorls. The shells are translucent, glossy, and with low spires of regularly increasing whorls. They are umbilicate with simple apertural lips. *Zonitoides* are most easily confused with species of *Oxychilus* and *Nesovitrea* of the same family.

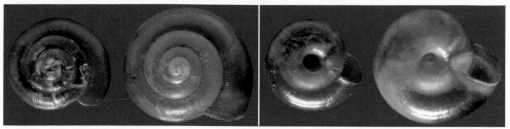

(L) *Z. arboreus*: Crum Canyon, Chelan Co., WA; (R) *Z. nitidus*: Sinlahekin Cr., Okanogan Co., WA

Key to the Species of *Zonitoides*

(a) Shell rather low with a low rounded spire. With 4½ whorls it measures 5 to 6 mm wide by 3 mm high. Body whorl flattened basally so aperture is ovally wider than high. Umbilicus narrow and symmetrical to the last whorl, then expands abruptly to about one-fifth the shell diameter . . . . *Zonitoides arboreus*

   (aa) Shell larger with a more conic spire, only slightly rounded. With 4½ whorls it measurers 6 to 7 (8) mm wide by 3.5 to 4 mm high. Body whorl well-rounded; the aperture roundly lunate. Umbilicus expanding rather uniformly to about one-fifth the shell diameter at the aperture. . . . . *Zonitoides nitidus*

Zonitoides arboreus (Say, 1816) Quick Gloss

Zonitoides arboreus: Crum Canyon, Chelan Co., WA; scale bar = 1 mm

Description: This snail is smaller than *Z. nitidus*, at 5 to 6 mm wide by as much as 3 mm high in 4½ to 4¾ whorls. The shell is translucent olive-buff, with a low, rounded spire of gradually increasing whorls. The umbilicus is narrow and symmetrical to the last ½ to 1 whorl, where it abruptly expands to about one-fifth or more of the shell diameter at the aperture. The body whorl is somewhat flattened basally, so the aperture is horizontally oval. The shell surface shows weak growth-wrinkles and may have very fine spiral striae. The animal is dark blue-gray dorsally in front of the shell, fading to whitish on the lower sides, the sole, and behind the shell.

Similar Species: *Z. nitidus* and *Oxychilus alliarius* are larger than *Z. arboreus*. *Z. nitidus* has a higher, more conic spire and a more roundly lunate aperture. The whorls of *Oxychilus alliarius*

are more flattened basally, giving it a more oval aperture. Looking at the apex, the first whorl of *O. alliarius* begins much narrower and hooks sharply inward to an oblique rather acute point, while those of the *Zonitoides* taper regularly to a mostly symmetrically rounded point. The sides and tail of *Z. arboreus* are white, while the other two animals are all black.

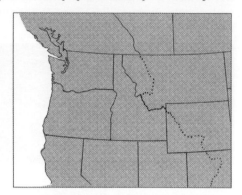

Distribution: *Z. arboreus* is native to North America and occurs throughout most of the continent and the West Indies. It has been introduced onto every other continent, the Hawaiian Islands, and Japan. It is common in forests of eastern Washington but occurs only sporadically west of the Cascades in that state. It is found in forested sites and wooded riparian zones, usually in and under logs and other woody debris.

Zonitoides nitidus   (Müller, 1774)   Black Gloss

Zonitoides nitidus: Sinlahekin Creek, Okanogan Co., WA; scale bar = 2 mm

Description: Another small snail, but larger than *Z. arboreus*, *Z. nitidus* has a translucent, olive-yellow or yellowish-brown shell, normally 6.1 to 7.0 (8.0) mm wide in 4½ to 4¾ whorls. The shell is 3.5 to 4 mm high, and the funnelform umbilicus expands rather uniformly from its origin to about one-fifth the shell diameter at the aperture. The spire is nearly conic, being only slightly rounded below the apex. The body whorl is well-rounded at the periphery. The aperture is roundly lunate, slightly wider than high. Shell sculpturing is lacking except for fine growth-wrinkles. The animal is wholly black.

Similar Species: *Oxychilus alliarius* is most similar in size and shape; it differs in having a lower spire and being less rounded basally, so the aperture is more oval. *Zonitoides arboreus* is smaller than *Z. nitidus*, with about the same number of whorls, and its spire is also lower. Its umbilicus is narrow and symmetrical to the last ½ to 1 whorl, where it expands rapidly to the aperture, and its body is pale ventrally, laterally and posteriorly. See also this discussion under *Z. arboreus* above.

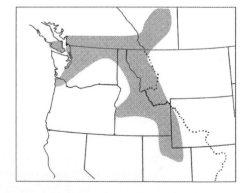

Distribution: *Z. nitidus* is nearly holarctic in distribution. In North America it occurs in southern Alaska; across the southern Canadian provinces, British Columbia to Quebec; and the

northern United States. It is scattered throughout the Pacific Northwest, and in the Pacific states it has been found as far south as San Diego, California. It is a forest species, but usually occurs near water, and can be found in and under woody debris in riparian areas and around wetlands.

Family: OXYCHILIDAE
Subfamily: OXYCHILINAE
Genus: *Oxychilus*

Oxychilus are exotic snails from Europe, the Near East, and North Africa. They have been introduced into North America and are often common around areas of human habitation. At least four species were reported by Pilsbry (1946) to occur within the United States. Hanna (1966) was uncertain of the validity of the record for *O. helveticus*. Kerney, Cameron, and Riley (1994) listed nine species of *Oxychilus* for Britain and northwest Europe. Records show three or four species introduced into the United States, all reported prior to 1946, although interceptions of three species on imported commodities continue (Robinson 1999). Therefore, the continued introduction of additional species should be expected.

Key to the Species of *Oxychilus*

(a)   With 4 to 4½ whorls, 5.5 to 7 mm wide. Spire very low, conic. Animal appears black. Strong garlic odor when disturbed. Umbilicus a little larger than one-sixth the shell width . . . . *O. alliarius*

    (aa)   With 5 or more whorls. Width greater than 7 mm . . . . [b/bb]

(b)   Width 8 to 10 mm with 5 whorls. Spire low convexly-conic with slightly angled periphery, and a black band behind the aperture in live animals. Umbilicus a little smaller than one-sixth the shell width . . . .*O. helveticus*

    (bb)   Width 9 mm or greater in 5 to 6 whorls. The umbilicus is about one-sixth the shell width . . . . [c/cc]

(c)   Width 9 to 12 mm, rarely to 14 mm. Spire very low-conic; shell glossy, transparent, whitish with a faint amber or yellow tint . . . . *O. cellarius*

    (cc)   Width 11 to 17 mm. Spire very low, convex; shell brown to yellowish-brown, rather opaque, not very glossy. . . . . *O. draparnaudi*

Distinguishing Characteristics of *Oxychilus* spp.

Species	Shell Width	Spire	Umbilical Width	Notes
O. alliarius	5.5 to 7 mm	very low-conic	wider	strong garlic odor
O. cellarius	9 to 12 mm	very low-conic	W/U=.6	lacks garlic odor
O. draparnaudi	11 to 17 mm	very low-convex	W/U=.6	largest *Oxychilus*
O. helveticus	8 to 10 mm	very low-conic	narrower	black apertural band

Oxychilus alliarius   (J. S. Miller, 1822)   Garlic Glass-snail

Oxychilus alliarius: Near Culdesac, Nez Perce Co., ID; scale bar = 1 mm

Description: A small snail, 5.5 to 7 mm wide by about 2.5 mm high in 4 to 4½ whorls. The shell is yellowish-brown to greenish, often with a white waxy sheen around the umbilicus. The spire is very low-conic. The whorls increase gradually but regularly. The aperture is roundly oval. The umbilicus is open, relatively broad for this genus, and increases rapidly in the last whorl. The animal is dark blue-gray, appearing black, and gives off a strong garlic odor when disturbed, hence its specific epithet referring to *Allium*, the genus of garlic.

Similar Species: For other *Oxychilus*, compare diameter per number of whorls, size of umbilicus, and body color. *Zonitoides nitidus* is difficult to distinguish from small *Oxychilus* specimens; it has a higher, more conic spire and rounder whorls than *O. alliarius*. The whorls of *O. alliarius* are somewhat flattened basally, giving it a more oval aperture. The characteristic garlic odor of *O. alliarius* is absent from these other species.

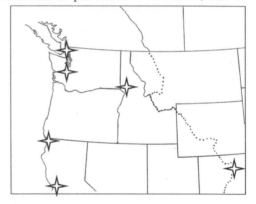

Distribution: Introduced into America from Europe, *O. alliarius* is usually found in suburban areas, around human habitation or where past human activity has occurred.

Oxychilus cellarius   (Müller, 1774)   Cellar Glass-snail

Description: This species is one of the larger *Oxychilus*, normally 9 to 12 mm wide (occasionally up to 14 mm) by 4.2 mm high in shells with 5 to 6 whorls. The umbilicus is symmetrical and is about one-sixth the width of the shell. The shell is translucent whitish, with a faint amber or yellowish tint. It has a very low conic spire and gradually but regularly increasing whorls with a rounded periphery. The aperture is roundly oval; the sutures are shallow. Sculpturing is indistinct or very fine spiral striae. The animal is usually pale blue-gray but sometimes darker. The sole is light-colored.

Similar Species: *O. cellarius* is most similar to *O. draparnaudi*, but it is smaller, its last whorl is relatively narrower, and it is without the gray mantle. It is larger than *O. alliarius* and lacks its strong garlic odor. It lacks the dark vertical stripes on the sides and the black apertural line of *O. helveticus*.

Distribution: Introduced into America from western Europe, *O. cellarius* is usually found in areas of human habitation, in gardens, greenhouses, cellars, and among debris. Mapped sites for this and other *Oxychilus* are locations for which I have records of their occurrence, but any of these species can be expected near any human habitation (with the exception of *Oxychilus helveticus*).

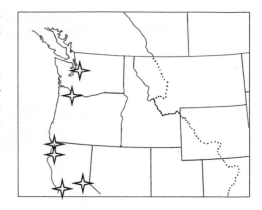

Oxychilus draparnaudi (Beck, 1837)   Dark-bodied Glass-snail
Synonym: *O. draparnaldi*

Oxychilus draparnaudi: Port Townsend, Jefferson County, WA; scale bar = 1 mm

Description: This is the largest *Oxychilus*, normally 11 to 16.5 mm wide by 6.5 to 8 mm high in 5½ to 6 whorls. The umbilicus is open but narrow, about one-sixth the width of the shell, which is only slightly glossy, rather opaque, brown dorsally and lighter, waxy-white basally. The spire is very low and convex, and the whorls increase slowly until the last one, which is much wider than the penultimate whorl. The aperture is widely oval and deeply lunate. The shell lacks sculpturing except for distinct growth-wrinkles. The animal is dark blue with a gray mantle.

Similar Species: Its larger size, brown shell, and more prominent white basal area distinguish *O. draparnaudi* from other *Oxychilus* and other snails of similar appearance. Its last whorl is relatively wider than that of *O. cellarius*, its nearest appearing relative.

Distribution: Native to Europe, the Middle East and North Africa, *O. draparnaudi* has been introduced throughout much of North America. It is usually found in areas of human habitation, in gardens, greenhouses, and under debris.

Oxychilus helveticus   (Blum, 1881)   Swiss Glass-snail

Description: Larger than *O. alliarius*, *Oxychilus helveticus* has approximately 5 whorls, with a diameter of 8 to 10 mm and a height of about 4.5 mm or more. Its umbilicus is narrow, symmetrical, and deep. The shell is brownish-yellow. The spire is low, slightly convex; the whorls increase regularly. The periphery is slightly roundly angled, and the aperture is roundly oval and deeply lunate. The animal is blue-gray with darker (zebra-like) stripes on the sides, and there is often a black collabral band at or near the apertural lip.

Similar Species: For other *Oxychilus* and *Zonitoides*, compare the diameter per number of whorls, size of umbilicus, and body color. Compared to other *Oxychilus*, the black band behind the aperture and the more elevated conic spire indicate *O. helveticus*. Its basal white area is confined to around the umbilicus. Its shell is more flattened basally, and its umbilicus is narrower. When disturbed, it may give off a lesser garlic odor than *O. alliarius*.

Distribution: Native to western Europe, the only record of *O. helveticus* from the western United States was a report by Hemphill of specimens in Oakland, California, cited by Pilsbry (1946). Hanna (1966) failed to find these specimens among Hemphill's collections or catalogs, so was uncertain of the validity of that report. Robinson (1999) listed the species as introduced in North America, but showed no interceptions by USDA APHIS Plant Protection and Quarantine inspectors on commodities entering the United States in 1993 through 1998. Thus, it is uncertain whether or not this species has been introduced into the western United States.

Subfamily: GODWINIINAE
Genus: *Nesovitrea*

The genus is represented in the Pacific Northwest by two small species, which, unmagnified, are similar in appearance to other snails of similar size and shape. The *Nesovitrea* are readily separable from the *Zonitoides* and *Oxychilus* by body whorls that are distinct in being much wider than the penultimate whorls and by radial grooves across the dorsal whorls that are distinct from the growth-wrinkles.

Key to the Species of *Nesovitrea*

(a)   Shell width 3.5 to 4.0 mm with 4 to 4½ whorls. Shell is translucent, white
       with green tint and with radial grooves and very fine spiral striae.
       . . . . *Nesovitrea binneyana occidentalis*

   (aa)   Shell width 4.5 to 5.2 mm with 3¾ to 4¼ whorls. Shell is clear,
           transparent with green or yellow tint and with radial grooves but
           lacks spiral striae. Animal is dark grayish-black dorsally, dark gray
           on the lower sides and foot. . . . . *Nesovitrea electrina*

Nesovitrea binneyana   (E. S. Morse, 1864)   Blue Glass
Synonym: *Retinella binneyana binneyana*

The typical variety of this species occurs in the northeastern and midwestern United States and adjacent Canada. In the Pacific Northwest it is represented by the subspecies *Nesovitrea binneyana occidentalis*, as described below.

Nesovitrea binneyana occidentalis   H. B. Baker, 1930

Nesovitrea binneyana occidentalis: Lopez Is., San Juan Co., WA; scale bar = 1 mm

Description: This small snail is about 3.5 to 4.0 mm wide with about 4 whorls. The shell is translucent, white with a greenish tint. It has a low rounded spire, the whorls increasing rather rapidly. The periphery is round, the aperture ovately-lunate and a little oblique. The umbilicus is funnelform and about one-fifth the shell diameter at its widest point. Sculpturing consists of radial grooves dorsally and very fine, spiral striae. The animal is dark blue-gray dorsally, pale ventrally.

Similar Species: *Nesovitrea electrina* is similar, but its shell is larger when compared to those of *N. binneyana* with the same number of whorls. It is clear and transparent, and the spiral striae of *N. electrina* are less distinct, discontinuous, or lacking. *Zonitoides arboreus* and some *Oxychilus* species are also similar in size and in transparent, glossy shells, but they are distinctly more tightly coiled and lack radial grooves on the dorsal shell surface.

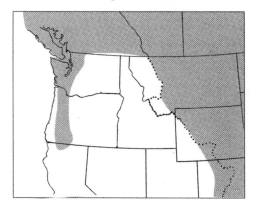

Distribution: *Nesovitrea binneyana binneyana* occurs from Maine to Michigan, Ohio, and Pennsylvania, and Ontario, Canada. *N. binneyana occidentalis* inhabits the Rocky Mountains from Alberta south to Colorado, and west to Vancouver Island, British Columbia, through the Pacific Northwest states and into northern California.

Nesovitrea electrina   (Gould, 1841)   Amber Glass
Synonym: *Retinella electrina*

Description: *Nesovitrea electrina* is small, but generally larger than *N. binneyana* with the same number of whorls. Adults are 4.5 to 5.2 mm wide in about 3¾ to 4¼ whorls. The whorls increase rapidly, the last, just behind the aperture, being more than 1½ times the width of the penultimate. The shell is transparent, with a yellow or greenish tint. The spire is low and slightly rounded. The periphery is round, the aperture roundly lunate, a little wider than high and a little oblique. The umbilicus is funnelform, about one-fifth to one-sixth the shell diameter at its widest point. The head, neck, and tentacles are grayish-black, fading to dark gray on the lower sides and foot.

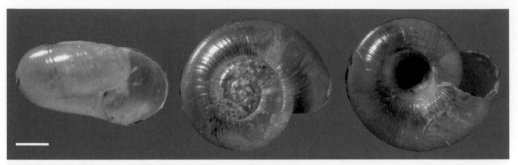

Nesovitrea electrina (subadult): Lewis Co.,WA; scale bar = 1 mm

Similar Species: *Nesovitrea binneyana* is very similar. Its shell with the same number of whorls is smaller than that of *N. electrina*, it is translucent white, and its spiral striations are more distinct (Pilsbry 1946). *Zonitoides arboreus* and some *Oxychilus* sp. are similar, but they are more tightly coiled and often have a brownish-colored shell.

Distribution: *Nesovitrea electrina* occurs in eastern Canada, across the northern United States and south in the Rocky Mountains to New Mexico and Arizona. In the Pacific Northwest, it occurs, at least, to the eastern Cascade Mountains and north into British Columbia, Canada, and Kodiak Is., Alaska.

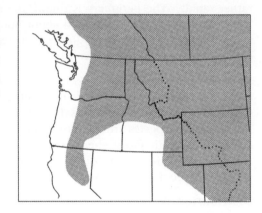

Family: PRISTILOMATIDAE

Current nomenclature places four genera of our Pacific Northwest snails into this family. *Hawaiia*, which is identified to species in the superfamily key to the families, appears first, then *Pristiloma* (including a key to its several species). *Ogaridiscus*, which is included in both of those keys because it is very similar to the *Pristiloma*, follows the *Pristiloma* species accounts, and *Vitrea* appears last.

Genus: *Hawaiia*

Hawaiia minuscula   (A. Binney, 1841)   Minute Gem

Hawaiia minuscula: Asotin Co., WA; width 2.4 mm; scale bar = 1 mm

Description: A minute snail, 2.0 to 2.5 mm wide by about 1.4 mm high with 4 whorls. The spire is low, but elevated and slightly rounded. The shell is widely umbilicate, with the umbilicus about one-third the shell width. The shell is thin and light gray. The whorls increase slowly, and the last is quite round. The aperture is simple and roundly lunate. Dorsally, the shell shows distinct growth-wrinkles, which smooth out ventrally. Spiral striae are lacking.

Similar Species: The shell of *Helicodiscus singleyanus* is nearly identical to that of *Hawaiia minuscula*, differing in its corneous or light-yellow color and fine spiral striae. It does not occur near the Pacific Northwest, however, but has been introduced into California (Hanna 1966).

Distribution: *Hawaiia minuscula* is generally found east of the Rocky Mountains in the United States and southern Canada, but it has been introduced locally in California and Oregon. Although not previously confirmed in Washington or Idaho, Frest (1999) found it at eight sites in the lower Salmon River watershed in Idaho, and Burke and Jepsen found it at two sites in Asotin Co., Washington. Branson (1977) reported a single dead shell from the Dungeness watershed, Clallam Co., Washington, a record that needs to be confirmed. Forsyth (2004) doubted the validity of reports from British Columbia.

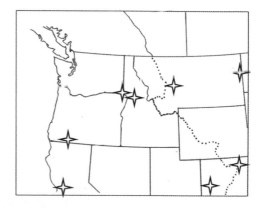

Genus: *Pristiloma*

Pristiloma is a group of minute to small snails. The shells of Pacific Northwest species are usually tightly coiled (except *P. johnsoni*), and imperforate or perforate, with low, conic spires. The aperture of most is narrowly lunate and obliquely positioned, extending from the outer palatal curve of the penultimate whorl, around the outer edge to the mid-ventral umbilical depression or perforation. The surface is usually translucent, smooth, and glossy. Two species are described as having radial grooves across the dorsal surface of the whorls, referred to as sulci. While the sulci are distinct on these two species, under magnification, faint or obsolete sulci-like creases can often be seen on most of the *Pristiloma* species.

Key to the Species of *Pristiloma* and *Ogaridiscus*

(a)   Shell imperforate (with no opening for the umbilicus) . . . .[b/bb]

    (aa)   Shell perforate or narrowly umbilicate. . . . . [i/ii]

(b)   Shell tightly coiled, increasing in size gradually . . . . [d/dd]

    (bb)   Whorls increasing in size regularly, the last whorl obviously wider
    than the adjacent penultimate whorl. Spire very low or nearly flat.
    Shell whitish or transparent . . . . [c/cc]

(c)   Shell diameter about 2.4 mm in 3¾ whorls, height about 1.0 mm.
Glossy-white. Width of the whorls increases rather rapidly . . . . . . *Pristiloma johnsoni*

    (cc)   Shell diameter 3 to 3.5 mm in 4 to 4½ whorls. Spire very low-conic and
    transparent; umbilicus normally perforate but sometimes closed. Width
    of the whorls increases regularly; aperture widely lunate. Northeastern
    Oregon, southern Idaho, Colorado, and Utah. . . . . *Ogaridiscus subrupicola*

(d)   Shells with, more or less, distinct radial grooves (sulci) dorsally . . . . [e/ee]

    (dd)   If faint sulci are visible dorsally, they are indistinct and inconsistent. . . . . [f/ff]

(e)   Relatively large, 3.4 to 3.8 mm wide by 2.4 to 2.7 mm high with 6⅓ to 7
whorls. Dorsal sulci shallow and rather closely and regularly spaced.
Periphery round or with very slight if any shoulder (immature shells
shouldered) . . . . *Pristiloma stearnsi*

    (ee)   Shell about 3.4 mm wide by 1.7 mm high with about 6½ whorls.
    Sulci deep, more widely spaced, forming a scalloped corona on the
    dorsal surface of the whorls. Shoulder rather abrupt, periphery
    somewhat vertical. . . . . *Pristiloma pilsbryi*

(f)   Shell glossy, brown, rather distinctly shouldered . . . . [g/gg]

    (ff)   Shell weakly or not at all shouldered. Head and neck transparent,
    light gray; tentacle retractors diffuse black; tail with darker gray area
    dorsally. No whitish granules in vicinity of mantle. . . . . [h/hh]

(g)   Shell transparent, measuring 2.3 to 3 mm wide with about 5½ whorls;
typically with a serrated rib inside of the aperture. Head, neck, and tentacles
gray, remainder of body white. A patch of whitish granules shows through
the shell in front of the mantle cavity. . . . . *Pristiloma lansingi*

    (gg)   Shell glossy, rather large, to 3.4 mm wide with 6 to 6¼ whorls.
    Tightly coiled and distinctly shouldered. Northeastern Washington,
    northeastern Oregon, and northern Idaho. . . . . *Pristiloma idahoense*

(h) Body whorl very slightly shouldered. With 5½ to 6 whorls it measures 2 to 2.7 mm wide. Color of shell rich amber-brown. Basal lip margin more roundly arched. . . . . *Pristiloma arcticum*

    (hh) Body whorl rounded, not at all shouldered. With 4½ to 5¼ whorls it measures 2.5 to 2.75 mm wide. Shell color pinkish-buff. Basal lip margin less roundly arched. *Pristiloma crateris*

(i) Shell diameter 3 to 3.5 mm in 4 to 4½ whorls. Spire very low-conic and transparent; umbilicus normally perforate but sometimes closed. Width of the whorls increases regularly; aperture widely lunate. Northeastern Oregon, southern Idaho, Colorado, and Utah. . . . . *Ogaridiscus subrupicola*

    (ii) Spire more elevated, convexly-conic; whorls rather tightly coiled; aperture narrowly lunate. Umbilicus perforate to very narrowly umbilicate . . . . . . [j/jj]

(j) Spire relatively high, somewhat domed. Periphery rounded, except angled a little at the aperture. With 5 to 5¼ whorls it is 3.25 to 3.55 mm wide by 1.85 to 2.1 mm high . . . . *Pristiloma chersinella*

    (jj) Spire moderately low, conic but slightly rounded. Periphery rounded, except slightly angled at the aperture. With 4½ to 5 whorls it measures 2.0 to 2.6 mm wide by 1.25 to 1.5 mm high . . . . *Pristiloma wascoense*

Genus: *Pristiloma*

For easier tracking to identify these snails, the species accounts for this genus have been organized by groups with similar shell characteristics rather than in alphabetical order. Thus we have *Pristiloma stearnsi* group: imperforate shells tightly coiled with distinct radial grooves or "sulci" (*P. pilsbryi* and *P. stearnsi*); *Pristiloma idahoense* group: tightly coiled without distinct sulci (*P. arcticum, P. crateris, P. idahoense,* and *P. lansingi*); *Pristiloma nicholsoni* group: rapidly increasing whorls (*P. johnsoni*); and subgenus *Priscovitrea*: perforate shells (*Pristiloma chersinella* and *P. wascoense*).

Subgenus: *Pristiloma*
Pristiloma stearnsi Group
Imperforate, tightly coiled with distinct sulci

Pristiloma pilsbryi (Vanatta, 1899) Crowned Tightcoil

Description: This is a minute snail, with a unique shell scalloped on the dorsal surface of the whorls by deep transverse impressions (sulci). With about 6½ whorls the shell measures 3.4 mm wide by 1.7 mm high. It is an imperforate, translucent, brown-colored shell with a low conic spire. It is tightly coiled and has a narrowly lunate aperture with a thin, simple lip. The periphery of the outer whorl is flattened vertically so the whorl is strongly shouldered. The sutures are deeply impressed, as are the sulci that form a corona-like scalloping around the top of the shell.

Similar Species: This beautiful little shell is unique among land snails in the Pacific Northwest. By description of the sulci, the novice might confuse *P. stearnsi* with this species, but the *P. stearnsi* shell is not distinctly shouldered like that of *P. pilsbryi*, nor does it have the prominent corona formed by the deep sulci of *P. pilsbryi*. By comparison the sulci of *P. stearnsi* are shallow and closely spaced.

1. *Pristiloma pilsbryi*: Pacific Co., WA
2. *Pristiloma pilsbryi*: Tillamook Co., OR

Distribution: *Pristiloma pilsbryi* occurs in Pacific County, Washington, and the northern Coast Range of Oregon. It is found among leaf litter under shrubs near the coast and other moist sites.

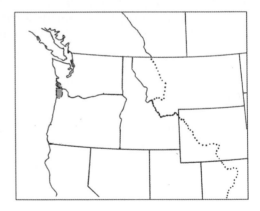

Pristiloma stearnsi   (Bland, 1875)   Striate Tightcoil

Pristiloma stearnsi: Quinault National Fish Hatchery, Grays Harbor Co., WA; scale bar = 1 mm

Description: Large for a *Pristiloma*, the shell measures 3.4 to 4.0 mm wide by 2.4 to 2.7 mm high in 6⅓ to 7 whorls. The shell is imperforate and has a low to moderately elevated, slightly rounded, conic spire. The spire is distinctly higher in full adults, but when immature the shell shape may be difficult to distinguish from *P. lansingi* or others. The periphery of adults is roundly and indistinctly shouldered, but in juveniles the shouldering is more distinct. The aperture is oblique and narrowly crescent-shaped. It extends from just above the periphery to the columellar insertion mid-ventrally. The lip is simple and sharp but expands a little near the columella. The first 1½ whorls are smooth, but the remainder are incised dorsally by rather distinct, fairly closely spaced, radial grooves (sulci), which may be variable for different individuals. The head, neck, and tentacles are transparent, grayish-white. Body similarly colored, but with a darker gray patch dorsally on the tail between the shell and the posterior end. Tentacle retractors are diffuse-black, this color extending back into the neck.

Similar Species: While this species might be thought to resemble *P. pilsbryi* because they are both described as being sulcate, in reality there is little or no similarity in their appearance (see discussion under *P. pilsbryi*). More confusion is likely to result over immature *P. stearnsi* and *P. lansingi* or *P. arcticum*. Adult *P. stearnsi* can usually be easily recognized by their larger size. *P. idahoense* is the only similar species of near the same size, but it does not occur in the same area, and it retains a strong shoulder into adulthood. The apertural rib of *P. lansingi* will distinguish that species, but it is often lacking even in adults.

Distribution: An endemic species, *P. stearnsi* may be found through most of western Oregon, Washington, and British Columbia, to Alaska. Occurring in moist forests and riparian areas, it is most common in the coastal mountain ranges, but it also occurs rarely in the Cascades.

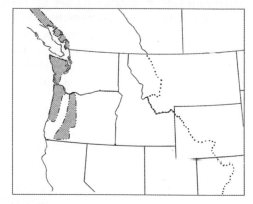

Pristiloma idahoense Group
Imperforate, tightly coiled, without distinct sulci

Pristiloma arcticum   (Lehnert, 1884)   Northern Tightcoil

Pristiloma arcticum: Cispus Watershed, Lewis Co., WA; scale bar = 1 mm

Description: This minute snail is about 2 mm wide by 1.5 mm high in 4¾ whorls. The whorls are tightly coiled; the shell is imperforate, and the spire is low-conic. The body whorl is nearly round at the periphery, but it is very slightly shouldered as it slopes inward more below the periphery than above it. The shell is yellowish-brown, smooth, and glossy. The aperture is oblique and narrowly lunate. The lip is simple, sharp, and slightly reflected at the columella. The basal margin is roundly arched. The body and tail are translucent, with a gray tint, and there is a darker gray patch on the tail dorsally between the shell and the posterior end.

Similar Species: Differences between *P. arcticum* and other species of *Pristiloma* are subtle. *P. crateris* is usually considered a subspecies of *arcticum*, but appears no more similar than some of the other species. *P. crateris* and *P. lansingi* are wider per number of whorls than *P. arcticum*. The periphery of *P. crateris* is rounder and its basal lip margin flatter than that of *P. arcticum*. *P. lansingi* is distinctly shouldered, while shouldering of *P. arcticum* is very slight. *P. lansingi* may or may not have a rib inside of the basal and outer aperture, but none of the others have that characteristic.

Distribution: *P. arcticum* occurs in Alaska southward through western Washington, including the Cascades. The Oregon distribution is questionable (Roth, personal communication). It may be found among shrubs, in forest floor litter, and associated with woody debris.

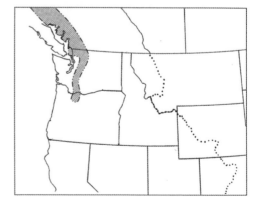

Pristiloma crateris   (Pilsbry, 1946)   Crater Lake Tightcoil
Synonym: *Pristiloma arcticum crateris*

P. crateris is usually considered a subspecies of *P. arcticum*, but it appears no more closely related to that species than other *Pristiloma* of this group; therefore, it is here considered a distinct species.

Pristiloma crateris: Deschutes Co., OR (OSAC #DES02-075); scale bar = 1 mm

Description: It is a minute species, with 4½ to 5¼ whorls measuring 2.5 to 2.75 mm wide by 1.45 to 1.6 mm high. The following description is of notes from observations of shells, with correlating statements from Pilsbry's (1946) description inserted in parentheses. The spire is low and subconical, being slightly convex ("depressed, with quite low, conoid spire"). The periphery is rounded, the widest point being at mid-whorl ("rounded periphery, median in position"). The shell is imperforate, and there is a small callus-like thickening around the columellar insertion, which extends down along the basal lip margin a short way and, in some specimens, forms a sinuosity along the edge of the inner basal lip outward from the insertion ("columellar margin slightly spreading, thickened within, reflected at the insertion in a small callus over the axis"). The basal arch of the apertural lip is flattened slightly. This is usually more apparent from about the mid-basal point toward the columella, but the flattening may be centered mid-basally ("very similar to *P. arcticum*, but the base is more flattened"). Growth-wrinkles are close set and regularly spaced, appearing somewhat like the sulci of *P. stearnsi* but faint by comparison ("sculpture of weak but subregular ripples of growth below the suture, soon disappearing, leaving the peripheral region and base smooth except for very weak lines of growth"). No spiral striae were seen ("very fine, close spirals are seen on the upper surface").

Similar Species: Subtle differences occur between *P. crateris* and other *Pristiloma* of this group. The periphery of *P. crateris* is distinctly rounded, while that of *P. arcticum* is slightly wider above the midpoint, giving it a very slight shoulder to the shell. The widest point of the body whorl of *P. lansingi* is distinctly above the midpoint.

Distribution: *P. crateris* occurs in the Oregon Cascade Mountains in mid- to higher elevation wet meadows and riparian areas.

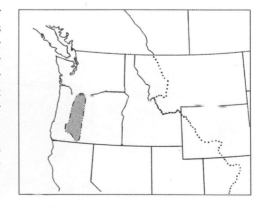

Pristiloma idahoense   Pilsbry, 1902   Thinlip Tightcoil

Pristiloma idahoense: Rocky Fork of Harvey Creek, Pend Oreille Co., WA; scale bar = 1 mm

Description: This Idaho species is relatively large and the only one of this group to inhabit its range. It measures 3.2 to 3.4 mm wide by 1.6 to 2.1 mm high in 6 to 6¼ whorls. Its shell is imperforate and brown, with a low conic spire. It is tightly coiled and distinctly shouldered. It has faint growth-wrinkles and also shows faint sulci.

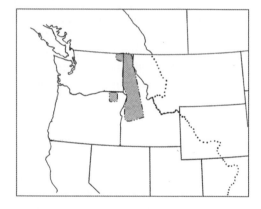

Similar Species: This is the only imperforate low-conic *Pristiloma* found in the Idaho Panhandle and northeastern Washington. Size and shape will separate it from other *Pristiloma* from outside of its range. It is even more strongly shouldered than *P. lansingi*.

Distribution: *P. idahoense* occurs in moist forest zones in the Pend Oreille River watershed in Pend Oreille County, Washington, throughout the Idaho Panhandle south to Boise County, and in the Blue Mountains, Umatilla Co., Oregon.

Pristiloma lansingi   (Bland, 1875)   Denticulate Tightcoil

Description: *P. lansingi* measures 2.3 to 3.0 mm wide by 1.3 to 1.5 mm high with a shell of about 5½ whorls. It is imperforate, with a rich brownish shell, having a low conic spire. The tightly coiled whorls are fairly distinctly shouldered, sloping roundly down and inward

Pristiloma lansingi: Woodard Bay, Thurston Co., WA; scale bar = 1 mm

from above the midpoint of the whorl. The aperture is obliquely, narrowly lunate, and has a whitish, irregularly serrated rib running inside of the lip from the columellar insertion to about the midpoint of the outer lip margin or into the curve of the shoulder. This rib may develop prior to full growth of the shell, so it is often a short way inside the aperture, or it may be farther inside and have a second rib forming nearer to the aperture. Many specimens lack the rib altogether. The head and neck are transparent, with a gray tint dorsally; the body, including the tail, is white. Tentacle retractors are diffuse-black. There is a patch of white square granules outlined in thin black lines in front of the mantle cavity.

Similar Species: There is no other species native to the Pacific Northwest with which to confuse typical specimens of *P. lansingi*, with its distinct *Pristiloma* form and serrated apertural rib. However, if the apertural rib is lacking, as it often is, other *Pristiloma* are very similar. The color of a live animal and the granulose pattern in front of the mantle cavity appear to be diagnostic for this species. With only shells available without the rib, the shouldered shell is the best indication. Immature *Pristiloma stearnsi* may be similar in shape to *P. lansingi*, but unless it is small, the closely-spaced sulci of the shell of *P. stearnsi* identifies that species. The similarly shaped *P. idahoense* is larger and is not found within the same region as *P. lansingi*. In other imperforate *Pristiloma* of this group, including *P. arcticum* and *P. crateris* (both of which have more rounded peripheries), the color patterns on the animals is more like that of *P. stearnsi* than *P. lansingi*.

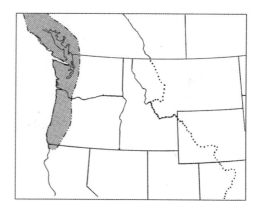

Distribution: *P. lansingi* occurs in southwestern British Columbia, through western Washington and Oregon to northern California.

Pristiloma nicholsoni Group

Pristiloma with shells that increase in size more rapidly. Most are umbilicate and occur only in California. Only one of this group is imperforate and occurs in the Pacific Northwest.

Pristiloma johnsoni (Dall, 1895) Broadwhorl Tightcoil

Pristiloma johnsoni: Patos Island, San Juan Co., WA; scale bar = 1 mm

Description: *P. johnsoni* is a tiny snail that is less tightly coiled than other *Pristiloma* in the Pacific Northwest. Its shell is 2.0 to 2.5 mm wide by 1.0 to 1.1 mm high with 3½ to 4 whorls. It is imperforate but may have a rather deep umbilical depression. The shell is translucent, waxy-white, with a nearly flat spire. The whorls increase rapidly in size; the periphery is round. The aperture is broadly lunate, rather oval, with a rounded outer margin, and it is somewhat flattened palatally. The body of the animal is white.

Similar Species: No other species of this size will fit the description of *P. johnsoni*. This species may be misidentified as a hatchling *Ancotrema* or *Megomphix*, with the misconception that the lack of an umbilicus is due to the whorls not yet having descended around the opening.

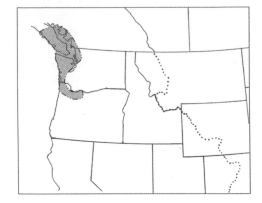

Distribution: This snail occurs from southwestern British Columbia through western Washington and into northwestern Oregon.

Subgenus: *Priscovitrea*

Perforate shells

This subgenus differs outwardly from the above *Pristiloma* by its open, although very narrow, umbilicus. Two species occur in the Mountains of Washington and Oregon, with disjunct populations in the Rocky Mountains in Montana.

Pristiloma chersinella   (Dall, 1886)   Blackfoot Tightcoil

Description: This species is rather large for a *Pristiloma*. With 4½ to 5¼ whorls it measures 3.25 to 3.55 mm wide by 1.85 to 2.1 mm high. The perforate umbilicus is less than one-ninth the width of the shell. The shell is translucent, yellow, with a moderately elevated, slightly rounded, conic spire. The whorls increase slowly in size. The periphery is round;

1. *Pristiloma chersinella*: Snoqualmie Pass, King Co., WA; scale bar = 1 mm
2. *Pristiloma chersinella*: Medford, Jackson Co., OR; scale bar = 1 mm

the aperture is oblique, narrowly lunate, and appears slightly shouldered. Sculpturing is of low growth-wrinkles and often very faint sulci-like lines. The animal has "considerable black pigmentation on the foot, [and] mantle edge" (Pilsbry 1946).

Similar Species: *P. chersinella* is similar to but larger than *P. wascoense* and, as an adult, the spire is higher and more conic. Its shape somewhat resembles *Zonitoides* species, but those species are larger and have a distinctly larger and more open umbilicus. There may be a superficial resemblance to *Euconulus fulvus*, but the spire is not as narrowly domed, and it does not have the radial striations of *Euconulus*.

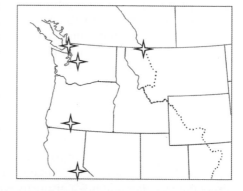

Distribution: *Pristiloma chersinella* is known from southern Oregon and northern California. It has also been reported from Glacier National Park, Montana, and we have specimens resembling it from the western Cascade Mountains and San Juan Islands of Washington.

Pristiloma wascoense (Hemphill, 1911)   Shiny Tightcoil

Pristiloma wascoense: Kittitas Co., WA; scale bar = 1 mm

Description: Similar to but smaller than *P. chersinella*, *P. wascoense* measures about 2.0 to 2.75 mm wide by 1.25 to 1.5 mm high with a shell of 4½ to 5 whorls. The shell is transparent. It is perforate, the last whorl intruding on the opening so the hole enters at an angle. It has a low-conic but somewhat rounded spire. The whorls increase slowly. The periphery is round. The aperture is oblique, narrowly lunate, and appears slightly shouldered. Sculpturing is of low growth-wrinkles and often very faint sulci-like lines.

Similar Species: *P. wascoense* is smaller than *P. chersinella* and has a lower spire. Species of *Zonitoides* are larger, with a more open umbilicus.

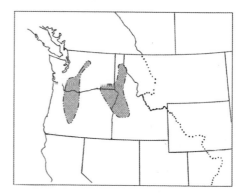

Distribution: Known from Wasco and Marion counties, Oregon, *Pristiloma wascoense* is also found in the Blue Mountains of Washington and Oregon, and on the east slope of the Cascade Mountains of Washington, primarily under deciduous trees such as quaking aspens.

Genus: *Ogaridiscus*

Ogaridiscus subrupicola (Dall, 1877)   Southern Tightcoil
Synonym: *Pristiloma (Ogaridiscus) subrupicola*

Description: This species is minute to small in size, measuring 2.9 to 3.5 mm wide by 1.4 to 1.6 mm high with 3⅔ to 4⅓ whorls. The umbilicus is very narrowly perforate, or may be covered by the reflected columellar lip margin. The shell is clear and transparent. The spire is very low-conic but distinctly elevated and with a rounded apex. Whorls increase rather slowly but regularly. The periphery is round and the aperture widely lunate or somewhat auriculate. The shell is smooth but may have some unevenly spaced grooves on the last half-whorl, and sometimes there may be very weak, irregular spiral striae on the base. The animal is translucent whitish to grayish.

Similar Species: *Pristiloma wascoense* and *P. chersinella* have higher spires and are more tightly coiled, with more narrow crescentic apertures, and wider perforate umbilici. *P. johnsoni* is smaller, less tightly coiled, and imperforate. The introduced *Vitrea contracta* is also a little smaller and has a distinctly wider umbilicus.

Distribution: *Ogaridiscus subrupicola* is known from northeastern Oregon, northern Idaho, and into Utah.

Genus: *Vitrea*

Vitrea contracta   (Westerlund, 1871)   Contracted Glass-snail

Vitrea contracta: Patos Island, San Juan Co., WA; scale bar = 1 mm

1.

Description: A minute snail introduced into North America from Europe, measuring about 2.5 mm wide by 1.1 mm high. Its shell has a low conic spire and is narrowly and deeply umbilicate. It is normally clear and transparent but sometimes clouded with white. The whorls increase rather slowly. The aperture is somewhat oblique, rather narrowly lunate, but mostly below the penultimate whorl. It is sculptured with faint, regularly spaced, radial grooves.

Similar Species: *Pristiloma wascoense* is the most similar in appearance to *Vitrea contracta* but has a relatively smaller, perforate umbilicus, a less transparent shell, and a higher, more convex spire. The shell of *Zonitoides arboreus* is nearly identical to that of *Vitrea contracta*, but is much larger and darker colored.

Distribution: *V. contracta* is a European species, introduced into local areas of British Columbia, western Washington, and California.

Selected references for the superfamily Gastrodontoidea: Forsyth 2004; Frest 1999; Hanna 1966; Henderson 1929a; Pilsbry 1946; Robinson 1999; Roth and Sadeghian 2006.

Slugs

The following taxa are those commonly known as slugs, the shells of which are reduced and completely enclosed within the mantle or, if visible, are too small for the animal to withdraw into. *Vitrina* is usually thought of as a snail, having a shell that covers most of the body, but it is included in the same superfamily as many of the old-world slugs; with its large aperture and body whorl, and small early whorls, the animal is not completely concealed when contracted.

Keys to the identification of the slugs begin on page 60, and a diagram of slug morphology appears on that page. Keys to the genera of the superfamily Arionoidea are found beginning on page 281. Keys to the species are presented with the following species accounts to their respective genera.

Directory to Taxonomy of the Slugs

Superfamily	Family	Genus	Page
LIMACOIDEA	VITRINIDAE	*Vitrina*	268
	BOETTGERILLIDAE	*Boettgerilla*	269
	LIMACIDAE	*Lehmannia*	270
		Limax	271
	AGRIOLIMACIDAE	*Deroceras*	274
PARMACELLOIDEA	MILACIDAE	*Milax*	279
TESTACELLOIDEA	TESTACELLIDAE	*Testacella*	280
ARIONOIDEA	ARIONIDAE	*Arion*	282
	ANADENIDAE	*Prophysaon*	289
		Kootenaia	300
		Carinacauda	302
		Securicauda	303
	ARIOLIMACIDAE	*Ariolimax*	305
		Gliabates	306
		Hesperarion	307
		Magnipelta	309
		Udosarx	310
		Zacoleus	311
	BINNEYIDAE	*Hemphillia*	314

Superfamily: LIMACOIDEA

Many of the members of this family group are old-world slugs introduced into North America. However, one snail, *Vitrina*, and some of the slug species of *Deroceras* are native to North America and the Pacific Northwest.

Family: VITRINIDAE
Genus: *Vitrina*

Vitrina pellucida   (Müller, 1774)   Western Glass-snail
Synonym: *Vitrina alaskana* Dall

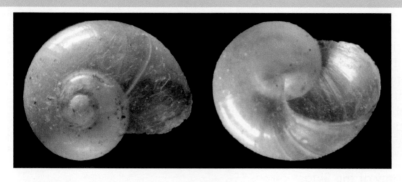

Vitrina pellucida:
Snake River,
Asotin Co., WA;
width 6-8 mm

Description: This small, uniquely shaped snail is 6 to 8 (rarely to 10) mm wide by 2+ mm high in about 2½ to 3 whorls. The shell is very thin, transparent, glossy, and light green. The spire is low and narrow, the greatest part of the width being in the body whorl, which increases in size very rapidly. It is imperforate. The aperture is larger than the rest of the shell, wider than high and a little oblique. The protoconch is sculptured with spiral rows of microscopic puncta (dots). At first sight of a living specimen, the transparent, glossy shell blends in with the body and the well-overlapping mantle edge, giving the animal the appearance of a small slug. A fresh, clean shell was so clear that it was barely visible in the photograph, so a soiled shell was used for the above illustration.

Similar Species: The unique shell of this snail effectively distinguishes it from all other snails in the Pacific Northwest.

Distribution: *Vitrina pellucida* is widespread from Alaska through British Columbia and the western and Rocky Mountain states, east to South Dakota and south into southern California, Arizona, and New Mexico. It is common under brush on open slopes, under aspen and other hardwood stands, and in riparian areas. In the southern part of its range, it occurs only at high elevations.

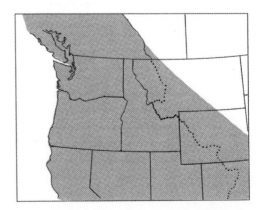

Slugs of the Families BOETTGERILLIDAE, LIMACIDAE,
AGRIOLIMACIDAE, MILACIDAE, and TESTACELLIDAE

Family: BOETTGERILLIDAE
Genus: *Boettgerilla*

Boettgerilla pallens   Simroth, 1912   Wormslug

Boetgerilla pallens: Vancouver Island, British Columbia (Kristiina Ovaska photo)

Description: A small to medium-sized slug, 30 to 60 mm long, pale gray or yellowish; darker blue-gray dorsally. Pneumostome in the third quarter back in the right side of the mantle. A distinct dorsal keel runs from the posterior edge of the mantle to the end of the tail. The sole is yellow, the mucus clear. The body is narrow and sinuous (worm-like) when extended and crawling.

Similar Species: The most similar species within our area is the Ryan Lake slug, *Zacoleus leonardi*, which is bluish-white and keeled the length of the tail. It can be distinguished from *Boettgerilla* by the position of its pneumostome, which is well back in the mantle, and a secondary pore even farther back, as in *Zacoleus idahoensis*, to which it appears closely related.

Distribution: *Boettgerilla pallens* is native to southeastern Europe but is rapidly spreading across the continent and through the British Isles. It has been reported on Vancouver Island, British Columbia, by Reise et al. (2000) and Forsyth (2004), but to date it has not been documented from the United States.

Family: LIMACIDAE

Genera: *Limax* and *Lehmannia*

Larger slugs, 50 to 200 mm long. Tail tapers gently to the acute end and is keeled for about the last half of its length. Concentric rings or wrinkles are centered on the midline of the mantle.

Key to the Species of *Limax* and *Lehmannia*

(a)   A large slug, 100 to 200 mm long, light brown or gray with 2 or 3 black
      or dark brown lateral bands that are often broken into rows of short
      stripes or spots. Mantle with irregular markings but not banded . . . . . *Limax maximus*

   (aa)   Smaller, less than 130 mm long. If greater than 75 mm, then
          without lateral bands. If smaller than 75 mm long, then the mantle,
          as well as the body, is banded . . . .[b/bb]

(b)   Length 70 to 130 mm. Small oval, yellow and gray tubercles on the back
      and sides give the animal a speckled appearance . . . . [c/cc]

   (bb)   Length 50 to 75 mm. Light-colored slugs, somewhat translucent
          or gelatinous appearing. . . . . [d/dd]

(c)   Length 70 to 100 mm. Yellow with gray or yellowish-gray marbling
      dorsally and on the tail, fading to light yellow on the lower sides and
      below and in front of the mantle. Tentacles blue. *Limax flavus*

   (cc)   Length 70 to 130 mm. Pale greenish-gray but mostly mottled
          with very dark gray or black. Tentacles gray. . . . . *Limax pseudoflavus*

(d)   Length 50 to 75 mm. Light brown dorsally, paler on the lower sides
      and appearing somewhat translucent. With a light but moderately
      distinct black band along each side of the mantle and back and a less
      distinct mid-dorsal line. May be found in trees. . . . . *Lehmannia marginatus*

   (dd)   Length 50 to 70 mm. Yellowish-gray or with a violet tint, darker
          on the head. Gelatinous appearing. Mantle with a darker mid-dorsal
          line and a pair of lateral bands. Body with one or two dorso-lateral
          bands. A ground dweller. . . . . *Lehmannia valentiana*

Genus: *Lehmannia*

> ### *Lehmannia marginatus*   Müller, 1774   Tree Slug
> **Synonyms:** *Limax (Lehmannia) marginata*   Müller;
> *Limax arborum*   Bouchart-Chantereaux, 1838

Description: This is a medium-sized slug, attaining a length of 50 to 75 mm. It is light brown dorsally and paler on the lower sides, appearing translucent. It is marked by a moderately distinct black band along each side of the mantle and back, and a less distinct mid-dorsal band. The pneumostome is about two-thirds back in the right side of the mantle.

Similar Species: See *Lehmannia valentiana*.

Distribution: *L. marginatus* is native to much of Europe and has been introduced into America (Robinson 1999). It occurs sporadically across North America and according to Hanna (1966) is widespread in California, although Roth (personal communications) believes that many of those records are more likely *Lehmannia valentiana*. It is a species of woodlands and is found under the bark of logs, among other woody debris, and in trees. In open country it may be found on walls and among rocks.

Lehmannia valentiana Férussac, 1821   Threeband Gardenslug
Synonym: *Limax valentianus*

Lehmannia valentiana: Corvallis, Benton Co., Oregon

Description: Medium-sized, 50 to 70 mm long, this slug is yellowish-gray or yellowish-violet with a slightly darker head. It is watery and gelatinous in appearance, usually with one dark band on each side of and near the midline, and sometimes a second, weaker band lower on the sides. The mantle has a median, dark band and a dark lateral band that forms a lyre shape on the right side. The sole is pale grayish; the mucus is clear. The posterior end has a short keel (Kerney, Cameron, and Riley 1994).

Similar Species: *Lehmannia marginatus* (which see) is very similar and may be considered a synonym by some authors. Turgeon et al. (1998) listed them separately; Kerney, Cameron, and Riley (1994) listed them as *Limax valentiana* and *Limax marginata*, and they illustrated distinct genitalia for each.

Distribution: *Lehmannia valentiana* is native to the Iberian Peninsula of southwestern Europe, but it is spreading through Europe and America. Kerney, Cameron, and Riley (1994) say it is "A terrestrial, not an arboreal slug, unlike *L. marginatus*." Roth (personal communication) says it is the most common slug in his garden in San Francisco. Although occurring in the Pacific Northwest, it does not appear to be abundant here yet.

Genus: *Limax*

Limax flavus Linnaeus, 1758   Yellow Gardenslug
Synonym: *Limacus flavus*

Description: This larger slug attains a length of 70 to 100 mm and is speckled with gray and yellow. It is yellowish with gray marbling dorsally and on the tail, and its tentacles are blue. The color fades on the sides, which become light yellow without the gray, anteriorly. The mantle is yellowish-gray with rather dense, irregular light yellow spots. The body below and behind the mantle is covered with rows of small oval tubercles, irregularly colored gray or yellow, which causes its speckled appearance. The sole is yellowish-white. The mucus is clear on the foot and yellow on the body. The short keel is confined to the posterior end. There is a long caecum on the hindgut.

Limax flavus: Seattle, King Co., WA

Similar Species: *Limax pseudoflavus* is similar in appearance but is darker colored, more distinctly greenish, and with gray tentacles.

Distribution: Native to southern and western Europe, *L. flavus* has been introduced with settlement over much of the world. In North America it occurs mostly in the eastern states and California, but it also occurs uncommonly in urban areas of western Oregon, Washington, and into British Columbia. It inhabits yards and gardens in urban and sub-urban areas.

Limax maximus Linnaeus, 1758   Giant Gardenslug

Limax maximus: Olympia, Thurston Co., WA

Description: The largest of our introduced slugs, the length of *L. maximus* normally exceeds 100 mm, and large specimens may attain a length of nearly 200 mm. It is usually light brown or grayish, light reflecting from the mucus often giving it a more gray or silvery appearance than its actual color. It normally has two or three black lateral bands on each side, most often broken into rows of elongated stripes or sometimes merely rows of spots. The mantle is relatively short, well forward on the back, the same color as the body, but the black spots or blotches are irregularly arranged instead of in rows. The sole is white; the mucus is clear and sticky. The tail is keeled for about the posterior one-third of its length.

Similar Species: Its size and color pattern are generally sufficient to distinguish *Limax maximus* from all other slugs, native or introduced, that are currently known within the Pacific Northwest.

Distribution: Native to Europe, Asia Minor, and North Africa, *L. maximus* has been widely introduced into most other continents. In the Pacific Northwest it is common in urban

1. *Limax maximus* (mating): Oregon (Joseph Furnish photo)
2. *Limax maximus*: Olympia, Thurston Co., WA
3. *Limax maximus* (hatchling): Olympic NF, Jefferson Co., WA

and suburban areas and is occasionally encountered in rural regions. It is common around buildings and is often seen on outer walls. When breeding, they hang from a string of mucus attached to the side of a building, tree, or shrub with their heads downward, their bodies entwined and their genitalia hanging entwined below them.

Limax pseudoflavus   Evans, 1978   Irish Gardenslug

Description: Another larger slug, this one attains a length of 70 to 130 mm. Its base color is pale greenish-gray and is predominantly covered with very dark gray or black mottling. The tentacles are gray. The sole is yellowish-white. The mucus is clear on the foot and tinted yellow or clear on the body. The short keel is confined to the posterior end.

Similar Species: *Limax flavus* is similar in appearance, but it is lighter colored, more distinctly yellowish (with green on only juveniles), and has blue tentacles instead of gray.

Distribution: Apparently native to the British Isles, *L. pseudoflavus* is most common in Ireland. Recently described, it may be more widely spread in western Europe but may have been mistaken for *L. flavus* in the past (Kerney, Cameron, and Riley 1994). Not previously documented in North America, a specimen found in a yard in Corvallis, Oregon, was tentatively identified as this species by Dr. Terrence Frest.

Family: AGRIOLIMACIDAE

Genus: *Deroceras*

The species of *Deroceras* are mostly small slugs; the largest seldom attains a length of more than 50 mm. Both native and introduced species occur in the Pacific Northwest. They are without bands or distinct stripes or markings, but they may have irregular black flecks or other faint irregular spots. There are concentric rings of wrinkles on the mantle resembling a fingerprint centered to the right of the midline. The tail has a short keel at its tip, which often slopes abruptly downward at its posterior end, lending it a truncated appearance from the lateral view.

Key to the Species of *Deroceras*

(a)   Small, about 16 to 22 mm long when preserved. Light brown with slightly darker spots. Mantle brown with small scattered light spots. Sole tripartite.
. . . .*Deroceras hesperium*

  (aa)   Length may be as little as 12 mm, but is generally longer than 15 mm as adults. If smaller than 15 mm, then closely and darkly speckled. . . . .[b/bb]

(b)   Animal mostly uniformly colored. It may have small flecks or pigments in the integument but without distinct black flecks or close speckling. . . . . [d/dd]

  (bb)   Animal fairly uniformly colored but with distinct dark flecks or close speckling. . . . . [c/cc]

(c)   Length of adults 35 to 50 mm. Color variable, whitish, buff, to light or dark gray, with or without irregular dark gray or black flecks on the mantle and body. Pneumostome well back in the mantle and with a pale border. Sole tripartite. Mucus clear, becoming thick and white when the animal is irritated . . . . *Deroceras reticulatum*

  (cc)   Length of adults 12 to 21 mm, to 32 mm in southern California. Gray or buff to brown, closely speckled with black or dark brown dorsally, except on the head and neck, and similarly speckled just above the pedal furrows. Pneumostome just behind the midpoint of the mantle, its border also somewhat speckled. Mucus clear. . . . . . *Deroceras monentolophus*

(d)   Length 15 to 25 mm. Varying shades of amber, reddish-brown, dark brown to blackish. Nearly uniformly colored, sometimes with gray flecks, but mantle may be lighter. Pneumostome with a light border and well behind the midpoint of mantle . . . .*Deroceras laeve*

  (dd)   Larger, adults usually longer than 28 mm. Colors vary . . . . [e/ee]

(e)   Length 28 to 32 mm. Brown with lighter shade on the mantle and neck. Pneumostome about two-thirds back in the mantle and encircled by a pale border. Neck well extended when crawling. . . . . *Deroceras panormitanum*

  (ee)   Very common slugs, adults 35 to 50 mm long. Colors varying but usually uniform per individual, nearly white to buff to light or dark gray. Neck not markedly long when crawling. Pneumostome well back in the mantle and encircled by a pale border. Mucus clear but turning thick and white when the animal is irritated. . . . . *Deroceras reticulatum*

Deroceras hesperium Pilsbry, 1944   Evening Fieldslug

Description: One of the smaller species; length in alcohol is about 16 to 22 mm. Color is light brown, paler below the mantle; may have some slightly darker brown spots. The mantle is brown with small, scattered light spots. The tail is keeled very shortly at the posterior end. The sole is tripartite; the middle field is a little darker than the outer ones. The shell is thin, rounded posteriorly and somewhat squared anteriorly, although the leading edge is still slightly curved. It measures about 3.3 mm long. The sides are roughly parallel, but the left side is slightly convex while the right side is straight. The apex is terminal, just left of center. It is concave interiorly; faint growth lines show exteriorly.

Similar Species: Generally smaller than the other *Deroceras*, with unique shell (see above) and genitalia. It is "most easily recognized by the enlarged duct of the spermatheca, which may be seen quite easily on opening the anterior end of the animal, without further dissection. The inner lining of this duct, formed of long crowded lamellae, is also peculiar to this species" (Pilsbry 1948).

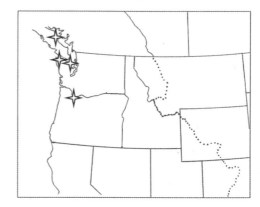

Distribution: An endemic species, *D. hesperium* has been found sparsely between the western Cascades and the Pacific Ocean in northwestern Oregon, western Washington, and on Vancouver Island, British Columbia.

Deroceras laeve (Müller, 1774)   Meadow Fieldslug

Description: A small slug; extended length is 15 to 25 mm. Color is of varying shades of amber, reddish-brown, dark brown to blackish, and it may have darker flecks in the integument. The mantle may be a little lighter. The pneumostome is located well back in the mantle and has a slightly lighter border around it. The posterior end is keeled very shortly. The sole is light brown, and the mucus is thin and clear. The shell is about 4 mm long. It is well rounded anteriorly, somewhat pointed posteriorly. It is strongly convex to the left side of the point, curving into the rather straight left edge, and rather straight to

Deroceras laeve: Olympia, Thurston Co., WA

Deroceras laeve: (L) Rimrock Lake, Yakima Co., WA; (R) Windy Point Campground, Yakima Co., WA

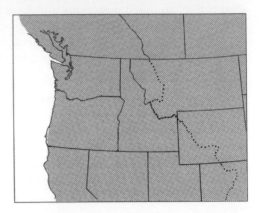

the right side, joining the straight right edge at an obtusely rounded angle. The apex is subterminal at the left, posterior end.

Similar Species: *D. monentolophus* and *D. panormitanum* (see below).

Distribution: *D. laeve* is a holarctic species, found generally throughout North and Central America and into South America. It is most commonly found in and around wetlands.

Deroceras monentolophus   Pilsbry, 1944   One-ridge Fieldslug

Description: This small slug attains a length up to 32 mm when extended. Southern animals are larger than northern ones. They are gray or buff to brown. The mantle is closely speckled with black or dark brown. The pneumostome is just behind the midpoint of the mantle, and its border is little, if at all, lighter and speckled like the rest of the mantle. The body behind the mantle is also closely speckled with black but the lower sides and anterior portion of the body lacks these markings. The pedal flange is speckled with fine gray dots. The head (dorsally) and tentacles are darker. The mucus is clear. The posterior end is shortly keeled and tapers downward to the tip. The shell measures about 4.5 mm long. It is rounded in front, skewed a little to the right. The posterior border is truncate between rounded corners; the apex is terminal at the left posterior angle. The sides are roughly parallel, the right slightly concave, the left more convex or very slightly sinuous. The plate is somewhat convex and thickened internally, more so on the left side.

Similar Species: *D. laeve* and *D. panormitanum* (see their descriptions).

Distribution: *D. monentolophus* occurs in southern California, but there is an 1896 report from Seattle, King County, Washington. Such an old, disjunct report causes one to question whether there was ever an established population in the Seattle area.

Deroceras panormitanum Lessona & Pollonera, 1882   Longneck Fieldslug
Synonym: *Deroceras caruanae* (Pollonera, 1891)

Description: A small slug, reaching 28 to 32 mm long, *D. panormitanum* is brown with a lighter-colored, somewhat orange-tinted mantle. Mantle and body behind it are speckled with fine, scattered blackish spots. The mantle is about 40% of the total length; the pneumostome, about two-thirds back in its right side, is encircled by a pale border. There are concentric, fingerprint-like wrinkles on the mantle. When crawling, the pale-colored neck is well extended in front of the mantle, which is also paler than the body of the animal. The keel is distinct but short, resembling a pinched area at the tip of the tail. The sole is gray. The mucus is clear on both body and sole.

Similar Species: *Deroceras reticulatum* is larger, the color more variable and the pneumostome is farther back in the mantle. See also *D. laeve*.

Distribution: *D. panormitanum* is native to southwestern Europe and the British Isles. Introduced in North America and apparently spreading. It is common in urban and suburban areas from California to southern British Columbia.

1

2

1. *Deroceras panormitanum*:
   Thurston Co., WA

2. *Deroceras panormitanum*:
   Olympia, Thurston Co., WA

3. *Deroceras panormitanum*:
   Thurston Co., WA

3

Deroceras reticulatum   (Müller, 1774)   Gray Fieldslug
Synonyms: *Agriolimax agrestis* L; *Agriolimax reticulatus* (Müller)

1& 2. *Deroceras reticulatum*: Olympia, Thurston Co., WA

Description: *D. reticulatum* is a small to medium slug, 35 to 50 mm long when extended. There is little difference between the color and/or markings on the mantle and back; the color ranges from whitish to buff, to light or dark gray. It may or may not be marked with irregular darker gray or black flecks on the mantle and the body behind it. The mantle has fine concentric wrinkles resembling a fingerprint. The mantle is a little longer than one-third the length of the animal; the pneumostome is located well back in its right side (about 80% back) and is encircled by a pale border. The keel is distinct but short, resembling a pinched area at the tip of the tail. The sole is tripartite, the outer fields whitish or yellowish, the median gray. The mucus is clear, but becomes white and thick when the animal is irritated.

Similar Species: *D. panormitanum* and *D. monentolophus* (see above).

Distribution: Native to western Europe and the British Isles, *D. reticulatum* has been introduced widely following European settlement. Found around human populated areas throughout most of the United States and southern Canada, and well into British Columbia. It is a common garden pest, found in lawns and in and under wood, stones, and vegetation.

Superfamily: PARMACELLOIDEA
Family: MILACIDAE
Genus: *Milax*

Milax gagates   (Draparnaud, 1801)   Greenhouse Slug

Description: This medium-sized slug extends to 50 or 70 mm. It is black to gray dorsally and laterally, fading near the foot; the sole is gray or brownish-white. The mantle has a characteristic groove in a horseshoe or diamond shape encircling the posterior two-thirds to three-fourths of its central area. The back is strongly keeled from the mantle to the posterior end, and there are distinct longitudinal grooves running parallel dorsally, some of them branching and becoming oblique laterally and anteriorly. The sole is tripartite.

Similar Species: Kerney, Cameron, and Riley (1994) list five species of *Milax* from Europe, but only *M. gagates* has been reported from the western United States. They should not be mistaken for species of any other family.

Distribution: Native to the Mediterranean, western Europe, and the British Isles, *M. gagates* has been recorded in the United States in Washington, Oregon, Idaho, California, Colorado, and some Atlantic states. They are typically found in greenhouses, hotbeds, and gardens.

1

2

1, 2, & 3. *Milax gagates*:
Central Point, Jackson Co., OR

3

Superfamily: TESTACELLOIDEA
Family: TESTACELLIDAE
Genus: *Testacella*

Testacella haliotidea Draparnaud, 1801   Earshell Slug

Testacella haliotidea: Vancouver, Clark Co., WA

Description: *Testacella* are unusual slugs with a small vestigial shell at their posterior ends. Since the shell covers the mantle, this slug lacks the usual anteriorly positioned mantle of other slugs. The animal is medium-sized, attaining a length of 60 to 120 mm when extended. It is usually creamy white or yellowish; the sole is normally whitish. A branching groove runs forward from under the mantle and extends along each side to the base of the tentacles. It is a subterranean species, in part, and is a predator of earthworms and other soil invertebrates.

Similar Species: Kerney, Cameron, and Riley (1994) describe three species of *Testacella*, only one of which we have found to have been reported in North America. The shell at the posterior end and the lack of a normal slug mantle easily distinguish the genus from other gastropods occurring here.

Distribution: Native to the British Isles and western Europe to North Africa, *Testacella haliotidea* has been introduced into North America and other continents. Records exist from California, the San Francisco area, western Oregon, and to Vancouver Island, British Columbia.

Selected references for the non-Arionid Slugs: Burch 1962; Forsyth 2004; Frest 1999; Frest and Johannes 1993, 2000; Hanna 1966; Kelley et al. 1999; Kerney, Cameron, and Riley 1994; Pilsbry 1948; Reise et al. 2000.

Superfamily: ARIONOIDEA

The Arionoidea are small to large slugs. The edge of the foot, between the pedal furrows and the angle of the sole, is relatively broad, forming a flange-like margin that widens and is most conspicuous in the tail region. Other characteristics are variable, but see the keys to the genera for distinctive combinations that differ from those of the Limacidae and other slug families.

Most of our native slugs are of this family, as are some exotics. The endemic species are usually found in native habitats and are generally not harmful to crops or other resources. Most of the slugs found in yards and gardens, including the pests, are exotic species. Although native slugs, which are seldom seen in urban or developed suburban sites, may be common around homes built in forested or remote settings, usually the most harm that they do is in repulsing the residents. Most people would rather not see a large olive-green bananaslug on their lawn, even though it is a common inhabitant in or near western Washington and Oregon forests.

Families and Genera of ARIONOIDEA

Under current taxonomy, four families and twelve genera within the superfamily Arionoidea occur in the Pacific Northwest. The status of *Gliabates* is uncertain at present; it has not been re-located since being described. The following list of genera includes those found in the Pacific Northwest, but the ranges of many also extend into California and British Columbia, Canada. The Arionidae are all exotic in North America, but most species of the other families are endemic to the Pacific Northwest.

Families and Genera of the Currently Known Arionoidea of the Pacific Northwest

ARIONIDAE
- *Arion* (the arionid slugs) not native to North America.

ANADENIDAE
- *Prophysaon* (taildroppers) endemic to the Pacific Northwest (PNW).
- *Kootenaia* (pygmy slug) from northern Idaho and western Montana.
- *Carinacauda* (Cascades axetail) northwestern Oregon Cascades.
- *Securicauda* (Rocky Mountain axetail) Northern Idaho.

ARIOLIMACIDAE
- *Ariolimax* (bananaslugs) native to the far-western United States.
- *Gliabates* (salamander slug) only one known specimen, from western Oregon.
- *Hesperarion* (westernslugs) endemic to western Oregon and California.
- *Magnipelta* (magnum mantleslug) northeastern Washington to northwestern Montana.
- *Udosarx* (lyre mantleslug) north Idaho and into adjacent Montana.
- *Zacoleus* (sheathed slug) endemic to Idaho and western Montana, with a new species in western Washington and northwestern Oregon.

BINNEYIDAE
- *Hemphillia* (jumping-slugs) endemic to the Pacific Northwest.

Family: ARIONIDAE

Genus: *Arion*

The *Arion* in America are all exotic species introduced from Europe. They vary in size from small to large, and they are generally broad with a relatively low profile. The pneumostome is in the front half (second quarter) of the right side of the mantle. The pedal flange, between the pedal furrows and angle of the foot, is relatively wide, spreading most conspicuously posteriorly. There is a caudal mucous pit at the posterior end where the pedal furrows meet.

Slugs of the genera *Prophysaon* and *Securicauda* also have their pneumostomes in the forward half of the mantle; in *Carinacauda* and *Kootenaia* the pneumostome is about centered in the right side of the mantle or is very slightly posterior. However, members of these other genera lack a caudal pit, and their posterior ends are more rounded in cross-section or keeled, unlike the dorsally flattened and broadly-rounded tail of the *Arion* species.

Key to the Species of *Arion*

(a)   Larger slugs, length exceeding 100 mm. Adults without darker bands on the mantle or tail, but there may be colored markings on the pedal flange. Color generally rusty or brownish, or sometimes black. Often with orange markings on the pedal flange. Most often found around human habitation, as in urban or suburban yards and gardens. . . . . *Arion rufus*

   (aa)   Smaller slugs, length less than 80 mm. Usually with 1 or 2 darker dorso-lateral bands. Pedal flange with or without markings . . . .[b/bb]

(b)   Length 50 to 70 mm. Color dark gray, brown, or reddish, changing abruptly on the sides to whitish with a tint of the dorsal color. Darker lateral bands, often indistinct because of the dark dorsal color, usually passing above the pneumostome. Sole yellowish, mucus yellowish on the body, clear on the sole. . . . . *Arion subfuscus*

   (bb)   Smaller slugs, less than 50 mm long. . . . . [c/cc]

(c)   Medium-sized, length 30 to 50 mm. Bell-shaped in cross-section when contracted. Pneumostome below lateral band on mantle. . . . . [d/dd]

   (cc)   Small slugs, length usually less than 30 mm, rarely as long as 40 mm. Mucus yellow or orange. When lateral band is present, it encloses the pneumostome . . . .[f/ff]

(d)   Color dark gray, or sometimes brown; paler on the sides; the sole and pedal flange nearly white. A thin dark dorso-lateral band on each side of the mantle and tail. Mantle with black speckles. Mucus thin and clear . . . . . .*Arion circumscriptus*

   (dd)   Gray dorsally with dark dorso-lateral band; the sides paler below. Without distinct black speckling on the mantle . . . .[e/ee]

(e)   Gray dorsally, with a yellow tint strongest just below the lateral band . . *Arion fasciatus*

   (ee)   Pale gray dorsally, whitish laterally. Lateral band broad and dark . *Arion silvaticus*

(f)   Length 25 to 30 mm. Dark gray or blue-gray, sometimes brownish, with darker lateral bands enclosing the pneumostome. Sole and mucus yellowish to orange; the mucus tinting the animal. Semicircular in cross-section when resting, and without mucous points on the tubercles. . . . . *Arion distinctus*

   (ff)   Length seldom greater than 20 mm. Color whitish or grayish, often with a light yellowish tint. Lateral bands gray when present and enclosing the pneumostome, but often faint or indistinct. Mucus yellow. When contracted, pointed mucous tips appear on the tubercles, thus the common name hedgehog slug . . . . *Arion intermedius*

Genus: *Arion*

Species accounts for this genus are organized by key characteristics instead of alphabetically as for most other genera. The arionids separate into groups naturally and this method of organization will assist in comparing them for identification. Refer to the subheadings between groups.

All species of the genus *Arion* are exotic in North America; they are most common in yards and gardens in suburban settings. Some species may also be found in uninhabited areas where they have been introduced by human activities such as forest management or recreation.

Large *Arion* (length > 100 mm); adults without dorso-lateral band

Arion rufus   (Linnaeus, 1758)   Chocolate Arion
Synonym: *Arion ater rufus*

Description: One of the largest species of the genus, *A. rufus* attains a length of 100 to 150 mm or longer. It is generally rusty red, brownish, orange, or sometimes black. The foot margin is usually orange and marked by black lines in obliquely vertical grooves. It has a low profile and broad rounded tail. The tubercles are relatively high, long, and narrow (5 to 10 times as long as wide). Very young juveniles have a lateral band.

Similar Species: *Arion ater* and *A. rufus* are often described as varieties or subspecies of each other, *A. ater* being the black color variety. However, color is not a distinguishing characteristic, and although black *A. rufus* are not uncommon, to date no *A. ater* has been confirmed in the Pacific Northwest.

Distribution: The *A. ater/A. rufus* complex is native to Europe. *Arion rufus* has been introduced into North America and is established and generally common in suburban and rural areas.

1

1 & 2: *Arion rufus*:
Olympia, Thurston Co., WA

2

Length 50 to 70 mm; often with a lateral band

Arion subfuscus   (Draparnaud, 1805)   Dusky Arion

Description: This medium-sized slug (50 to 70 mm long) is dark gray or brown to reddish, fading laterally to whitish on the lower sides. The lateral band is not always distinct, but it has a rather abrupt outer edge between the darker band and lighter sides. On the mantle, the pneumostome is located below the band or darker-colored area. When contracted, the lateral profile is low, skewed

1, 2, & 3. *Arion subfuscus*: Olympia, Thurston Co., WA

forward—not rounded to form a distinct hemisphere or semicircular arch. The sole is pale yellowish, the pedal flange with vertical black lines.

Similar Species: *Arion distinctus* is smaller than *A. subfuscus,* but it might resemble juveniles of that species. *A. distinctus* has a yellow or orange sole, but it lacks the vertical lines on the edge of the foot and has a more distinct lateral band delineated by a narrow lighter area between the band and the dorsal color on both the mantle and the tail.

Distribution: Native to Europe, *A. subfuscus* is introduced in North America with scattered distribution in Washington, Oregon, and Montana. It is often seen in and around human habitation but may also be found some distance away in areas where less concentrated activities (e.g., timber management or recreation) have occurred.

Small to medium slugs (20 to 50 mm long) with dorso-lateral band, which may be indistinct. Lateral band runs above the pneumostome; when contracted, cross-section is bell-shaped.

Arion circumscriptus   Johnston, 1828   Brown-banded Arion

1 & 2. *Arion circumscriptus*: Olympia, Thurston Co., WA

Description: A small to medium-sized slug (30 to 40 mm long), *A. circumscriptus* is dark gray or sometimes brown dorsally and distinctly paler laterally. There is a rather thin but distinct, dark dorso-lateral band on the mantle and tail. The mantle is speckled with black, and the pneumostome is below the band. The head and tentacles are blackish, and the sole is white. Its mucus is thin and colorless. Its resting posture in lateral view is hemispherical, and when contracted its cross-section is bell-shaped.

Similar Species: *A. fasciatus* has a yellowish tint below the lateral band. *A. silvaticus* is somewhat smaller and pale gray dorsally with white sides.

Distribution: *A. circumscriptus* is a slug of northwestern Europe. It is introduced in North America and is locally common in suburban and inhabited rural settings throughout the Pacific Northwest.

Soles of *Arion circumscriptus* (A)
and *A. distinctus* (B)

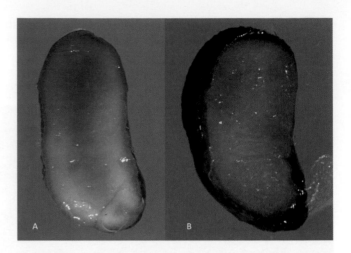

Arion fasciatus   Nilsson, 1823   Orange-banded Arion

Description: This small to medium-sized slug (40 to 50 mm long) is gray dorsally, paler laterally, and has a yellow tint most conspicuously below the lateral band. The band on the right side of the mantle passes above the pneumostome. The pedal flange is whitish. The sole is whitish with a gray cast; the mucus is colorless. When contracted, its lateral profile is hemispherical, and its cross-section is bell-shaped.

Similar Species: *A. circumscriptus* has a speckled mantle, and *A. silvaticus* is smaller and more distinctly pale gray and whitish. With *A. fasciatus*, these three species have been considered synonyms in the past (Burch 1962; Forsyth, 2004; Hanna 1966; Kerney, Cameron, and Riley 1994; Pilsbry 1948).

Distribution: Kerney, Cameron, and Riley (1994) retained *Arion circumscriptus, A. fasciatus*, and *A. silvaticus* as separate species. Since Quick (1960) had placed these species in synonymy, it is difficult to interpret which species have been recorded as occurring in North America. Burch (1962) included *fasciatus* but not *circumscriptus* or *silvaticus* and listed California as a location. Hanna (1966) included both *circumscriptus* and *fasciatus* as synonyms and listed them as occurring in California and Oregon. Pilsbry (1948) included only *circumscriptus*, showing *fasciatus* as a synonym, and listed it as occurring in California and British Columbia as well as other states and provinces. At present there is no confirmed record of *Arion fasciatus* in the Pacific Northwest.

Arion silvaticus   Lomander, 1937   Forest Arion

Description: A small to medium-sized slug (30 to 40 mm long), *A. silvaticus* is pale gray dorsally and whitish laterally below a broad, dark lateral band. Its sole is white; mucus is colorless. Its lateral profile is hemispherical, and it is bell-shaped in cross-section when contracted.

Similar Species: *A. silvaticus* is similar in appearance to *A. circumscriptus* and *A. fasciatus*, but it is somewhat smaller and is more distinctly pale gray and white.

Distribution: *A. silvaticus* is common in northwestern Europe, and has been introduced into North America. It has been found in flowerbeds in Colville, Stevens Co., Washington, and in the forests of the Blue Mountains in southeastern Washington and northeastern Oregon.

Small to medium slugs (20 to 50 mm long) with dorso-lateral band, which may be indistinct. Lateral band on the mantle encloses the pneumostome; when contracted, cross-section is semi-circular.

Arion hortensis Group

This group contains at least 3 closely related species (Kerney, Cameron, and Riley 1994). The species currently found in the Pacific Northwest is considered to be *A. distinctus* Mabille, 1868 (Forsyth 2004) or *A. hortensis* Férussac, 1819 "form A" (Kerney, Cameron, and Riley 1994).

Arion distinctus   Mabille, 1868   Darkface Arion

Arion distinctus: Olympia, Thurston Co., WA

Description: This is a small to medium-sized slug, usually 25 to 30 mm or sometimes longer. Its color is slate to bluish-gray or brownish (often tinted by the mucus), darkest mid-dorsally and becoming lighter on the sides. There is a distinct black dorso-lateral band, separated from the dark dorsal color by a narrow light band. The pneumostome is enclosed in the band on the right side of the mantle. The head and tentacles are dark blue-gray; the neck is lighter. The sole is yellow to orange, not noticeably tripartite. The mucus is yellow-orange and tints the body an orangish color. When contracted the cross-section (back and sides) is cylindrical to the pedal furrows.

Similar Species: Similar species include *A. circumscriptus*, *A. fasciatus*, and *A. silvaticus*, which have white soles and are bell-shaped in cross-section when at rest. *A. subfuscus* is larger as an adult but has a yellowish sole and mucus. The pneumostomes of all three of these are below the lateral band on the mantle.

Distribution: *A. distinctus* is native to Europe and introduced into North America. It is found locally in the PNW, rather commonly in suburban areas west of the cascades.

Quite small slug with faint markings; tubercles with pointed mucous tips.

Arion intermedius   (Normand, 1852)   Hedgehog Arion

Description: A small slug (normally less than 20 mm long), *A. intermedius* is generally yellowish-gray or whitish. There is usually a faint gray band, which is sometimes difficult to see. When present and distinct, the band encloses the pneumostome. There is a row of black dots above the anterior half of the pedal furrows. The head and tentacles are dark. The sole is light yellowish; mucus is lemon yellow. When contracted, mucous points form on the tubercles, hence the common name "hedgehog arion."

Similar Species: The small size and usually light yellowish color, faint lateral bands, and prickly appearance when contracted, distinguish this slug from other arions.

Distribution: *A. intermedius* is a European slug, introduced in North America and common locally in areas of western Washington and Oregon. It may be found in yards and gardens, in undeveloped grassy areas or forest edge.

1

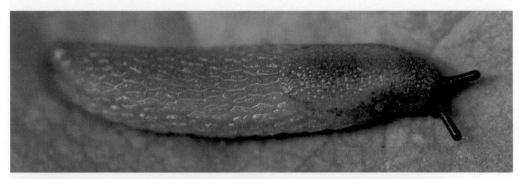

2

1. *Arion intermedius*: Olympia, Thurston Co., WA
2. *Arion intermedius*: Corvallis, Benton Co., OR

Family: ANADENIDAE

Four genera of this family of slugs occur in the Pacific Northwest: *Prophysaon* (taildroppers), *Kootenaia* (pygmy slug), *Carinacauda* (Cascade axetail), and *Securicauda* (Rocky Mountain axetail). Of these genera, only *Prophysaon* contains several species; the other three have only one known species each. They are mostly small slugs, although some *Prophysaon* grow to 60 or 80 (rarely to 100) mm long. The pneumostomes of all four genera are located a little forward of the midpoint in the right side of the mantle or near to or slightly posterior to the midpoint.

Genus: *Prophysaon*

The *Prophysaon* are named for their pneumostome placement in the front half of the mantle, a characteristic shared with the *Arion*. Unlike the *Arion*, *Prophysaon* lack a caudal mucous pit, and when attacked or harassed, they are able to self-amputate their tails, hence their common name, taildroppers. There is usually a visible line at the point of abscission, which may be seen as an oblique indented line, or just a whitish line, across the sole; or it is inconspicuous in some species. A specimen that has recently dropped its tail will often have a vertical crease in its posterior end that may be mistaken for a caudal pit. The sole is not tripartite.

Key to the Species of *Prophysaon*

(a)   Without dark-colored dorso-lateral bands behind the mantle, but they may
       be present on the mantle . . . .[b/bb]

  (aa)   With dark-colored dorso-lateral bands on the tail as well as the mantle,
          but they are sometimes inconspicuous because of a lack of contrast
          between the markings and base colors . . . .[g/gg]

(b)   Rather large for the genus (30 to 100 mm long), with a diamond or elongated
       hexagon pattern of thin, black lines on the sides of the tail and usually with a
       light mid-dorsal line behind the mantle . . . . [c/cc]

  (bb)   Smaller (10 to 40 mm, seldom longer). Tail with longitudinal or oblique,
          deeply impressed grooves or, if conspicuously spotted, then with fine,
          oblique, posteriorly descending, lateral lines joined regularly by fainter,
          forwardly descending oblique lines. . . . .[e/ee]

(c)   Length 30 to 60 mm. Tail behind the line of abscission relatively short, color
       and texture differing little, if at all, from the body in front of that line. Color
       variable but normally without a distinct yellow border around the mantle. . . [d/dd]

  (cc)   Length 50 to 80 mm or longer. Color variable but normally with a yellow
          border around the mantle. Tail behind the line of abscission relatively
          long, laterally compressed, and slightly translucent, usually differing
          conspicuously from the area in front of that line . . . . *Prophysaon foliolatum*

(d)   Mantle usually with distinct or faint lateral bands (sometimes lacking). Tail
       with light mid-dorsal stripe and with regular, impressed pattern of diamond
       or elongated hexagon shapes. Widespread . . . .*Prophysaon andersoni*

  (dd)   Body light brown with black spots. Lateral bands on mantle composed
          of irregular black blotches. Light dorsal tail stripe and diamond pattern
          present but not as conspicuous. Northern Oregon Coast Range.
          . . . . *Prophysaon andersoni* (spotted variety)

(e)   Small to medium slug, 20 to 40 mm long, light brown with black spots scattered
       over the mantle and tail. Southwestern Washington and northwestern Oregon.
       . . . . *Prophysaon vanattae pardalis*

(ee) Small to medium-sized slugs, less than 40 mm long. Without conspicuous spots. If irregular brown marbling on mantle, then the animal is distinctly papillose on the mantle, sides, and tail . . . .[f/ff]

(f) Small (15 to 40 mm long), blue-gray with whitish flecks in the integument but no other colored markings. Parallel longitudinal ridges and grooves on the tail . . . . *Prophysaon coeruleum*

    (ff) Smaller, seldom more than 25 mm long. Light brown or sometimes bluish, with irregular darker brown markings on the mantle. Thread-width brown or black indented lines on the tail, irregularly parallel dorsally but oblique laterally. Body covered with distinct papillae . . . . *Prophysaon dubium*

(g) Typical markings behind the mantle consist of a black lateral band and a dorsal wedge of distinct, sometimes rather bright color (orange, red, various shades of brown or blue-gray). Between the dorsal wedge and lateral band is a narrow border of a lighter hue of the dorsal color. . . . . [h/hh]

    (gg) Markings similar but less obvious because the markings lack contrast between the colors. Colors are dark olive-brown or dark gray to near black with a dark gray sole. Columbia Gorge through southwestern Washington . . . .*Prophysaon obscurum*

(h) Lateral lines relatively thin with edges sharply delineated; the light-colored border between the lateral line and the dorsal wedge distinct. Dorsal color orange, red, varying shades of brown or blue-gray. Mantle markings variable (lateral band or irregular markings). Sides below the lateral band contrasting, whitish. Western Washington through northwestern Oregon . . . . *Prophysaon vanattae*

    (hh) Lateral lines broader, fading toward the edges, often extending forward on the mantle (though not continuous at the border). Sides light-colored to whitish. Eastern Washington and northern Idaho . . . . . .*Prophysaon humile*

Genus: *Prophysaon*
Subgenus: *Prophysaon*

Prophysaon andersoni (J. G. Cooper, 1872)   Reticulate Taildropper

Description: A medium-sized slug, 30 to 60 mm long; its mantle is about one-third the body length, with the pneumostome located slightly forward of the middle in the right side. The front edge of the mantle is free nearly back to the pneumostome. The color is variable, reddish-gray, light brown, or buff; the mantle usually a little lighter and with darker lateral bands confined to the mantle and above the pneumostome on the right side. There is a light, mid-dorsal stripe behind the mantle. The sides are marked with narrow, oblique, indented lines crossing to form a diamond-shaped or elongated hexagon pattern. The head is light brown with darker tentacles. The short tail behind the line of abscission appears to be a normal extension of the body except for an oblique, impressed line that delineates it from the forward portion. The shell, enclosed in the mantle, is thick and solid.

    There is a spotted variety in Tillamook Co., Oregon, that appears to be *P. vanattae pardalis*, but its genitalia proves it to be *P. andersoni*. Morphologically, this form of *P. andersoni* can be recognized by its larger size, the sparse diamond pattern on the sides of its tail, and the faint, light dorsal stripe.

Similar Species: Several varieties of *P. andersoni* have been described, based primarily on color variations, but these are not recognized as subspecies. *P. foliolatum*, the most similar species, is somewhat larger than *P. andersoni*. It is often of the same color and with similar

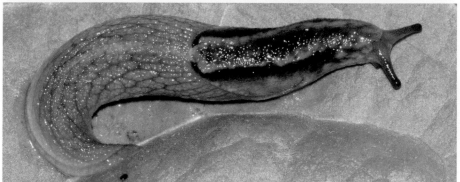

1. *Prophysaon andersoni*: Riverside State Park, Spokane Co., WA
2. *Prophysaon andersoni*: Touchet River, Umatilla National Forest, Columbia Co., WA
3. *Prophysaon andersoni* (spotted form): Tillamook Co., OR

markings, but usually has a yellow border around the edge of the mantle; it varies most by the distinctive difference between the laterally compressed portion of the tail behind the line of abscission and the area forward of that line. In *P. andersoni*, there is no obvious distinction between these areas.

Distribution: *P. andersoni* has been reported from the Pacific Slope, from Alaska through western Washington and Oregon to Central California and eastward, including localized areas in northeastern Washington, the Blue Mountains of Washington and Oregon, the Idaho Panhandle, and northwestern Montana. It is a species of moist conifer forests, is usually associated with logs and other debris, and may be seen in the open during wet weather.

Prophysaon coeruleum   Cockerell, 1890   Blue-gray Taildropper

Description: The blue-gray taildropper is a small to medium slug, 20 to 40 mm in length when extended. The mantle is about one-third the overall length, with the pneumostome just forward of the middle, and it may be covered with fine granules (small rounded bumps). The color is usually a uniform blue-gray with tiny white flecks in the integument. Sometimes it may be lighter on the sides and the sole may be white. Prominent grooves and ridges on the tail are horizontal and generally parallel, although occasionally interconnecting. The grooves become obliquely angled forward to nearly vertical below the mantle and slope forward anteriorly.

1. *Prophysaon coeruleum*: Gifford Pinchot NF, Lewis Co., WA
2. *Prophysaon coeruleum*: Chatcolet Lake, Benewah Co., ID
3. *Prophysaon coeruleum*: Benewah Co., ID

Similar Species: *P. coeruleum* is a species complex rather than a single species. The typical form is found in western Washington and Oregon, although color variations to nearly black occur in the Columbia Gorge in Oregon and at higher elevations in the southern parts of its range. A blue-white variety (discovered by Dr. J. Applegarth) occurs in the Central Coast Range in Oregon, and a bluish-brown variety (Klamath taildropper of Frest and Johannes) is found in the more open habitats of the southern Oregon Cascades. These two varieties appear to be separate species. Wilke and Duncan, using mitochondrial DNA analysis, reported three major clades and eight subclades among specimens collected from throughout the range of the species.

Bluish-colored *P. dubium* are occasionally found, causing *P. dubium* to be grouped with *P. coeruleum* by some authors, but that species differs so distinctly that there is really little similarity. Its body is mostly covered by papillae, and the posterior-lateral grooves fan out.

In northern Idaho, the pygmy slug, *Kootenaia burkei*, is similar in color and appearance, but it is not a taildropper, its pneumostome is posterior to the middle of the mantle, and the ridges on its tail end in a unique pattern.

Distribution: The type locality for *Prophysaon coeruleum* is Olympia, Thurston Co., Washington. The range extends from Vancouver Island, British Columbia, Canada, south through western Oregon, to the western slopes of the southern Oregon Cascades, and into northern California. A disjunct population inhabits the area around Lake Coeur d'Alene and as far south as Lapwai Creek in southern Nez Perce Co., Idaho. It is a species of moist conifer forests, found in forest floor litter in association with Douglas-fir logs and other woody debris.

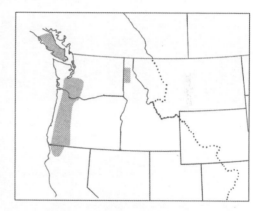

Prophysaon dubium   Cockerell, 1890   Papillose Taildropper

Description: This small slug is usually 15 to 25 mm long when extended, but large specimens may reach 30 mm. It is generally light brown to olive-brown with irregular dark brown markings on the mantle. Occasionally bluish or blue-gray colored specimens are encountered. Tentacles are sometimes dark blue-gray. Prominent conical papillae cover the body, including the mantle, sides, and tail, but not the head and neck. Narrow, thread-width, dark brown or black indented lines form a specific pattern on the tail and sides. The most dorsal two lines run irregularly parallel from the mantle to about halfway back to the tail, then branch out laterally. The lines run laterally from under the mantle down the sides of the tail and become progressively steeper nearer the mantle, to vertical below the anterior end of the mantle, and slant anteriorly farther forward.

Similar Species: *P. dubium* was originally considered a variety of *P. coeruleum*, the type specimen apparently being found with that species and bluish in color. Pilsbry (1948) separated them based on genitalia. The pattern of impressed lines and ridges on the tails of the two are unique for each species, and while *P. coeruleum* may have low rounded bumps on its mantle, these are nothing like the distinct papillae covering most of the body of *P. dubium*.

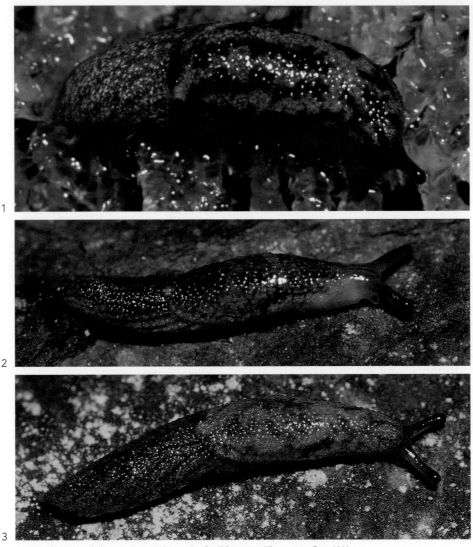

1. *Prophysaon dubium*: Priest Point Park, Olympia, Thurston Co., WA
2. *Prophysaon dubium*: Siuslaw NF, Tillamook Co., OR
3. *Prophysaon dubium*: Idaho Panhandle National Forest, Kootenai Co., ID

Distribution: *P. dubium* occurs in Washington and Oregon, into northern California. It is found in conifer forests west of the Cascades and in some riparian areas in the eastern Cascades, usually associated with hardwood litter (logs, leaves, etc.). Disjunct populations are found in northern Idaho and northeastern Washington, and a population in the Tieton River drainage in south-central Washington was found among ponderosa pine slash, an unusual habitat for this slug.

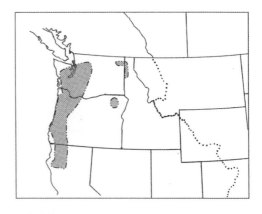

Prophysaon foliolatum   Gould, 1851   Yellow-bordered Taildropper

Description: A medium-sized to fairly large slug, *P. foliolatum* is usually about 50 to 80 mm long but may attain a length of over 100 mm. The mantle is about one-third the body length, and the pneumostome is noticeably forward of the middle in the right side. The color varies but is usually reddish-brown or light brownish, often clouded with black. The mantle is lighter colored and has a diffuse black band that passes above the pneumostome, and usually a distinct yellow border. All-yellow animals are not rare, and localized populations of other colored (gold, black, or whitish) animals may be found. The sides of the body and tail are lined with thin, thread-width, indented, oblique lines, darkly colored, crossing to form a foliate pattern of diamond shapes or elongated hexagons. There is usually a light mid-dorsal line behind the mantle. The line of abscission is distinct, and the tail behind it is laterally compressed and somewhat translucent, so it is distinctly different in appearance from the body in front of that line. The shell is of thin calcareous granules held together by the periostracum.

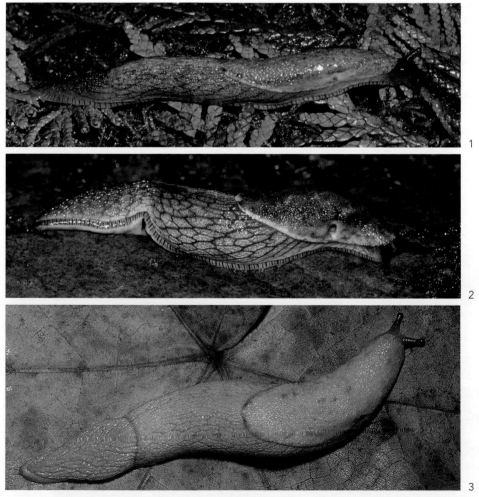

1. *Prophysaon foliolatum*: Naselle, Pacific Co., WA
2. *Prophysaon foliolatum*: Rialto Beach, Olympic NP, Jefferson Co., WA
3. *Prophysaon foliolatum*: Wishkah R., Grays Harbor Co., WA

Similar Species: *P. andersoni* is the most similar species. From outward appearances, the size, the yellow border around the mantle, and the obvious distinction between the body and the tail behind the line of abscission indicate *P. foliolatum*. Pilsbry (1948) stated that "one of the best specific characteristics" to distinguish these species may be the radular teeth of *P. andersoni*, which "differ quite perceptibly in the shorter, blunter cusps of the outer laterals and inner marginal" of *P. foliolatum*. Color variations in individuals along the Pacific Coast of Washington are distinct enough that further study may yield additional evidence of scientific value.

Distribution: *P. foliolatum* occurs mostly in western Washington along the Pacific Coast, where it is most common, and inland to the west slopes of the Cascade Mountains and around Puget Sound. Its range extends south into western Oregon, and Forsyth (2004) reported it from north to the Queen Charlotte Islands, British Columbia, Canada. This slug is often found under loose bark on logs or in other woody debris, in moist forest stands.

Subgenus: *Mimetarion* Pilsbry, 1948

Prophysaon humile   Cockerell, 1890   Smoky Taildropper

Description: A small slug (to about 24 mm, preserved), with mantle covering about two-fifths of the length. The mantle and dorsal body are smoky-gray or brownish, darkest mid-dorsally and behind the mantle. Markings are similar to that of *P. vanattae* but more diffuse—the bands are dark but wider than those of *P. vanattae* and not as sharply delineated at the edges. The mid-dorsal wedge is not distinct, as the lighter line between it and the lateral band is wider. The pneumostome is in the second quarter of the mantle, distinctly forward of the middle. The mantle is free anteriorly, back to near the pneumostome.

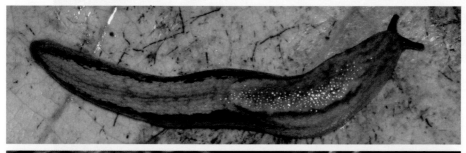

1. *Prophysaon humile*: Benewah Co., ID
2. *Prophysaon humile* (point of excision visible): Lochsa River, Idaho Co., ID

Anatomically, the epiphallus is rather thick anteriorly, with one rounded pilaster inside. Length of the epiphallus is about 9 mm.

Similar Species: *P. humile* appears much like *P. vanattae* but differs in its distribution and, exteriorly, by its wider bands with less sharply defined edges. Internally it differs in having only one pilaster inside of the epiphallus, while *P. vanattae* has two.

Distribution: The type locality is the woods around Lake Coeur d'Alene, Idaho. *P. humile* is endemic to the Idaho Panhandle, and its range extends eastward into northwestern Montana. Externally, specimens in Chelan County, Washington, on the east slopes of the Cascades, appear to be this species. They can be found in forest stands in and under logs and other woody debris, and around springs and seeps in more open areas.

Prophysaon obscurum Cockerell, 1890   Mottled Taildropper
Synonym: *P. fasciatum* var. *obscurum* Cockerell, 1893

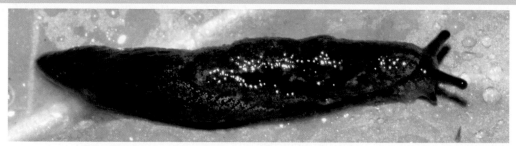

Prophysaon obscurum: Olympia, Thurston Co., Washington

Description: A small to medium-sized slug (20 to 40 mm long), with body gray to dark olive-brown to black. It is very dark dorsally, with obscure dark lateral bands. In alcohol the mantle is black, similar to the body color when live, with pale marbling at the edges. The sole is dark gray.

Similar Species: Originally considered a variety of the banded taildropper (*P. fasciatum* Cockerell, 1890, from northwestern California), this slug is well-removed from the range of that species. *P. obscurum* has "the penis-sac narrower and more tapering, and coloration different." (Pilsbry 1948). Outwardly it appears to be a darkly colored *P. vanattae* with obscure markings, and there appears to be some convergence of colors between certain specimens of brown-colored *vanattae* and *obscurum*.

Distribution: South Puget Sound southward in western Washington and into the Columbia Gorge and eastward along the Oregon/Washington border. The type locality is Chehalis, Lewis Co., Washington. *P. obscurum* is generally found in moist areas under the bark of logs or other woody debris, or crawling in the open during rains.

Prophysaon vanattae   Pilsbry, 1948   Scarletback Taildropper

Description: *P. vanattae* is a small to medium-sized slug, usually 25 to 35 mm long but rarely to 50 mm. Its color is variable, but basically it is orangish, light brown, or buff-colored, or sometimes bluish-gray; its common name is derived from specimens with red backs. The mantle is variable, usually some shade of the dorsal tail color with or without lateral bands or irregular markings. While the color is quite variable, the markings are more distinct to the species; however, see *P. vanattae pardalis*. Other than the *pardalis* variety, there is always a distinct, dark, lateral band running from the edges of the mantle to the posterior end. Around the inner edge of the band is a narrow band of a light shade of the dorsal color, and inside that band is a dorsal wedge of a dark shade of the same color. Below the lateral band, the sides are very light colored to whitish, with oblique, thread-width, indented, dark lines sloping posteriorly, and with fainter, less regular lines sloping forward. Although auto-amputation of the tail occurs, the droppable portion is short, and the line of abscission is not conspicuous. Anatomically, the epiphallus contains two high ridges (pilasters) inside.

Similar Species: *Prophysaon obscurum* is similar, but its color, usually a dark olive-brown or gray-brown, is so dark that the markings are not obvious. Some brownish *P. vanattae* are rather similar in appearance to *obscurum*. *P. humile* is similar in appearance, but has wider, less sharply delineated bands and occurs east of the Cascade Crest. *P. vanattae* var. *pardalis* (see below) is distinctly different in appearance.

Distribution: *P. vanattae* is a common slug endemic to the forests of western Washington and south to Corvallis, Oregon. The type locality is Seattle, King County, Washington. It has also been found in the Blue Mountains of Oregon. It can often be seen crawling on the forest floor, on logs or in the branches of shrubs, especially during rains. Logs and other woody debris provide cover for these slugs.

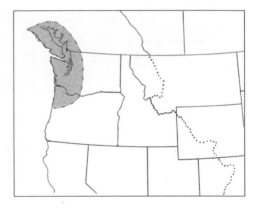

Prophysaon vanattae (golden form)

Description: A unique form, golden-colored except for a dark gray sole and pedal flange, gray or brown tentacles, and gray oblique indented lateral lines on the tail.

Similar Species: This slug is unlike others in outward appearance because of its color, but anatomically it is identical to *P. vanattae*.

Distribution: *P. vanattae* (golden form) was collected from Umatilla County, Oregon, by Leonard and Richart in 2009.

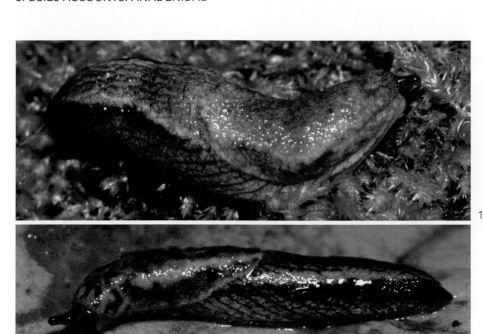

1

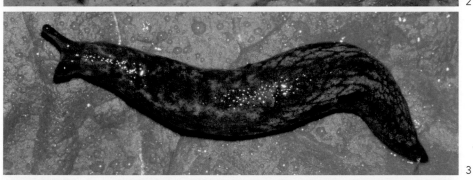

2

3

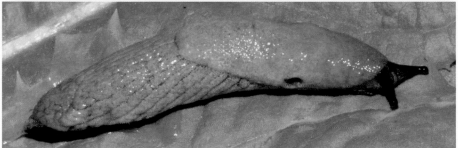

4

1. *Prophysaon vanattae*: Skokomish River, Mason Co., WA

2. *Prophysaon vanattae*: Gifford Pinchot National Forest, Lewis Co., WA

3. *Prophysaon vanattae*: Tiger Cr., Umatilla Co., OR

4. *Prophysaon vanattae* (golden variety): Tiger Cr., Umatilla Co., OR

5. *Prophysaon vanattae* (mating): W of White Pass, Gifford Pinchot National Forest, Lewis Co., WA

5

Prophysaon vanattae pardalis Pilsbry, 1948   Spotted Taildropper

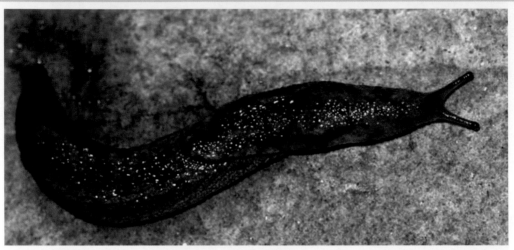

Prophysaon vanattae pardalis: Knappton, Pacific Co., WA

Description: Pilsbry (1948) named this as a color variety because it "is so different in color pattern that a name for it seems desirable." However, "the genitalia . . . appear to be entirely typical of the species." About the same size or a little smaller than typical *P. vanattae*, but lacking the typical markings; it is light brown and its mantle and tail are covered with black spots of irregular size and shape. The spots are sparse or lacking only in front of the mantle and on the sides below it.

Similar Species: This is a spotted form of *P. vanattae*. A similar-appearing spotted form of *P. andersoni* occurs in Tillamook Co., Oregon, that is larger and retains the light dorsal stripe and some of the lateral diamond pattern of that species.

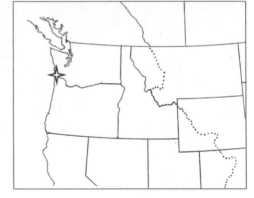

Distribution: This color variety is found in southwestern Washington and northwestern Oregon. The type locality is Long Beach, Pacific County, Washington.

Genus: *Kootenaia*

Kootenaia burkei Leonard, Chichester, Baugh & Wilke, 2003   Pygmy Slug

Description: This is a small slug, 10.2 mm long when extended. The roundly elliptic mantle measures about two-fifths of the body length. It is nearly uniformly blue-gray or dark gray to nearly black with tiny bluish-white flecks in the integument. The head and neck are lighter, as are the medial half of the ocular tentacles. The lower tentacles are black medially, with a transparent tip. The sole and pedal flange are transparent gray, lighter than the dorsal color, and with small whitish flecks. The lower sides become increasingly lighter as a result of the increasing density of the light flecks. Wrinkles on the mantle radiate as rounded ridges and valleys, giving it a scalloped appearance around the edges. The pneumostome is slightly posterior to the middle of the right edge of the mantle. There are about 10 flat, rather broad ridges running parallel on the tail, so that they spread as the tail narrows. Posteriorly, a line crosses these ridges, forming a scalloped junction in a daisy-petal pattern, and leaving a second set of very short ridges or polygons at the end of the tail.

Similar Species: The blue-gray taildropper, *Prophysaon coeruleum*, is most similar to *Kootenaia burkei*, and occurs in the same areas but with a much broader range. However, the taildropper lacks the small polygons following the caudal ridges at the end of the tail, and its pneumostome is anterior to the midpoint of the mantle.

Distribution: *K. burkei* occurs around Lake Coeur d'Alene and southwestern Shoshone County, Idaho, northward in the Idaho Panhandle into adjacent British Columbia and eastward to include Mineral and Sanders counties in northwestern Montana. It can be found in closed moist forest and riparian areas in litter and woody debris.

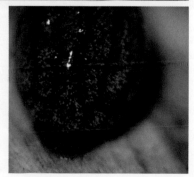

1. *Kootenaia burkei*: Idaho Panhandle National Forest, Shoshone Co., ID
2. *Kootenaia burkei*: Idaho Panhandle National Forest, Shoshone Co., ID
3. *Kootenaia burkei* (posterior end of tail showing branching of grooves): Shoshone Co, ID

Genera: *Carinacauda* and *Securicauda*

Although of different genera, the following two species are very similar in outward appearance; however, they differ distinctly in their internal anatomy. *Carinacauda* was originally discovered in Oregon by Dr. John Applegarth while an employee of the US Bureau of Land Management at Eugene, Oregon; subsequently the similar-appearing *Securicauda* was discovered by Bill Leonard in Idaho in August 1999. Dr. Applegarth dubbed his find the axetail slug because when contracted the keel on its short tail protrudes conspicuously above the posterior edge of the mantle. It was suggested that the animal might be *Gliabates oregonius* Webb, 1959, and although John expressed doubts as to that identification, that is what it has been called for the past several years. Collaboration between Bill Leonard and other researchers has shown, through anatomical and DNA analyses, that these slugs are of two different genera; a paper has recently been published by those researchers documenting that the Cascade axetail is not Webb's *Gliabates* and describing these slugs as *Carinacauda stormi* and *Securicauda hermani*.

Genus: *Carinacauda*

Carinacauda stormi Leonard, Chichester, Richart & Young, 2011   Cascade Axetail

Description: This very small slug is about 12 to 15 mm long, with a relatively long mantle and short tail. The mantle covers 58% to 64% of the total length when live and in motion. The tail extends behind the mantle 20% to 27% of the extended length of

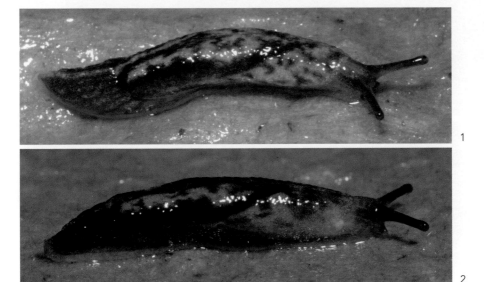

1. *Carinacauda stormi*: Willamette National Forest, Lane Co., OR

2. *Carinacauda stormi*: Willamette National Forest, Lane Co., OR

3. *Carinacauda stormi* (close-up of tail showing dorsal keel and ridges and grooves): Willamette National Forest, Lane Co., OR

the animal. The mantle is pale gray or light tan with dark brown to nearly black irregular markings and relatively broad lateral stripes. The pneumostome is nearly centered on or slightly posterior to the middle of the right side of the mantle, and it is positioned below the oblique lateral stripe. The head and neck are generally lighter gray than the rest of the animal and a little translucent; the tentacles are dark brown or gray. Below the mantle the sides are pale gray with or without dark markings. The tail has high keeled ridges (a central keel and several slightly oblique laterals on each side). The mid-dorsal keel has a pale gray crest and the grooves between the ridges are marked with dark lines. The central keel, and to a lesser extent the lateral ridges, are higher than the posterior edge of the mantle when the slug contracts. The gray sole is flecked with white. The mucus is clear.

Similar Species: The small size, distinctly long mantle, and short keeled tail separate this slug from all but *Securicauda hermani*, which is only known from the Rocky Mountains of northern Idaho.

Distribution: *Carinacauda stormi* was originally found in the Cascade Mountains on the north side of the McKenzie watershed, Linn Co., Oregon, at about 1200 meters (4000 ft) elevation (Applegarth, personal communication). Since then, additional specimens have been found in Marion, Linn, and Lane counties, Oregon, all on the west slope of the Cascade Mountains between 610 and 1190 meters (2000 and 3900 ft) elevation. They occur in Douglas-fir/western hemlock forests in Douglas-fir needle litter and around woody debris where understory vegetation is lacking.

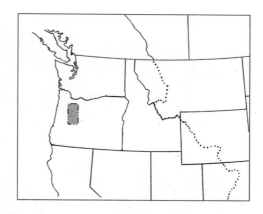

Genus: *Securicauda*

Securicauda hermani   Leonard, Chichester, Richart & Young, 2011
   Rocky Mountain Axetail

Description: This very small slug is about 8 to 10 mm long with a relatively long mantle and short tail. The mantle covers 64% to 80% of the total length when live and in motion. The tail extends behind the mantle 7% to 20% of the extended length of the animal. The mantle is tan or gray with dark gray or brown blotches and relatively broad but somewhat indistinct lateral stripes. The position of the pneumostome is slightly anterior to the mid-line on the right side of mantle and below the lateral stripe. The head is tan; the tentacles dark gray or brown. The neck is somewhat translucent. Below the mantle the sides are gray or tan. The tail is light brown with or without dark brown markings. The tail is with high keeled ridges (a central keel and slightly oblique laterals). The central keel, and to a lesser extent the lateral ridges, are higher than the posterior edge of the mantle when the slug contracts. The sole is gray or tan with white flecks. The mucus is clear.

Similar Species: The small size, distinctly long mantle, and short keeled tail separate this slug from all but *Carinacauda stormi*, which is known only from the Cascade Mountains of Oregon.

Distribution: *Securicauda hermani* occurs in Benewah and Shoshone counties, Idaho, in western hemlock and western red cedar forests at 1060 to 1300 meters (3480 to 4265 ft) elevation. It has been found in or near riparian areas and seeps under woody debris or among mosses.

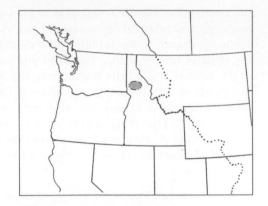

1. *Securicauda hermani*: Idaho Panhandle National Forest, Shoshone Co., ID
2. *Securicauda hermani*: Idaho Panhandle National Forest, Shoshone Co., ID
3. *Securicauda hermani*: Benewah Co., ID

Family: **ARIOLIMACIDAE**
Genus: *Ariolimax*

Ariolimax columbianus   (Gould, 1851)   Pacific Bananaslug

Description: *Ariolimax columbianus* is our largest slug, reaching 180 to 260 mm long when extended. Its usual color is rather uniformly olive green, but some are of varying shades of brown, greenish brown, yellow, or white. It is sometimes spotted or mottled with black, which may appear as just a small spot to irregularly blotched to mostly or nearly all black. The mantle is rather short, one-quarter to one-third of the body length. The pneumostome is large and is in the third quarter of the right side of the mantle, distinctly behind the middle. The back is keeled for most of its length behind the mantle, and there is a large caudal pit at the juncture of the pedal furrows.

1. *Ariolimax columbianus*: Naselle River, Pacific Co., WA
2. *Ariolimax columbianus*: Woodland Cr., Thurston Co., WA
3. *Ariolimax columbianus* (juvenile): McAllister Cr., Thurston Co., WA

Similar Species: Its large size, keeled tail, and caudal pit separate this species from all but other species of *Ariolimax*. The other known species and subspecies occur in west-central and southern California. Other varieties of the bananaslug occur on some of the San Juan Islands of Washington, and islands of British Columbia. Forsyth (2004) included a photograph of a white *Ariolimax* from Graham Island, British Columbia, and Leonard (this work) has a photograph of one from Thurston County, Washington. Since this is not a common occurrence on the Washington mainland, Leonard's photograph may be of an albino. A yellow variety occurs on Patos Island, Washington. It is smaller than typical *A. columbianus* from the

Ariolimax columbianus (eggs and hatchling): Vancouver Island, British Columbia

mainland and from other islands in the San Juan group. It is a uniformly banana-yellow color, and all of the adults observed appear to be of consistent size and color. No black markings were seen among the more than 140 Patos Island specimens observed. *Ariolimax steindachneri* Babor, 1900, was described from "Puget Sound." Mead (1943) believed it to be an abnormal form of *A. columbianus*, and Pilsbry (1948) considered it a "species dubia."

Distribution: *Ariolimax columbianus* is a species of the western sides of the Pacific states, found from southern Alaska, through western British Columbia, Washington, Oregon, and to west central California. In Washington it also occurs sparsely on the east slopes of the Cascades in riparian areas or other moist forest sites. Disjunct populations are also known from northern Idaho and from Umatilla Co., Oregon. It is a species of moist forest habitats, regularly seen on the forest floor or on or under logs or other woody debris.

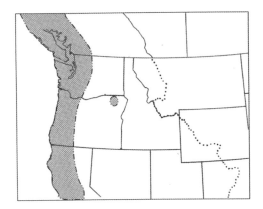

Genus: *Gliabates*

Gliabates oregonius   Webb, 1959   Salamander Slug

Description: Paraphrasing the description by Webb (1959): Very small, about 10 mm long. The mantle is about half the body length. It is gray dorsally and with a darker area on the strong keel on the back and tail. The mantle is darker gray than the sides of the body, which lack other markings.

Webb had originally thought, when this specimen was collected, that it was an immature example of *Prophysaon coeruleum*. The shell plate is moderately solid, broadly oval, the left margin nearly straight and less curved than the right margin. It is noteworthy that Webb's description was based on a single specimen in the early stages of decomposition (Webb 1977), which may explain why mention of so many basic characteristics were omitted.

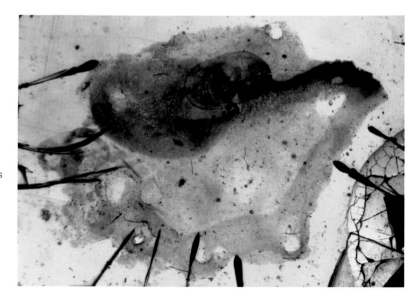

Gliabates oregonius (slide-mounted type specimen [FMNH #308279]): Long Tom R., adjacent to Alderwood State Park, Lane Co., OR (Jochen Gerber photo)

Similar Species: Webb's statement that he had at first though it to be a *Prophysaon coeruleum* indicates a similarity in appearance between that slug and his new species. Webb also compared his new species to *Hesperarion hemphilli*. Since the mantle of the *Hesperarion* species is relatively short, one would not think them similar in outward appearance, but Webb may have been confining this comparison to the genitalia.

Distribution: The only specimen of *Gliabates oregonius* was from Lane Co., Oregon, the type locality on the east bank of Long Tom River adjacent to Alderwood State Park. The slug was found in leaf litter under bushes in a mature conifer forest. It has not been documented again since Webb's original find over fifty years ago.

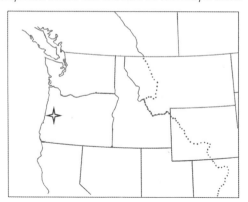

Genus: *Hesperarion*

Hesperarion Simroth, 1891, is a genus of rather small slugs in which the mantle is small, with the pneumostome located in the third quarter. The genital pore is located behind the right tentacle below the leading edge of the mantle. There is a caudal mucous pit. Of four described species: *H. hemphilli* (Binney, 1875) occurs in southwestern California; *H. niger* (J. G. Cooper, 1872) is found in west-central California; *H. plumbeus* Roth, 2004, from Shasta County, California; and *H. mariae* Branson, 1991, the only species in the Pacific Northwest, occurs in northwestern Oregon.

Hesperarion mariae Branson, 1991   Tillamook Westernslug

Description: This small to medium-sized slug is 20 mm long preserved. The mantle is about one-third the overall length. The color is light reddish-tan, the mantle marked with irregularly spaced black spots, flecks, or blotches. The reddish-tan on the body pales on

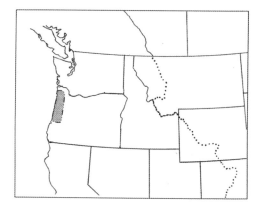

1 & 2. *Hesperarion mariae*: Hebo Ranger District, Siuslaw NF, Tillamook Co., OR

the sides, and rows of black flecks follow the grooves on the back and sides behind the mantle. The pneumostome is located in the third quarter of the right side, just posterior to the middle of the mantle. The dorsal ridge behind the mantle is roundly angled. The foot margin is light yellowish; the sole is white. At the meeting of the pedal furrows is a distinct caudal pit, below which the posterior end of the sole is deflected downward.

Similar Species: Other *Hesperarion* species are found well south of the range of *H. mariae* in west-central and southern California. Juvenile *Ariolimax columbianus*, with their caudal pit and similarly positioned pneumostome, often have black or gray flecks similar to those of *Hesperarion*, but they also have a distinct long, sharp keel on their posterior half, which *Hesperarion* lacks.

Distribution: *Hesperarion mariae* was first described by Branson (1991) from two specimens collected from decaying wood in a Sitka spruce forest in the Oregon Coast Range, Tillamook Co., in June 1981. It has since been found in the Oregon Coast Range as far south as Douglas County.

Genus: *Magnipelta*

Magnipelta mycophaga   Pilsbry, 1953   Magnum Mantleslug

Description: A medium-sized slug, attaining a length of 60 to 80 mm or more. Its long, elliptical mantle covers about two-thirds to three-quarters of the animal's total length, and the leading edge of the mantle is free for about 10% of its length. When contracted the slug is completely beneath its mantle. The tail behind the mantle is less than one-fourth of the total length, and it has an angled dorsal ridge. The pneumostome is located approximately 60% of the length of the mantle back in its right side and is overhung by an angular protuberance of the mantle surface. The head and neck are light brown and textured with low rectangular tubercles. Base color of the mantle is light buff, and it is mottled with dark grayish-brown. A blackish-brown lateral line runs nearly the length of

1. *Magnipelta mycophaga*: Pend Oreille Co., WA (TEB photo)
2. *Magnipelta mycophaga*: Rand Creek, Flathead Co., MT
3. *Magnipelta mycophaga* (defensive posture): Pend Oreille Co., WA (TEB photo)

the mantle on each side, passing above the pneumostome on the right. Ground color of the body and tail is very light gray-buff. Blackish-brown impressed lines radiate obliquely from under the mantle and from the indistinctly defined, light mid-dorsal line of the tail to the pedal furrows and, less distinctly, to the angle of the foot. These lines fade anteriorly but continue as a row of dots between the pedal furrows to below the head. The sole is dirty white, the center portion more gray than the outer areas and displays a distinct forward wave action during locomotion. In preserved specimens a groove can be seen separating the longitudinal sections. At rest, the head and nearly all of the tail are concealed under the mantle. The shell is white, 5.2 by 2.9 mm, porous to granular, most dense at the nucleus and held together by a thin membrane. When first disturbed, *Magnipelta* will suddenly flare its mantle out, like popping open an umbrella.

Similar Species: The exceptionally large mantle, combined with the size of the animal, separates *M. mycophaga* from all other slugs in the Pacific Northwest.

Distribution: *M. mycophaga* is found under logs and woody debris in the upper subalpine fir zones in the Selkirk Mountains of northeastern Washington, northern Idaho, northwestern Montana, and into southeastern British Columbia. Also in the Selway-Bitterroot Range of Idaho and Montana.

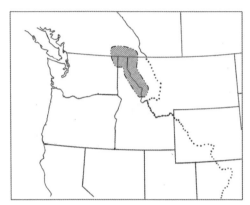

Genus: *Udosarx*

Udosarx lyrata   Webb, 1959   Lyre Mantleslug

Description: A very small slug, about 10 to 12 mm long. Its color is variable shades of grayish or bluish-gray, lighter on the keel. The mantle is 45% of the total length, and is also somewhat lighter and has black flecks across its mid-portion. Near its posterior end the mantle is marked with heavy black lateral bands, the right one curving over the pneumostome so that the bands form the shape of a lyre, and the area between the bands is light buff-colored. There are 7 to 8 lateral grooves on the tail between the mid-dorsal keel and the pneumostome. The pneumostome is located nearly three-quarters of the way back in the right side of the mantle, and there is a secondary notch in the right posterior margin of the mantle.

Similar Species: The conspicuous lyre-markings on the mantle stand out, making this slug unique among others with which it may be found.

Distribution: *Udosarx lyrata* is found in forested sites in the northern Idaho Panhandle and adjacent parts of Mineral and Ravalli counties, Montana.

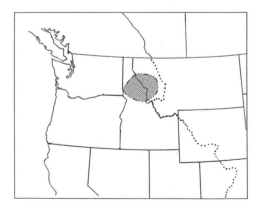

1. *Udosarx lyrata*: Shoshone Co., ID
2. *Udosarx lyrata*: W of Darby, Bunkhouse Cr., Ravalli Co., MT
3. *Udosarx lyrata*: Shoshone Co., ID

Genus: *Zacoleus*

Zacoleus is a small slug with the pneumostome well back in the posterior half of the mantle and a notch in the right posterior edge of the mantle over a second pore. It is keeled behind the mantle, has a tripartite sole, and lacks a caudal pit. Only one species of *Zacoleus* has been described to date, but a newly discovered, closely related species appears here as *Zacoleus leonardi* new species. The two differ in their color and body shape, and in having widely separated ranges.

Zacoleus idahoensis   Pilsbry, 1903   Sheathed Slug

Description: This small slug, 20 to 24 mm long, has a mantle that is about 40% of its length in live, extended specimens. The pneumostome is more than two-thirds back in the right side of the mantle, and there is a second opening under a notch in the right posterior edge of the mantle. The animal is light brown with tiny black specks that give it a grayish appearance. The mantle is smooth, generally the same color as the rest of the animal, but the concentration of black specks gives its posterior border and sometimes the anterior quarter a blackish appearance. On the head and neck are parallel longitudinal grooves, and fine, close-set, white pigments in the neck give it a bluish-silver cast. The bluish cast is most apparent in specimens from the southern portion of its range. The body is distinctly keeled behind the mantle, but most prominently at the posterior end. The tip is rather sharply tapered laterally. Long, low tubercles form parallel horizontal rows on the tail. The sides are translucent yellowish with a gray cast. The sole is dirty white, slightly yellowish, and tripartite. The genital pore is located behind and a little lower than the right tentacle, about two-thirds of the way back to the mantle. Pedal furrows are distinct; there is no caudal pit.

Similar Species: The Ryan Lake slug (*Z. leonardi*) is a new species of *Zacoleus* that differs in its lighter color and range. *Deroceras laeve* is similar in size and somewhat in coloration,

1. *Zacoleus idahoensis*: Meadows, Adams Co., ID
2. *Zacoleus idahoensis*: Meadows, Adams Co., ID (TEB photo)
3. *Zacoleus idahoensis*: Idaho Panhandle National Forest, Benewah Co., ID
4. *Zacoleus idahoensis*: Salmon River, Idaho Co., ID

but the keel on its tail is confined to the posterior end while that of *Zacoleus* is lower and
extends from the mantle to the posterior end.
In addition, *Deroceras laeve* lacks the notch in
the right posterior margin of the mantle.

Distribution: *Zacoleus idahoensis* is found in
the Idaho Panhandle and adjacent western
Montana. Type locality is Meadows, Idaho. It
is found in moist forest stands and riparian
zones.

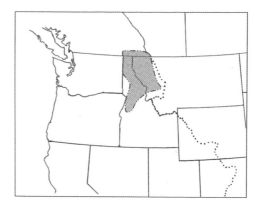

Zacoleus leonardi new species   Ryan Lake Slug

Description: This small slug measures to 22 mm extended. It is generally a uniform light
bluish-white color (sometimes with a pinkish hue), with white specks of pigment embed-
ded in the integument. The pneumostome is located well back in the right side of the
mantle, and there is a second pore under a notch in the right posterior edge of the mantle.

1. *Zacoleus leonardi* ("Ryan Lake Slug"): Ellsworth Cr., Pacific Co., WA
2. *Zacoleus leonardi* ("Ryan Lake Slug"): Wynoochee River, Grays Harbor Co., WA
3. *Zacoleus leonardi* ("Ryan Lake Slug"): Snow Cr., Olympic National Forest, Jefferson Co., WA

The tail is keeled, from the mantle to its tip, the narrow ridge of which is adorned with tiny white or black pigments. Specimens included as genus B in DNA analysis (Wilke and Ziegltrum, unpublished) showed this slug to be most closely related to *Zacoleus idahoensis*, borderline to being the same or a separate genus. Being so close to *Zacoleus idahoensis* in appearance, I have kept it within the same genus. A type specimen was deposited in the Carnegie Museum of Natural History as CM126221. The species name recognizes William Leonard for his extensive study of slugs of the PNW, and his works with and photography of animals of the lower orders.

Similar Species: This species' color is light blue-gray, while that of *Z. idahoensis* is brown or grayish-brown. These two species are also widely separated geographically. *Boettgerilla pallens*, with its blue-gray color, long worm-like appearance, and long keel on the tail, is very similar in appearance, but its pneumostome, while in the third quarter of the mantle, is farther forward, and it lacks the second pore and the notch in the posterior, right edge of the mantle.

Distribution: This little slug is generally found at moderate elevations, and is usually associated with large woody debris. It was originally discovered in 1995 by T. Burke at the type locality, Ryan Lake, Skamania County, Washington, within the Mount St. Helens blast zone. *Zacoleus leonardi* has since been found in other sites in the southwest Washington Cascades, on the Olympic Peninsula, and in northwestern Oregon.

Family: BINNEYIDAE
Genus: *Hemphillia*

Slugs of the family Binneyidae have a shell that is not completely covered by the mantle. The viscera are elevated into a hump under the mantle and the visceral cavity does not extend into the tail behind the mantle as in other slugs. Of the two genera, *Binneya*, which has an exposed, sub-spiral shell, is found in coastal California and some of the offshore islands. *Hemphillia*, which has a slightly convex shell-plate, partially exposed through a slit in the mantle in back of the visceral hump, is endemic to the Pacific Northwest. A third, recently described genus, *Staala gwaii* Ovaska et al., 2010, is similar to the *Hemphillia* in external morphology, but differs in having its shell completely enclosed within the mantle. This unique little slug is not found within the Pacific Northwest, but inhabits areas in the Queen Charlotte Islands, British Columbia, Canada.

Hemphillia, the jumping-slugs, are so called because when irritated by handling or attack, they writhe and flop around like a fish out of water. There are two basic forms of *Hemphillia*. The *Hemphillia glandulosa* group, including *H. glandulosa*, *H. burringtoni*, and *H. pantherina* consists of small slugs, about 20 mm long. Their bodies are compressed under the visceral hump, and their tails behind the hump are relatively short, arched, and distinctly keeled. The *Hemphillia camelus* group, including *Hemphillia camellus*, *H. danielsi*, *H. malonei*, and *H. dromedarius* consists of larger slugs, often 30 to 50 mm or longer. Their tails are relatively long with a flat dorsal line. In cross-section, the tail is somewhat triangular but with an indented dorsal line behind the mantle, separating a herringbone pattern of oblique lateral grooves.

Key to the Species of *Hemphillia*

(a)   Small, about 20 mm long or less. Tail behind the visceral hump relatively short (about one-half the length of the mantle) with a conspicuous dorsal keel, arched for all or a part of its length . . . . *Hemphillia glandulosa* group [b/bb]

    (aa)   Medium-sized, 30 to 60 mm long. Tail behind the visceral hump relatively long (more or less the same length as the mantle). Dorsal profile of tail straight. Tail laterally compressed but not raised into an arched keel . . . .*Hemphillia camelus* group[c/cc]

(b)   Posterior one-third of visceral pouch not covered by the granulose-textured mantle. Head, tentacles, and tail with a white dorsal stripe. . . . *Hemphillia pantherina*

    (bb)   Mantle completely covers the visceral pouch, except where shell shows through the slit as in all of our *Hemphillia*. Mantle usually distinctly papillose, seldom otherwise . . *Hemphillia burringtoni* and *Hemphillia glandulosa*

(c)   Found in the Cascades and westward . . . . [d/dd]

    cc)   Found in northeastern Washington, northern Idaho, and western Montana . . . . [e/ee]

(d)   Primarily dark gray markings over dishwater-white ground color. No light-colored, mid-dorsal, tail stripe. . . . . *Hemphillia dromedarius*

    (dd)   Primarily brown or brownish markings, or some black, over buff or light-gray ground color. Mantle variously patterned or merely light-colored with a few black spots. Normally with a distinct, buff-colored, mid-dorsal line on the tail. . . . . *Hemphillia malonei*

(e)   Dark brown to blackish or gray markings over buff or dishwater-white ground color. Distinct light-colored mid-dorsal stripe on tail . . . . . *Hemphillia camelus*

    (ee)   Lacking conspicuous mid-dorsal stripe on tail. Crest of the tail may lack dark flecks or mottling of the sides, leaving a clear line of the ground color, but it is not brighter or of different shade . . . . [f/ff]

(f)   Mantle distinctly papillose. Brown or gray markings over buff ground color. Southern Idaho Panhandle (new taxon). . . . . *Hemphillia danielsi* ssp.

    (ff)   Gray, black, or bluish-black markings over dishwater-white or bluish-gray ground-color. With irregular lateral black stripe on posterior half of the mantle. Western Montana . . . . *Hemphillia danielsi*

Hemphillia glandulosa Group

Small jumping-slugs with relatively short, high keeled tails.

Hemphillia burringtoni   Pilsbry, 1948   **Burrington Jumping-slug**

H. glandulosa   Bland & W. G. Binney, 1872   Warty Jumping-slug

Description: These two little slugs are very similar in appearance and nearly impossible to distinguish as one or the other. The descriptions by Pilsbry (1948) and Branson (1972) separated the two species by the papillose mantle of *H. glandulosa*, and by the strong black oblique lines on the tail and a row of gray dots above the pedal furrows of *H. burringtoni*, characteristics that have not been found to be consistent. Pilsbry (1948) also described and illustrated the genitalia of *H. burringtoni* as quite different from that of *H. glandulosa*, characteristics which have not been found through recent dissections. Branson (1972) elevated *H. burringtoni* to a full species, "guided by the consistent differences of genitalia and external pigmentation patterns." Branson also described the distribution of the two species as approximately the same areas, as determined by Wilke and Ziegltrum through molecular (DNA) analyses.

During surveys by the US Forest Service and Bureau of Land Management, *H. glandulosa* and *H. burringtoni* were found to be indistinguishable by field biologists; therefore, a study, "Genetic and Anatomical Analyses of the Jumping Slugs" (Wilke and Ziegltrum, unpublished), was initiated by the Olympic National Forest. Under this study, the researchers found that "The morphological characters previously used to distinguish between *H. glandulosa* and *H. burringtoni* (notably the papillosity of the mantle and black spots above the pedal furrow) do not allow for reasonable differentiation between the two species."

As far as distinguishing between the species *H. glandulosa* and *H. burringtoni*, the study of Wilke and Ziegltrum was inconclusive. They found a species complex that they were unable to separate by anatomical or morphological characteristics. The group may be a single species, but neither the historic literature (Pilsbry 1948; Branson 1972) nor the findings of Wilke's genetic work indicated that. Therefore, at this time we consider this a species complex composed of at least two species that are not separable by our current knowledge of their morphological or anatomical characteristics; the geographical range is currently the best species indicator other than DNA analysis.

1, 2, & 3. *Hemphillia burringtoni*: Dickey River, Olympic National Park, Jefferson Co., WA

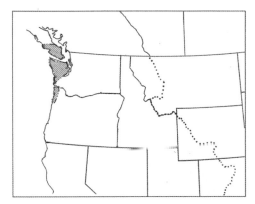

1. *Hemphillia glandulosa*: Woodard Bay, Thurston Co., WA

2. *Hemphillia glandulosa* "species complex": Kennedy Cr., Mason Co., WA

Similar Species: Genetically, Wilke and Ziegltrum found that "the 118 specimens . . . studied cluster in two major groups." They felt that the two "clades" were better separated by their geographic distribution than by the morphological characteristics used to distinguish their species.

Distribution: As determined by Wilke and Ziegltrum, Clade I (*H. burringtoni*) occurs in the northern and western part of the known range; that is, the vicinity of Vancouver Island, British Columbia, and the Olympic Peninsula and Pacific Coast area between Willapa Bay and Tillamook, Oregon. Clade II (*H. glandulosa*) occurs in the central-eastern part of their known range, which is Astoria, Oregon, eastward to include the Cascade Mountains of southwestern Washington.

Hemphillia burringtoni   *Hemphillia glandulosa*

Hemphillia pantherina   Branson, 1975   Panther Jumping-slug

Description: Dr. Branson's type specimen is the only known record for this species; therefore, the following description is paraphrased from his original paper: It is a small jumping-slug, the preserved specimen measuring 14.2 mm long. The mantle is off-white with a black stripe on the right side above and behind the pneumostome, which is located about center in the right side. Dorsally the mantle is marbled with dark gray flecks, but unmarked around the edges. It is granulose-textured and covers only the anterior two-thirds of the visceral pouch. Tubercles on the tail are similar to those of the other species of this group. The tail is grayish laterally, but the keel is white. The head and tentacles are also white dorsally but pale gray ventrally. Laterally the anterior half of the body is white.

Similar Species: *H. pantherina* is similar to *H. glandulosa* and *H. burringtoni* in size and external morphology. It differs by the white dorsal color pattern, and in its mantle not completely covering the visceral pouch.

Distribution: The holotype is from "beneath deep forest litter" near the Miller Creek Crossing in the Lewis River watershed, Skamania Co., Washington. The area has been re-searched several times for additional specimens, without success, leading one to consider *pantherina* a dubious species, but Branson's description and photograph indicate a unique specimen. Knowing that there are other undescribed species and varieties in the Pacific Northwest, it seems prudent to continue to include *H. pantherina* among the fauna of the area, at least for the time being.

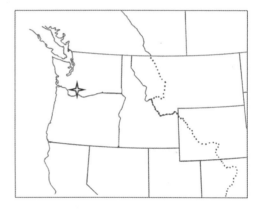

Hemphillia camelus Group
Medium-sized, large for jumping-slugs; tails relatively long with straight dorsal profiles.

Hemphillia camelus   Pilsbry & Vanatta, 1897   Pale Jumping-slug

Description: A medium-sized slug, 30 to 50 mm long or more. Its base color is light tan to buff, and it is clouded with gray. There is a light tan or buff-colored mid-dorsal stripe the length of the tail. The mantle is mottled or speckled with gray and has a blackish lateral band most prominent on the posterior half to three-quarters, or sometimes faint. The sides

Hemphillia camelus: Ross Cr., Lincoln Co., MT

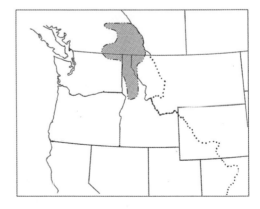

1. *Hemphillia camelus*: Kaniksu NF, Pend Oreille Co., WA
2. *Hemphillia camelus* (eggs and hatchling): Kaniksu NF., Pend Oreille Co., WA

below the mantle are light-colored, with little or no speckling. The pneumostome is in the third quarter of the right side of the mantle. The genital opening is behind the right tentacle.

Similar Species: While *H. malonei* is most similar to *H. camelus* in outward appearance, the ranges of the two species are widely separated. Also see *H. danielsi*, which occurs within the eastern portion of the range of *H. camelus*.

Distribution: *H. camelus* occurs in northwestern Montana and northern Idaho into northeastern Washington (westward into the Kettle Mountain Range of eastern Ferry Co.), and north into southeastern British Columbia.

Hemphillia danielsi   Vanatta, 1914   Marbled Jumping-slug

Description: A rather dark gray slug, 30 to 40 mm or a little longer. The tail is about the same length as the mantle. The pneumostome is in the third quarter of the right side of the mantle, which is gray with an irregular black lateral band. There is no light mid-dorsal caudal stripe.

In Idaho, a second subspecies (or possibly a new species) occurs. There, the species that compares anatomically to *H. danielsi* is more similar to *H. camelus*. A light-brown slug, with the sides of the tail a light brown with darkly pigmented oblique indented lines. There is a mid-dorsal caudal stripe that is devoid of darker clouding or maculae, but not lighter than the ground color. The tentacles are dark brown. The sides below the mantle are white or light buff. The mantle is more distinctly light brown with sparse, irregular markings and a faint or irregular lateral band. The mantle is covered with rather dense, small, but distinct papillae.

Similar Species: In Montana, *H. danielsi* is generally darker-colored than *H. camelus* and it lacks the light dorsal stripe on the tail. In Idaho it is more like *H. camelus* in color, but it lacks the contrasting light mid-dorsal tail stripe, that line being merely of the ground

1. *Hemphillia danielsi*: Bitterroot Mountains, Bitterroot National Forest, Ravalli Co., MT
2. *Hemphillia danielsi*: Slate Cr., Idaho Co., ID
3. *Hemphillia danielsi* (note papillose mantle): South Fork Clearwater River, Idaho Co., ID

color of the animal without the darker flecks or mottling seen on the sides of the tail; and its mantle is distinctly papillose.

Distribution: The type locality for *H. danielsi* is Camas Creek, Bitterroot Mountains, Montana. It is known from the Bitterroot Range in Mineral and Ravalli counties, Montana. The Idaho variety occurs in the Clearwater and Salmon River watersheds, Clearwater and Idaho counties, Idaho.

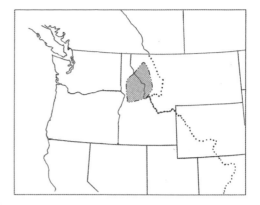

Hemphillia dromedarius Branson, 1972   Dromedary Jumping-slug

Description: This is a medium-sized slug, about 30 mm or more in length when extended. Its base color is whitish with dark gray to blue-gray markings. Sometimes the sides and sole are an orangish color. It is generally dark dorsally and on the sides of the tail, and more white below the mantle. The head and tentacles are dark gray. The tail is without a light dorsal stripe. Rather evenly spaced impressed lines run obliquely-posteriorly down the sides of the tail from a mid-dorsal groove that extends two-thirds or more from the posterior edge of the mantle toward the end of the tail.

1. *Hemphillia dromedarius*: Lake Kachelus, Kittitas Co., WA
2. *Hemphillia dromedarius*: Olympic National Forest, Clallam Co., WA

Similar Species: *H. malonei*, the only other jumping-slug of this group occurring in the same range, normally has a distinct light mid-dorsal line on the tail.

Distribution: The range of *H. dromedarius* is primarily in the Washington Cascades and Olympic Mountains. It extends into western Oregon and north to Vancouver Island, British Columbia.

Hemphillia malonei   Pilsbry, 1917   Malone Jumping-slug

Description: This medium-sized slug measures up to 60 mm long but is usually smaller. The mantle is light brown or buff (a little darker around the shell) with a few irregularly scattered, small black spots or flecks. The body is light tan, clouded with dark gray or brown on the sides of the tail, which becomes darker as it nears a distinct buff, light brown, or orange-brown, mid-dorsal stripe running the length of the tail. The lower sides and the pedal flange are lighter with little darker mottling. The head and neck are light to medium brown dorsally, and the tentacles are a medium shade of gray. There are two color variations. Within the Columbia Gorge and at lower elevations of northern Oregon, most are of the typical color as described above. In Washington, north of the Columbia Gorge and at some higher elevations in Oregon, the mantle is marked with dark brown, somewhat bilaterally symmetrical, marbling. Some of them are rather striking in appearance. All have the light, mid-dorsal tail stripe.

Similar Species: Very much like *H. camelus*, some are nearly identical in appearance, with some variations in color patterns. However, their ranges are widely separated. *H. dromedarius* occurs within the same range, but is generally dark gray and lacks the light-colored dorsal stripe on its tail.

Distribution: *H. malonei* occurs in the northern Oregon Cascades, Columbia Gorge, Washington Cascades, and Olympic Peninsula. It may be found in logs and stumps in moist forests, but is more abundant in riparian and other wet areas. Young are commonly found on skunk cabbage leaves.

Selected references for the family Arionidae: Branson 1972, 1975a, 1991; Brunson and Kevern 1963; Burch 1962; Forsyth 2004; Frest and Johannes 1996; Hanna 1966; Harper 1988, Hendricks et al. 2007; Kerney, Cameron, and Riley 1994; Kozloff and Vance 1958; Leonard et al. 2003, 2011; Mead 1943; Pilsbry 1917, 1948, 1953; Pilsbry and Brunson 1954; Quick 1960; Robinson 1999; Roth 2004; Turgeon et al. 1998; Vanatta 1914; Webb 1959, 1977; Wilke and Duncan 2004; Wilke and Ziegltrum (unpublished report).

1. *Hemphillia malonei*: Tilton River, Lewis Co., WA
2. *Hemphillia malonei*: Cowlitz River, Lewis Co., WA
3. *Hemphillia malonei*: Gifford Pinchot National Forest, Cowlitz Co., WA

Glossary of Terms

acute—an angle less than 90 degrees; a sharper or more pointed angle.

antepenultimate—"before the penultimate"; the whorl before the penultimate whorl.

apertural Teeth—protrusions into the aperture from the surrounding shell. Also called denticles or, collectively, dentition.

aperture—the opening of a snail's shell through which the body of the snail protrudes.

aperture lip—the edge of the shell around the aperture, the configuration of which may be used in identification of a species.

apical—at, of, or on the apex. The top or peak of the shell.

armature—the apertural teeth or dentition.

auriculate—ear-shaped, most often referring to the aperture.

axis—the central line around which the whorls of a spiral shell revolve. The columella of some shells form the axis, or the umbilicus may form a cylinder or funnel around it.

basal—the bottom of a snail shell when referring to the position of a structure, or characteristic (e.g., basal tooth, basal sculpturing, etc.).

body whorl—the last whorl of a snail shell.

bristles—hair-like projections formed by the periostracum on the surface of a shell (also called hairs).

calcareous—composed of calcium carbonate as the substance forming a shell; a chalky texture.

callus—a thickened area on a shell, such as a callus ridge within the apertural lip, a transparent thickening on the parietal surface, etc.

capacious—of large capacity (i.e., in relation to the overall size of the shell).

carina (*pl* carinae)—a keel or raised ridge-like projection such as around the periphery of some shells.

caudal fossa (= caudal pit or mucous pit)—a pit or slit at the posterior end of a gastropod where the pedal furrows come together. It is usually filled with mucus and sometimes bits of debris.

collabral—oriented at the same angle as the aperture.

columella (columellar)—the central axis of a snail shell. Columellar—structures on or attached to the columella (e.g., columellar tooth).

concave—indented or curved inward causing a depression.

conic (conical)—cone-shaped; relatively straight-sided tapering toward a point.

contractile tentacles—tentacles that are shortened or lengthened by contraction and extension. Compare with **retractile tentacles**.

convex—inflated or curved outward, as a rounded or domed spire.

crest—a wave-like ridge in the last whorl of the shell just behind the outer or palatal lip of the aperture.

cusp—a point on a structure forming a tooth-like protuberance.

cuticular—pertaining to skin or skin-like in origin. Cuticular riblets or other sculpturing are formed from the periostracum.

dentition—referring to apertural teeth. Denticle—an individual tooth-like structure. Denticulate—having tooth-like structures.

dextral—whorled to the right—recognized by the aperture on the right side of the shell when viewed from the front with the apex upward.

diameter (width)—Greatest or major diameter is measured from the outer edge of the aperture directly across the shell with the axis vertical and the shell turned to measure its maximum width. Lesser or minor diameter is measured with the axis vertical and the shell turned so that the outer edge of the last whorl just behind the aperture or peristome is parallel to the measuring lines of the scale.

discoidal (disk-shaped)—shells with a flat or nearly flat spire.

disjunct—being separated from similar occurrences. A disjunct population is well separated from other populations of the same species.

distal—located away from the center (i.e., farthest from the central axis).

elliptical—an oval shape with the ends similarly curved.

elongate—of snail shells, those that are distinctly higher than wide.

epiphragm—a membrane or seal formed from mucus secreted over the aperture of the shell to protect the snail from desiccation during hibernation or aestivation.

flammules—irregular markings, usually reddish or orange, resembling flames.

foot—the underside and lower body of a gastropod (stomach-foot).

funnelform—narrowing inward from the opening like a funnel (i.e., an umbilicus that expands with the growth of the shell diameter). Compare with **symmetrical**.

fusiform (spindle-shaped)—a relatively long narrow shell that tapers toward both ends.

globose—spherical or ball-shaped. Globose shells are inflated to be somewhat spherical in shape.

granules (granulose)—small grain-like bumps. Granulose—surface appearing as if covered with small seeds or scattered grains of sand.

growth lines (growth-wrinkles)—waves or wrinkles in the shell surface resulting from various rates or periods of growth.

hairs—fine bristles formed by the periostracum on the surface of a shell (also called bristles).

height—measured from the apex to the base of the lip with the axis of the shell (columella) oriented vertically, or perpendicular to the lines of the measuring scale.

heliciform—the helix-shape of snail shells that are wider than high.

imperforate—lacking any open or perforate umbilicus.

impressed line (also a **stria** or *pl.* striae)—an indented line or fine groove or furrow.

keel—a narrow, raised ridge; examples can be seen on the shell peripheries of some snails or on the tails of some slugs.

lamella (*pl.* lamellae)—a structure formed from a fold in the shell material forming, for example, a thin ridge or laterally flattened denticle.

laterally compressed—pertaining to the sides; pinched in or flattened from the sides.

lip—the edge of the aperture formed by the termination of the last whorl. See also **simple lip**.

lira (*pl.* lirae)—raised rib-like or thread-like sculpturing on the shell surface.

lunate—shaped somewhat like a crescent moon.

lyrate—shaped like a lyre (a harp of ancient Greece), as the pattern of dark lines on the mantle of some slugs.

malleation (malleated)—dents, as if repeatedly struck with a hammer. Malleated—dents spread over a surface.

mantle—the tissue lining the shell and from which the shell is secreted. On a slug it is the saddle-like structure on the anterior portion of the back, usually just behind the head or neck.

measurements: see **diameter (width)**; **height**; **width of the umbilicus**.

minute—referring to the size of a snail, one that is tiny—less than about 3 mm in the greater of its height or diameter.

mucus (n) **mucous** (adj)—a viscid fluid secreted by mucous glands in the skin of gastropods, on which they glide during locomotion. Mucus covers the animals' bodies and provides some protection against desiccation. For identification, it may be thick or watery; it may be clear, yellow, or white; in some species it changes color and/or consistency when the animal is irritated or attacked.

nacre—the hard iridescent material that lines the inside of some shells; mother-of-pearl.

obovate—reverse-ovate. A structure (shell) that is wider toward the apex than around the apertural end. An aperture that is wider at the distal than at the proximal end.

obtuse—an angle greater than 90 degrees; more blunt or less sharp than a right angle.

opaque—blocking the passage of light; neither translucent nor transparent.

operculum—a chitinous or horny disk, attached to the back of some snails, behind the shell, that closes the aperture when the animal withdraws into its shell. In this work, snails in the subclass Eugastropoda (or Prosobranchia) have an operculum, but it is lacking in the subclass Pulmonata to which most land snails belong.

outer lip—on snail shells that are wider than high, that portion of the apertural lip margin between the palatal and the basal margins, on the opposite side of the aperture from the columellar margin.

ovate (ovoid)—egg-shaped.

oviparous—reproducing by eggs that hatch after being deposited by the parent.

ovoviviparous—reproducing by eggs that are retained in the body of the parent until they hatch or are about to, so the young are born as baby snails.

palatal—referring to the aperture or apertural lip, that area adjacent to the upper or outer connection of the aperture with the penultimate whorl. For shells that are higher than wide, the palatal margin is the outer portion (between the previous whorl and the basal margin). For shells that are wider than high, the palatal margin is the dorsal area of the aperture (between the penultimate whorl and the "outer margin").

palatal insertion—the joining of the palatal lip margin with the penultimate whorl.

papilla (*pl.* papillae) (**papillose**)—small nipple-like projections. Papillose—partially or wholly covered with papillae.

parietal—a location inside of the aperture on the penultimate whorl (e.g., parietal tooth).

patelliform—kneecap-shaped (i.e. dish or cap-shaped as a limpet shell).

pedal flange—the edge of the gastropod foot between the angle of the sole and the pedal furrows in the infraorder or division Aulacopoda. It is also called the skirt by some authors when describing slugs.

pedal furrows—two indented lines (one often obscure) between the edge of the foot (pedal flange) and the sides of a snail or slug.

penultimate whorl—the whorl preceding the last or body whorl.

perforate—referring to the umbilicus, being very small (i.e., pinhole-sized).

periostracal—a structure formed from the periostracum (i.e. periostracal hairs). See also **cuticular**.

328 LAND SNAILS AND SLUGS OF THE PACIFIC NORTHWEST

periostracum—a thin membrane over the shell. Some shell sculpturings, such as hairs and some riblets, are formed from the periostracum.

periphery—the outer edge of the last whorl of a snail shell, which may be rounded, shouldered, angled, or sculptured in various ways.

peristome—the reflected or thickened rim or lip around the aperture.

protoconch—the first or embryonic whorls. Usually thought of as those whorls present on the embryonic or hatchling snail.

pupiform (pupilliform)—shaped like an insect pupa, as the shells of snails of the family Pupillidae or Vertiginidae.

radial—referring to the orientation of shell sculpturing, radiating outward or across the whorls (also transverse).

recurved—referring to the reflected peristome, the edge of which is reflected more narrowly a second time.

reflected lip—in which the aperture lip turns abruptly outward, causing a relatively wide border around the aperture. The lip may also be flared, widening less abruptly or only slightly, or merely thickened.

reniform—kidney or bean-shaped.

reticulated—marked with lines that cross in a network pattern.

retractile tentacles—tentacles that are essentially a tube and are shortened by pulling the distal end, containing the eye-spot, down through its center. Compare with **contractile tentacles**.

revolute—rolled or curled back, as an apertural lip that is rolled back instead of reflected to a flat peristome.

riblets—very small or thin ribs, usually formed by the periostracum of the shell. See also **cuticular**.

ribs—raised ridges (lirae) running radially across the whorls, forming rib-like structures, usually formed from the solid shell.

rimate—a slit-like umbilicus, caused by a sharp turning of the inner edge of tightly coiled whorls or by the close covering of the umbilicus by the apertural lip margin.

sculpture—raised or indented lines or areas forming patterns on the surface of a shell.

simple lip or aperture—a thin and straight lip, not reflected, flared, or conspicuously thickened.

sinistral—whorled to the left. Recognized by the aperture being on the left side of the shell when viewed from the front with the apex up.

sinulus—a wave, usually in the palatal lip margin of some pupillid snails, in which the margin is indented, forming an angle that points inward.

sinuous—wavy or undulating lines.

skirt—see **pedal flange**.

spiral—oriented with or parallel to the whorls when referring to shell sculpturing or markings.

spire—the usually elevated part of the shell between the body whorl and the apex, including the first through the penultimate whorls.

striae—indented or impressed lines or grooves in a shell.

succiniform—a distinctive elongated shell shape typical to members of the Family Succineidae, in which there is a large body whorl and aperture and a relatively very small spire.

sulci—relatively broad grooves forming radial ridges and grooves across the dorsal whorls of the shells of certain snails (e.g., some *Pristiloma*).

supernumerary—in excess of the normal or usual number, as in supernumerary bands on the shells of some snails (e.g., *Oreohelix strigosa*).

suture—the line (usually indented) where the whorls of a spiral shell are joined.

symmetrical—similar in size on both sides. A symmetrical umbilicus is nearly uniform in its diameter throughout its depth. Compare **funnelform**.

teeth—see **apertural teeth** and **dentition**.

teleoconch—the shell following the protoconch. The whole shell of a snail, except the protoconch.

threads—fine raised thread-like lines or lirae on a shell surface.

translucent—semitransparent so that light may pass through but not clear enough to see through. Translucent shells are usually clouded with white or some faint coloration.

transparent—clear enough to see through, although not necessarily without distortion. Transparent shells are normally colorless or clear with only a very faint tint of color.

transverse—see **radial**.

tripartite—divided into three parts or sections. A tripartite sole is divided into three longitudinal sections.

tubercles—raised blocks or short ridges on the body of a snail or slug.

turbinate—top-shaped; a shell with a broad base, tapering to a pointed apex.

umbilicate—having an open umbilicus larger than just pin-sized; may be narrowly to widely umbilicate.

umbilicus—the opening in the central base of a shell around which the whorls revolve.

viviparous—bearing live young that have been nourished by and developed within the parent.

whorl—one or each complete turn or revolution of a spiral shell.

width of the umbilicus—is measured in a line directly across the umbilicus from the aperture.

References and Literature Cited

Anderson, R. C. 1965. "Cerebrospinal nematodiasis (*Pneumostrongylus tenuis*) in North American cervids." Transcript of the North American Wildlife Conference 30:156–167.

Anderson, R. C., and A. K. Prestwood. 1981. "Lungworms," chap 2 in *Diseases and Parasites of White-tailed Deer*, W. R. Davidson, F. A. Hayes, V. F. Nettles, and F. E. Kellogg, 266–317. Tall Timbers Research Station, Misc. Publication no. 7, Tallahassee, FL.

Arnold, Winifred H. 1965. "A Glossary of a Thousand-and-One Terms Used in Conchology." *The Veliger* 7 (supplement). Reprinted and distributed with the agreement of the California Malacozoological Society, Inc., by *The Shell Cabinet*, Falls Church, Virginia.

Baker, H. B. 1930. "New and Problematic West American Land Snails." *The Nautilus* 43 (3): 95–101.

Baker, H. B. 1932. "New Land Snails from Idaho and Eastern Oregon." *The Nautilus* 45 (3): 82–87.

Barnes, R. D. 1968. *Invertebrate Zoology*. Second edition. Philadelphia, London, Toronto: W. B. Saunders Co.

Berman, J., and J. T. Carlton. 1991. "Marine Invasion Processes: Interactions between Native and Introduced Marsh Snails." *Journal of Experimental Marine Biology and Ecology* 150 (1991): 267–281.

Berry, S. S. 1932. "Three New Mountain Snails from Idaho and Nevada. *Journal of Entomology and Zoology* 24 (4): 57–63.

Berry, S. S. 1937. "Some Lesser Races of *Monadenia fidelis* (Gray)." *The Nautilus* 51 (1): 28–33.

Berry, S. S. 1940. "Nine New Snails of the Genus *Monadenia* from California." *Journal of Entomology and Zoology* 32 (1): 1–17.

Berry, S. S. 1955. "An Important New Land Snail from the Mission Range, Montana." *Bulletin of the Southern California Academy of Sciences* 54 (1): 17–19.

Binney, W. G. 1873. "Catalogue of the Terrestrial Air-Breathing Mollusks of North America with Notes on their Geographical Range." *Harvard University Museum of Comparative Zoology Bulletin* 3 (9): 191–220.

Binney, W. G. 1885. *A Manual of American Land Shells*. USDI, U.S. National Museum Bulletin no. 28. Smithsonian Institution, Government Printing Office, Washington, D. C., 57–528.

Borror, D. J., and R. E. White. 1970. *A Field Guide to the Insects of America North of Mexico.* Boston: Houghton Mifflin.

Bouchet, P., and J-P Rocroi. 2005. "Classification and Nomenclature of Gastropod Families." *Malacologia* 47 (1-2): 1–397.

Branson, B. A. 1972. "*Hemphillia dromedarius*, a New Arionid Slug from Washington." *The Nautilus* 85 (3): 100–106.

Branson, B. A. 1975a. "*Hemphillia pantherina*, a New Arionid Slug from Washington." *The Veliger* 18 (1): 93–94.

Branson, B. A. 1975b. "*Radiodiscus hubrichti* (Pulmonata: Endodontidae), a New Species from the Olympic Peninsula, Washington." *The Nautilus* 89 (2): 47–48.

Branson, B. A. 1977. "Freshwater and Terrestrial Mollusca of the Olympic Peninsula, Washington." *The Veliger* 19 (3): 310–330.

Branson, B. A. 1980. "Collections of Gastropods from the Cascade Mountains of Washington." *The Veliger* 23 (2): 171–176.

Branson, B. A. 1991. "*Hesperarion mariae* (Gastropoda: Arionidae: Ariolimacinae), a New Slug Species from Oregon." *Transactions of the Kentucky Academy of Sciences* 52 (3-4): 109–110.

Branson, B. A., and R. M. Branson. 1984. "Distributional Records for Terrestrial and Freshwater Mollusca of the Cascade and Coast Ranges, Oregon." *The Veliger* 26 (4): 248–257.

Brown, H. A., R. B. Bury, D. M. Darda, L. V. Diller, C. R. Peterson, and R. M. Storm. 1995. *Reptiles of Washington and Oregon.* Seattle: Seattle Audubon Society.

Brunson, R. B. 1956. "The Mystery of *Discus brunsoni.*" *The Nautilus* 70 (1): 16–21.

Brunson, R. B., and N. Kevern. 1963. "Observations of a Colony of *Magnipelta.*" *The Nautilus* 77 (1): 23–27.

Brunson, R. B., and U. Usher. 1957. "Haplotrema from Western Montana." *The Nautilus* 70 (4): 121–123.

Burch, J. B. 1962. *How to Know the Eastern Land Snails.* Dubuque, Iowa: William C. Brown Co.

Burch, J. B., and T. A. Pearce. 1990. "Terrestrial Gastropoda," chap. 9 in *Soil Biology Guide,* edited by D. L. Dindal, 201–309. New York: J. Wiley.

Burke, T. E. 2006. "Mollusk Survey on Patos Island Final Report." Report to the US Dept. Interior, BLM, Wenatchee, WA.

Burke, T. E. 2008. "Mollusk Survey on Lopez Island San Juan County, Washington." Report to the US Dept. Interior, BLM, Spokane, WA.

Capizzi, J. 1962. "A Land Snail (*Succinea campestris*)—Oregon— Cooperative Economic Insect Report." US Dept. Agriculture, Agricultural Research Admin. No. 49: p.1237.

Chamberlin, R. V., and D. T. Jones. 1929. *A Descriptive Catalog of the Mollusca of Utah.* Biological Series vol. 1, no. 1. Bulletin of the University of Utah, 19 (4): ix, 1–203.

Coan, E., and B. Roth. 1987. "The Malacological Taxa of Henry Hemphill." *The Veliger* 29 (3): 322–339.

Dougherty, E. C. 1945. "The Nematode Lungworms (Suborder Strongylina) of North American Deer of the Genus *Odocoileus.*" *Parasitology* 36:199–208.

Emberton, K. C. 1991. "Polygyrid Relations: A Phylogenetic Analysis of 17 Subfamilies of Land Snails (Mollusca: Gastropoda: Stylommatophora)." *Zoological Journal of the Linnean Society* 103:207–224.

Fairbanks, H. L. 1980. "Morphological notes on *Oreohelix amariradix* Pilsbry, 1934 (Pulmonata: Oreohelicidae)." *The Nautilus* 94 (1): 27–30.

Fairbanks, H. L. 1984. "A New Species of *Oreohelix* (Gastropoda: Pulmonata: Oreohelicidae) from the Seven Devils Mountains, Idaho." *Proceedings of the Biological Society of Wash.* 97 (1): 179–185.

Forsyth, R. G. 2004. *Land Snails of British Columbia.* Victoria, British Columbia, Canada: Royal BC Museum.

Frest, T. J. 1999. "A Review of the Land and Freshwater Mollusks of Idaho." Report prepared for Idaho Conservation Data Center, Idaho Dept. Fish and Game, Boise, ID.

Frest, T. J., and E. J. Johannes. 1993. "Mollusk Species of Special Concern within the Range of the Northern Spotted Owl." Final report prepared for the Forest Ecosystem Management Working Group, USDA Forest Service, Pacific Northwest Region, Portland, OR; Seattle, WA: Deixis Consultants.

Frest, T. J., and E. J. Johannes. 1995. "Interior Columbia Basin Mollusk Species of Special Concern." Final report prepared for the Interior Columbia Basin Ecosystem Management Project. Seattle, WA: Deixis Consultants.

Frest, T. J., and E. J. Johannes. 1996. "Additional Information on Certain Mollusk Species of Special Concern Occurring within the Range of the Northern Spotted Owl." (Supplement to 1993 report). Seattle, WA: Deixis Consultants.

Frest, T. J., and E. J. Johannes. 1997. *Land Snail Survey of the Lower Salmon River Drainage, Idaho.* Idaho Bureau of Land Management Technical Bulletin no. 97-18. Seattle: Deixis Consultants.

Frest, T. J., and E. J. Johannes. 2000. "An Annotated Checklist of Idaho Land and Freshwater Mollusks." *Journal of the Idaho Academy of Sciences* 36 (2): 1–51.

Glidden, Helene. (1951) 2002. *The Light on the Island: Tales of a Lighthouse Keeper's Family in the San Juan Islands.* Woodinville, WA: San Juan Publishing.

Grimm, F. W., R. G. Forsyth, F. W. Schueler, and Aleta Karstad. 2009. *Identifying Land Snails and Slugs in Canada: Introduced Species and Native Genera*. Canadian Food Inspection Agency. Ottawa Plant and Seed Laboratories, Ottawa.

Hanna, G. D. 1966. *Introduced Mollusks of Western North America*. Occasional Papers, California Academy of Sciences no. 48. San Francisco.

Hanna, G. D., and A. G. Smith. 1939. "Notes on some forms of *Oreohelix strigosa* (Gould)." *Proceedings of the California Academy of Sciences* 23 (25): 381–392.

Harper, Alice Bryant. 1988. *The Banana Slug: A Close Look at a Giant Forest Slug of Western North America*. Aptos, CA: Leaves Press.

Harris, S. A., and L. Hubricht. 1982. "Distribution of the Species of the Genus *Oxyloma* (Mollusca, Succineidae) in Southern Canada and the Adjacent Portions of the United States." *Canadian Journal of Zoolology* 60:1607–1611.

Hemphill, H. (1890). "Descriptions of New Varieties of North American Land Shells." *The Nautilus* 3 (12): 133–135.

Henderson, J. 1924. "Mollusca of Colorado, Utah, Montana, Idaho, and Wyoming." *University of Colorado Studies* 13 (2): 65–223.

Henderson, J. 1929a. "Non-marine Mollusca of Oregon and Washington." *University of Colorado Studies* 27 (2): 47–190.

Henderson, J. 1929b. "Some notes on *Oreohelix*." *Proceedings of the California Academy of Sciences* 18 (8): 221–227.

Henderson, J. 1931. "Molluscan Provinces in the Western United States." *University of Colorado Studies* 18 (4): 177–186.

Henderson J., and L. E. Daniels. 1916. "Hunting Mollusca in Utah and Idaho." *Academy of Natural Sciences of Philadelphia Proceedings* 68:315–339.

Hendricks, P. 2003. *Status and Management of Terrestrial Mollusks of Special Concern in Montana*. Helena, MT: Montana Natural Heritage Program.

Hendricks, P., B. A. Maxell, Susan Lenard. 2006. *Land Mollusk Surveys on USFS Northern Region Lands*. Helena, MT: Montana Natural Heritage Program.

Hendricks, P., B. A. Maxell, Susan Lenard, and C. Currier. 2007. *Land Mollusk Surveys on USFS Northern Region Lands: 2006*. Helena, MT: Montana Natural Heritage Program.

Hoagland, K. Elaine, and G. M. Davis. 1987. "The Succineid Snail Fauna of Chittenango Falls, New York: Taxonomic Status with Comparisons to Other Relevant Taxa." *Proceedings of the Academy of Natural Sciences Philadelphia* 139:465–526.

IUCN. 1993. *1994 IUCN Red List of Threatened Animals Compiled by the World Conservation Monitoring Centre*. Gland, Switzerland, and Cambridge, UK: IUCN World Conservation Union.

Jones, D. T. 1944. "*Oreohelix howardi*, new species." Proceedings of the Utah Academy of Sciences, Arts, and Letters. (Symposium; author's copy).

Kay, E. Alison, ed. 1995. "The Conservation Biology of Molluscs: Proceedings of a Symposium Held at the Ninth International Malacological Congress, Edinburgh, Scotland, 1986." Occasional Paper of the IUCN Species Survival Commission 9.

Kelley, Rachel, S. Dowlan, Nancy Duncan, and T. Burke. 1999. *Field Guide to Survey and Manage Terrestrial Mollusk Species from the Northwest Forest Plan*. US Bureau of Land Management, Oregon State Office, Salem.

Kerney, M. P., R. A. D. Cameron, and G. Riley. (1979) 1994. *Land Snails of Britain and North-West Europe*. Collins Field Guide Series. Hong Kong: Harper Collins Publishers.

Kozloff, E. N. 1993. *Seashore Life of the Northern Pacific Coast: An Illustrated Guide to Northern California, Oregon, Washington, and British Columbia*. Seattle and London: University of Washington Press.

Kozloff, Eugene N. 1996. *Marine Invertebrates of the Pacific Northwest with Additions and Corrections*. Seattle and London: University of Washington Press.

Kozloff, E. N., and Joann Vance. 1958. "Systematic status of *Hemphillia malonei*." *The Nautilus* 72 (2): 42–49.

Lankester, M. W., and R. C. Anderson. 1968. "Gastropods as Intermediate Hosts of *Pneumostrongylus tenuis* Dougherty of White-tailed Deer." *Canadian Journal of Zoology* 46:373–861.

Leonard, W. P., H. A. Brown, L. L. C. Jones, K. R. McAllister, and R. M. Storm. 1993. *Amphibians of Washington and Oregon.* Seattle, WA: Seattle Audubon Society.

Leonard, W. P., L. Chichester, J. Baugh, and T. Wilke. 2003. "*Kootenaia burkei*, a New Genus and Species of Slug from Northern Idaho, United States (Gastropoda: Pulmonata: Arionidae)." *Zootaxa* 355:1–16.

Leonard, W. P., L. Chichester, C. Richart, and Tiffany A. Young. 2011. "*Securicauda hermani* and *Carinacauda stormi*, Two New Genera and Species of Slug from the Pacific Northwest of the United States (Gastropoda: Stylommatophora: Arionidae), with Notes on *Gliabates oregonius* Webb 1959." *Zootaxa* 2746:43–56.

Leopold, A. 1949. *A Sand County Almanac and Sketches Here and There.* New York: Oxford University Press.

McGraw, R., Nancy Duncan, and E. Cazares. 2002. "Fungi and Other Items Consumed by the Blue-gray Taildropper Slug (*Prophysaon coeruleum*) and the Papillose Taildropper Slug (*Prophysaon dubium*)." *The Veliger* 45 (3): 261–264.

Mead, Albert R. 1943. Revision of the Giant West Coast Land Slugs of the Genus *Ariolimax* Moerch (Pulmonata: Arionidae)." *American Midland Naturalist* 30:675–717.

Nedeau, E. J., A. K. Smith, Jen Stone, and Sarina Jepsen. 2009. *Freshwater Mussels of the Pacific Northwest.* Portland, OR: The Xerces Society.

Neil, Kerry M. 2001. "Microhabitat Segregation of Co-existing Gastropod Species." *The Veliger* 44 (3): 294–300.

Nekola, J. C., and B. F. Coles. 2010. "Pupillid Land Snails of Eastern North America." *American Malacological Bulletin* 28 (1-2): 29–57.

Ovaska, Kristiina, L. Chichester, and L. Sopuck. 2010. "Terrestrial Gastropods from Haida Gwaii (Queen Charlotte Islands), British Columbia, Canada, Including Description of a New Northern Endemic Slug (Gastropoda: Stylommatophora: Arionidae)." *The Nautilus* 124 (1): 25–33.

Pennak, R. W. 1978. *Freshwater Invertebrates of the United States.* Second ed. John Wiley & Sons, New York.

Pilsbry, H. A. 1896. "The Aulacopoda: A Primary Division of the Monotremate Land Pulmonata." *The Nautilus* 9 (10): 109–111.

Pilsbry, H. A. 1917. "A New *Hemphillia* and Other Snails from Near Mt. Hood, Oregon." *The Nautilus* 30:117–119.

Pilsbry, H. A. 1933. "Notes on the Anatomy of *Oreohelix*, III, with Descriptions of New Species and Subspecies." *Proceedings of the Academy of Natural Sciences of Philadelphia* 85:383–410.

Pilsbry, H. A. 1939. *Land Mollusca of North America (North of Mexico).* Academy of Natural Sciences of Philadelphia, monograph no. 3, I (1): i–573.

Pilsbry, H. A. 1940. *Land Mollusca of North America (North of Mexico).* Academy of Natural Sciences of Philadelphia, monograph no. 3, I (2): i–ix, 575–994.

Pilsbry, H. A. 1946. *Land Mollusca of North America (North of Mexico).* Academy of Natural Sciences of Philadelphia, monograph no. 3, II (1): iv–520.

Pilsbry, H. A. 1948. *Land Mollusca of North America (North of Mexico).* Academy of Natural Sciences of Philadelphia, monograph no. 3, II (2): i–xlvii, 521–1113.

Pilsbry, H. A. 1953. "*Magnipelta*, a New Genus of Arionidae from Idaho." *The Nautilus* 67 (2): 37–39.

Pilsbry, H. A., and R. B. Brunson. 1954. "The Idaho-Montana slug *Magnipelta* (Arionidae)." *Notulae Naturae of the Academy of Natural Sciences of Philadelphia* 262:1–6.

Quick, H. E. 1960. "British slugs (Pulmonata): Testacelladae, Arionidae, Limacidae." *Bulletin of the British Museum (Natural History) Zoology* 6 (3): 103–326.

Reise, H., J. M. C. Hutchinson, R. G. Forsyth, and T. J. Forsyth. 2000. "The Ecology and Rapid Spread of the Terrestrial Slug *Boettgerilla pallens* in Europe with References to Its Recent Discovery in North America." *The Veliger* 43 (4): 313–318.

Robinson, D. G. 1999. "Alien Invasions: The Effects of the Global Economy on Non-marine Gastropod Introductions into the United States." *Malacologia* 41 (2): 413–438.

Roth, B. 1980. "Shell Color and Banding in Two Coastal Colonies of *Monadenia fidelis* (Gray)." *The Wasmann Journal of Biology* 38 (1-2): 39–51.

Roth, B. 1981. "Distribution, Reproductive Anatomy, and Variation of *Monadenia troglodytes* Hanna and Smith (Gastropoda: Pulmonata) with the Proposal of a New Subgenus." *Proceedings of the California Academy of Sciences* 42 (15): 379–407.

Roth, B. 1985. "A New Species of *Punctum* (Gastropoda: Pulmonata: Punctidae) from the Klamath Mountains, California, and First Californian Records of *Planogyra clappi* (Valloniidae)." *Malacological Review* 18:51–56.

Roth, B. 1987. "Identities of Two California Land Snails Described by Wesley Newcomb." *Malacological Review* 20:129–132.

Roth, B. 1990. "New Haplotrematid Land Snails, *Ancotrema* and *Haplotrema* (Gastropoda: Pulmonata), from the Sierra Nevada and North Coast Ranges, California." *The Wasmann Journal of Biology* 47 (1-2): 68–76.

Roth, B. 1991. "A Phylogenetic Analysis and Revised Classification of the North American Haplotrematidae (Gastropoda: Pulmonata)." *American Malacological Bulletin* 8 (2):155–163.

Roth, B. 2001. "Identity of the Land Snail *Monadenia rotifer* Berry, 1940 (Gastropoda: Pulmonata: Bradybaenidae)." *American Malacological Bulletin* 16 (1-2): 61–64.

Roth, B. 2003. "*Cochlicopa* Férussac, 1821, not *Cionella* Jefferys, 1829; Cionellidae Clessin, 1879, not Cochlicopidae Pilsbry, 1900 (Gastropoda: Pulmonata: Stylommatophora)." *The Veliger* 46 (2): 183–185.

Roth, B. 2004. "Observations on the Taxonomy and Range of *Hesperarion* Simroth, 1891 and the Evidence for Genital Polymorphism in *Ariolimax* Morch, 1860 (Gastropoda: Pulmonata: Arionidae: Ariolimacinae)." *The Veliger* 47 (1): 38–46.

Roth, B., C. Hertz, and R. Cerutti. 1987. "White Snails (Helicidae) in San Diego County, California." *The Festivus* 19 (9): 84–88.

Roth, B., and W. B. Miller. 1992. "A New Genus of Polygyrid Land Snail (Gastropoda: Pulmonata) from Oregon." *The Veliger* 35 (3): 222–225.

Roth, B., and W. B. Miller. 1993. "Polygyrid Land Snails, *Vespericola* (Gastropoda: Pulmonata), 1. Species and populations formerly referred to as *Vespericola columbianus* (Lea) in California." *The Veliger* 36 (2): 134–144.

Roth, B., and W. B. Miller. 1995. "Polygyrid Land Snails, *Vespericola* (Gastropoda: Pulmonata), 2. Taxonomic status of *Vespericola megasoma* (Pilsbry) and *V. karokorum* Talmadge." *The Veliger* 38 (2): 133–144.

Roth, B., and W. B. Miller. 2000. "Polygyrid Land Snails, *Vespericola* (Gastropoda: Pulmonata), 3. Three New Species from Northern California." *The Veliger* 43 (1): 64–71.

Roth, B., and T. A. Pearce. 1984. "*Vitrea contracta* (Westerlund) and Other Introduced Land Mollusks in Lynnwood, Washington." *The Veliger* 27 (1): 90–92.

Roth, B., and P. H. Pressley. 1986. "Observations on the Range and Natural History of *Monadenia setosa* (Gastopoda: Pulmonata) in the Klamath Mountains, California, and the Taxonomy of Some Related Species." *The Veliger* 29 (20): 169–182.

Roth, B., and Patricia S. Sadeghian. 2006. *Checklist of the Land Snails and Slugs of California.* Second edition. Santa Barbara Museum of Natural History Contributions in Science, no 3.

Severinghaus, C. W., and L. W. Jackson. 1970. "Feasibility of Stocking Moose in the Adirondacks." *New York Fish and Game Journal* 17 (1): 18–32.

Simms, B. T., C. R. Donham, J. N. Shaw, and A. M. McCapes. 1931. "Salmon Poisoning." *Journal of the American Veterinary Medical Association* 13 (2): 181–195.

Smith, A. G. 1937. "The Type Locality of *Oreohelix strigosa* (Gould)." *The Nautilus* 50 (3): 73–77.

Smith, A. G. 1943. "Mollusks of the Clearwater Mountains, Idaho." *California Academy of Sciences* 23 (36): 537–554, pl.48.

Smith, A. G. 1959. "Note on *Trilobopsis tehamana* (Pilsbry), a Rare Northern California Land Snail." *The Veliger* 2 (4): 97.

Smith, Allyn G. 1960. *A New Species of Megomphix from California.* Occasional Papers of the California Academy of Sciences no. 28.

Smith, Allyn G. 1970. 6. "American Malacological Union Symposium: Rare and Endangered Mollusks: 6— Western Land Snails." *Malacologia* 10 (1): 39–46.

Solem, A. 1974. *The Shell Makers: Introducing Mollusks*. New York: John Wiley & Sons.

Solem, A. 1975. "Notes on Salmon River Oreohelicid Land Snails, with Description of *Oreohelix waltoni*." *The Veliger* 18 (1): 16–30.

Sport Fishing Institute. 1981. "Snail Fever Cure." *SFI Bulletin* (July 1981) 326:2.

Talmadge, R. R. 1960. "Color Phases in *Monadenia fidelis* (Gray)." *The Veliger* 2 (4): 83–85, pl. 18.

Taylor, D. W. 1981. "Freshwater Mollusks of California: A Distributional Checklist." *California Fish and Game* 67 (3): 140–163.

Thomas, L. (1974) 1980. *The Medusa and the Snail: More Notes of a Biology Watcher*. New York: Bantam Books .

Turgeon, Donna D., J. F. Quinn Jr., A. E. Bogan, E. V. Coan, F. G. Hochberg, W. G. Lyons, Paula M. Mikkelsen, R. J. Neves, C. F. E. Roper, G. Rosenberg, B. Roth, Amelie Scheltema, F. G. Thompson, M. Vecchione, and J. D. Williams. 1998. *Common and Scientific Names of Aquatic Invertebrates from the United States and Canada: Mollusks*. Second edition. American Fisheries Society Special Publication 26. Bethesda Maryland: AMS.

US Department of Agriculture. 1956. "Animal Diseases." *Yearbook of Agriculture 1956*. USDA: Washington, D.C.

USDA Forest Service and USDI Bureau of Land Management. 1994. (The Northwest Forest Plan). "Record of Decision for Amendments to Forest Service and Bureau of Land Management Planning Documents within the Range of the Northern Spotted Owl, and Standards and Guidelines for Management of Habitat for Late-successional and Old-growth Forest Related Species within the Range of the Northern Spotted Owl." Portland and Washington, D.C.

USDI Bureau of Land Management and USDA Forest Service. 2003. Survey Protocol for Survey and Manage Terrestrial Mollusk Species from the Northwest Forest Plan, Version 3.0. Portland, OR.

Vanatta. 1914. "Montana Shells." *Proceedings of the Academy of Natural Sciences of Philadelphia* 66:367–371.

Webb, G. R. 1959. "Two New North-western slugs, *Udosarx lyrata* and *Gliabates oregonia*." *Gastropodia* 1 (3): 22–23, 28 (pl. 14).

Webb, G. R. 1977. "Additional Data on *Gliabates oregonia* Webb, 1959." *Gastropodia* 1(10): 108.

Webb, G. R., and R. H. Russell. 1977. "Anatomical Notes on a *Magnipelta*: Camaenidae?" *Gastropodia* 1 (10): 107–108.

White, R. E. 1983. *A Field Guide to the Beetles of North America*. Boston: Houghton Mifflin Co.

Wilke, T., and Nancy Duncan. 2004. "Phylogeographical Patterns in the American Pacific Northwest: Lessons from the Arionid Slug *Prophysaon coeruleum*." *Molecular Ecology* 13:2303–2315.

Wilke, T., and Joan Ziegltrum. "Genetic and Anatomical Analyses of the Jumping Slugs." Unpublished. Final Report Contract # 43-05G2-1-10086, USDA Forest Service.

Index

Page numbers in bold type are those on which the species or subspecies is described in the species accounts.